Graphics and GUIs
with
MATLAB®
THIRD EDITION

Graphics and GUIs
with
MATLAB®

THIRD EDITION

PATRICK MARCHAND

NVIDIA

O. THOMAS HOLLAND

The Naval Surface Warfare Center Dahlgren Division

CHAPMAN & HALL/CRC

A CRC Press Company
Boca Raton   London   New York   Washington, D.C.

Library of Congress Cataloging-in-Publication Data

Marchand, Patrick.
    Graphics and GUIs with MATLAB / by Patrick Marchand and O. Thomas Holland.—
3rd ed.
        p. cm.
    Includes bibliographical references and index.
    ISBN 1-58488-320-0
    1. Computer graphics. 2. Graphical user interfaces (Computer systems) 3. MATLAB.
I. Holland, O. Thomas. II. Title.

T385 .M3634 2002
006.6'6—dc21   2002034769

Visit the CRC Press Web site at www.crcpress.com

© 2003 by Chapman & Hall/CRC

No claim to original U.S. Government works
International Standard Book Number 1-58488-320-0
Library of Congress Card Number 2002034769
Printed in the United States of America 1 2 3 4 5 6 7 8 9 0
Printed on acid-free paper

PREFACE

First, I must say that it was quite an honor to be asked to update Patrick's seminal work. The original "Graphics and GUIs with MATLAB" was my introduction to the graphics capabilities of MATLAB®, and it was by that book that I came to a working knowledge of handle graphics. That was way back with MATLAB 4. Now we are at MATLAB 6 (release 13 is in beta release at the time of this writing) and MATLAB is more capable, powerful, and user friendly than ever – a far cry from MATLAB 4!

As with Patrick's earlier text, this book is intended to present a comprehensive discussion of the MATLAB graphics system. This third edition builds on the earlier editions by including the objects and properties new to MATLAB version 6 and includes the new features of the MATLAB environment. The organization of this edition is a little different as well. In teaching MATLAB, I have observed that not everyone wants to be a handle graphics guru (but they don't know what they are missing!). Many just want to be able to plot their data quickly and effectively. MATLAB has addressed this desire by expanding, for instance, the Figure Window tools, and providing the more casual user with a tool to modify many figure properties. Consequently, most of the first half of this book requires little or no knowledge of handle graphics. The second half thoroughly covers the concept of handle graphics, and how to create graphical user interfaces.

As with the earlier editions, this book has been written to be useful to anyone, regardless of their level of expertise with MATLAB. If you know nothing about MATLAB programming, you can learn much by starting at the beginning and working through the examples in this book. If you are already conversant with the MATLAB programming language, you will find a great deal of information here that is not readily apparent in the MATLAB documentation. However, I must point out that MATLAB's documentation has improved with the product and you are encouraged to delve into the documentation – but be aware, there is a lot of it!

The folks at the MathWorks continue to improve MATLAB, and its capabilities have grown well beyond the scope of a single text. New objects have been created for the latest versions, and the integrated development environment is more capable and customizable than ever.

MATLAB is a registered trademark of
The MathWorks Inc.
For production information, please contact:
The MathWorks, Inc.
3 Apple Hill Drive
Natick, MA 01760-2098, USA
Phone: (508) 647-7000
Email: info@mathworks.com
www.mathworks.com

The code in this text is written with version 6 in mind, so some of it will not work with earlier versions. The code has been written for clarity, not necessarily efficiency, and the functions kept as simple as possible so that you can focus on the graphics aspects. You can download any of the code in this text by going to

http://www.infinityassociates.com

and following the links for this book. You will also find the solutions to the end of chapter exercises.

If you are familiar with the earlier editions, you will see some familiar things here. Good is good and Patrick did such a fine job that much of what he presented then is still quite applicable and educational. Some things have been dropped, such as Patrick's GUI builder. In many respects, modern MATLAB doesn't need that any more. There are many new ideas in this edition, along of course with the new functions, features, and properties that the latest MATLAB has to offer. Perhaps one of the best aspects of MATLAB is that The MathWorks has continued to develop the product with very little compatibility problems. Although there were some major changes from version 4 to version 5 that led to a few problems for some extreme users, the transition from 5 to 6 has been smooth and has yielded a version that is more user friendly and more capable than ever.

So why am I writing this preface instead of Patrick? Patrick is very busy with new challenges in his career that have taken him a little out of the MATLAB world, at least as a regular user. My little consulting company, Infinity Technology Associates, has used his text for some time as a complement to our teachings, and I use MATLAB extensively for modeling and analysis in my position as director of a modeling and simulation facility for the Department of the Navy. Through one turn or another, I was contacted about a follow-up to Patrick's second edition. It has been exciting updating and expanding Patrick's original work and I know you will find this book a valuable tutorial and resource as you grow in your knowledge and skill of programming MATLAB Graphics and GUIs. However, don't stop with this book. Use MATLAB as much as you can, read the abundance of documentation that comes with MATLAB, and by all means experiment. Soon you will wonder how you ever got your work done without MATLAB.

Best wishes and happy programming!

Thomas Holland

DEDICATION

This book is first dedicated to the Creator, who has made us with inquisitive minds able to fathom the mysteries of the universe.

And secondly to Linda, Katy, and Danny, the best family I could ever have.

I also would like to acknowledge the influence of Wendy Martinez and Ronald Gross – two GUI gurus with whom it is a pleasure to work with, and of course, Patrick Marchand, who started it all.

CONTENTS

1 INTRODUCTION   1

1.1 OVERVIEW 1
1.2 ORGANIZATION OF THIS BOOK 2
1.3 TERMINOLOGY AND THE MATLAB PROGRAMMING LANGUAGE 4
    1.3.1 Getting Started 5
    1.3.2 Getting Help 6
1.4 OTHER REFERENCES 9

2 VISUALIZATION CONSIDERATIONS   11

2.1 WHY VISUALIZE? 11
2.2 CHARACTERISTICS OF GOOD DATA VISUALIZATION 12
2.3 DATA QUANTITY AND DIMENSION 13
2.4 COLOR, LIGHT, AND SHADING 14
2.5 MOTION 17
2.6 INTERACTION 17

3 PLOTTING IN TWO DIMENSIONS   19

3.1 SOURCES OF DATA 19
    3.1.1 Function Data 19
    3.1.2 Measured Data 20
3.2 IMPORTING DATA 21
    3.2.1 MATLAB Data Formats 21
    3.2.2 Importing High-Level Data 22
    3.2.3 Importing Low-Level Data 24
3.3 ELEMENTARY 2-D PLOTS 25
    3.3.1 A General Overview of the Plot Command 26
    3.3.2 Logarithmic Plots 35
3.4 SIMPLE 2-D PLOT MANIPULATION 37
    3.4.1 Generating Plots with Multiple Data Sets 37
    3.4.2 Using Axis to Customize Plots 41
    3.4.3 Creating Supporting Text and Legends 51
    3.4.4 Text Placement 57
    3.4.5 Special Text Character Formats 60
    3.4.6 Using Subplot to Create Multiple Axes 63
3.5 SPECIALIZED 2-D PLOTTING 67
    3.5.1 Bar Graphs 67
    3.5.2 Histograms 72
    3.5.3 Stairstep Graphs 74
    3.5.4 Stem Plots 75
    3.5.5 Plots with Error Bars 77
    3.5.6 Pie Charts 78
    3.5.7 Area Plots 83
    3.5.8 Working with Complex Data 84

3.5.9   Using the Polar Coordinate System 87
3.5.10   Plotting Functions with MATLAB 91
3.5.11   Creating Filled Plots and Shapes 93
3.6   PLOT EDITING IN THE MATLAB FIGURE WINDOW 95
3.6.1   Plot Editing Mode 96
3.6.2   The Property Editor 97
3.6.3   Zooming and Rotating 98
3.6.4   Exporting, Copying, and Pasting 99
3.7   ILLUSTRATIVE PROBLEMS 100

4 PLOTTING IN THREE DIMENSIONS   101

4.1   ELEMENTARY 3-D PLOTTING 101
4.1.1   Using Plot3 101
4.1.2   Creating 3-D Meshes and Surfaces 104
4.1.3   Waterfall Plots 109
4.1.4   3-D Plots of Non-Uniformly Sampled Data 110
4.1.5   Creating Shaded Surface Plots 112
4.1.6   Removing Hidden Lines 113
4.1.7   Contour Plots 115
4.1.8   Quiver Plots 121
4.1.9   Combination Plots 122
4.1.10   3-D Stem Plots 127
4.1.11   Generating Surfaces with Triangles 129
4.1.12   Polygons in a 3-D Space 131
4.1.13   Built-In Surface Functions 132
4.2   SIMPLE 3-D PLOT MANIPULATION 136
4.2.1   The Camera Toolbar 136
4.2.2   Generalizing the Axis for 3 Dimensions 138
4.2.3   3-D Plot Rotation 140
4.2.4   Using the View Command 142
4.3   VOLUME VISUALIZATION 145
4.3.1   Scalar Volume Data 145
4.3.1.1   Slice Planes 147
4.3.1.2   Contour Slices 149
4.3.1.3   Isosurfaces and Isocaps 151
4.3.2   Vector Volume Data 153
4.3.2.1   Stream Plots 156
4.3.2.2   Stream Lines 157
4.3.2.3   Stream Particles 159
4.3.2.4   Stream Ribbons 160
4.3.2.5   Stream Tubes 161
4.3.2.6   Cone Plots 162
4.4   A WORD ABOUT ANNOTATING 3-D PLOTS 164
4.5   ILLUSTRATIVE PROBLEMS 165

5 IMAGE GRAPHICS   167

5.1   IMAGE FILES AND FORMATS 167

5.1.1   Common Image File Types...168
5.2   IMAGE I/O...170
5.2.1   Reading a Graphics Image 172
5.2.2   Displaying a Graphics Image 173
5.2.3   Writing a Graphics Image...175
5.3   IMAGE TYPES AND PROPERTIES 176
5.3.1   Indexed Images 176
5.3.2   Intensity Level Images..178
5.3.3   Truecolor Images 181

6 GENERATING OUTPUT   183

6.1   THE QUICKEST WAY TO PAPER...183
6.1.1   Page Setup 184
6.2   PRINTING COLORED LINES TO BLACK & WHITE PRINTERS 185
6.3   ELECTRONIC OUTPUT...186
6.3.1   Using File Export 186
6.3.2   Using the Windows Clipboard 186
6.4   USING THE PRINT COMMAND...187
6.4.1   Creating Hardcopy with PRINT...187
6.4.2   Creating Graphics Files Using Print...................................187
6.4.3   Adding Additional Figures to a File 188
6.4.4   Publishing Using 4-Color Separation.................................188
6.4.5   EPS with a Preview Image 188
6.4.6   Rendering Method with -zbuffer or -painters.....................189
6.4.7   Indicating Which Figure Window to Print 189
6.4.8   Saving Figures for Future Use 190

7 HANDLE GRAPHICS   191

7.1   GRAPHICS OBJECTS 191
7.2   GRAPHICS OBJECTS HIERARCHY..193
7.3   GRAPHICS OBJECTS HANDLES 198
7.3.1   Determining Handles at Creation......................................199
7.3.2   Getting Handles of Current Objects...................................200
7.4   PROPERTIES...202
7.4.1   The Property Editor 203
7.4.2   Manipulating Properties...204
7.4.3   Universal Object Properties 206
7.4.3.1   ButtonDownFcn, BusyAction, and Interruptible 207
7.4.3.2   Children and Parent...208
7.4.3.3   Clipping...208
7.4.3.4   CreateFCN and DeleteFCN 210
7.4.3.5   HandleVisibility 211
7.4.3.6   HitTest...212
7.4.3.7   Selected and SelectionHighlight 212
7.4.3.8   Tag and Type 213
7.4.3.9   UserData...214
7.4.3.10   Visible...214

7.5   OBJECT SPECIFIC PROPERTIES 214
   7.5.1   Root Properties 215
      7.5.1.1   Display Related Root Properties 216
      7.5.1.2   Root Properties Related to the State of MATLAB 217
      7.5.1.3   Behavior Related Properties of the Root 219
   7.5.2   Figure Properties 220
      7.5.2.1   Figure Properties Affecting Position 222
      7.5.2.2   Style and Appearance Properties of the Figure Object 223
      7.5.2.3   Figure Properties that Control the Colormap 224
      7.5.2.4   Figure Properties that Affect Transparency 225
      7.5.2.5   Properties that Affect How Figures are Rendered 225
      7.5.2.6   Properties Related to the Current State of a Figure 226
      7.5.2.7   Figure Properties that Affect the Pointer 229
      7.5.2.8   Figure Properties that Affect Callback Execution 230
      7.5.2.9   Figure Properties that Control Access to Objects 234
      7.5.2.10 Figure Properties that Affect Printing 235
   7.5.3   Axes Properties 236
      7.5.3.1   Axes Properties Controlling Boxes and Tick Marks 238
      7.5.3.2   Properties Affecting Axes Character Formats 245
      7.5.3.3   Axes Properties Determining Axis Location and Position ...245
      7.5.3.4   Axes Properties Affecting Grids, Lines, and Color 248
      7.5.3.5   Properties Affecting Axis Limits 256
      7.5.3.6   Axes Properties Related to Viewing Perspective 265
   7.5.4   Line Properties 266
   7.5.5   Rectangle Properties 272
   7.5.6   Patch Properties 273
      7.5.6.1   Properties Defining Patch Objects 275
      7.5.6.2   Properties Specifying Lines, Color, and Markers 277
      7.5.6.3   Properties Affecting Lighting and Transparency 280
   7.5.7   Surface Properties 281
   7.5.8   Image Properties 287
   7.5.9   Text Properties 289
7.6   SETTING DEFAULT PROPERTIES 295
7.7   UNDOCUMENTED PROPERTIES 296
7.8   USING FINDOBJ 297
7.9   ILLUSTRATIVE PROBLEMS 300

8 USING COLOR, LIGHT, AND TRANSPARENCY   301

8.1   SIMPLE COLOR SPECIFICATIONS 301
8.2   COLOR MAPS 301
   8.2.1   Effects of Color Maps in General 304
   8.2.2   Color Axis Control 305
      8.2.2.1   Color Control with Direct Mapping 305
      8.2.2.2   Color Control with Scaled Mapping 306
   8.2.3   Color Maps as they Relate to Graphics Objects 307
      8.2.3.1   Color Maps and the Surface Object 307
      8.2.3.2   Patch Objects and the Color Map 313
      8.2.3.3   Images and the Color Map 315
   8.2.4   Color Shading 319

8.2.5   Brightening and Darkening Color Maps 319
8.2.6   Spinning the Color Map...322
8.2.7   Making Use of the Invisible Color with NaN 323
8.2.8   Creating Simple Color Bars 328
8.2.9   The Pseudocolor Plot 329
8.2.10   Texture Mapping...334
8.3   MODELING OBJECT LIGHTING...338
8.3.1   Light Properties 338
8.3.2   Functions that Make Use of Light................................339
8.3.2.1   Lighting Commands 343
8.3.3   Lighting Models...344
8.3.3.1   The Diffuse Lighting Model....................................344
8.3.3.2   The Ambient Lighting Model 346
8.3.3.3   The Specular Lighting Model 347
8.3.3.4   Combining Lighting Models....................................349
8.3.3.5   A Final Word on Light Objects 350
8.3.4   Creating Color Varying Lines with Surface Objects.................350
8.4   OBJECT TRANSPARENCY 352
8.4.1   Alpha Properties..352
8.4.1.1   AlphaData 353
8.4.1.2   Alphamap 353
8.4.1.3   ALim..353
8.4.1.4   ALimMode 353
8.4.1.5   AlphaDataMapping..354
8.4.1.6   FaceAlpha 354
8.4.1.7   EdgeAlpha 354
8.4.1.8   FaceVertexAlphaData...354
8.4.2   Alpha Functions...355
8.4.2.1   alpha 355
8.4.2.2   alphamap...355
8.4.2.3   alim 356
8.4.3   Setting a Single Transparency Value 357
8.4.4   Mapping Data to Transparency 357
8.5   ILLUSTRATIVE PROBLEMS..359

9 ANIMATION   361

9.1   FRAME-BY-FRAME CAPTURE AND PLAYBACK...................................361
9.1.1   Taking a Snapshot..363
9.1.2   Playing a Movie 366
9.1.3   Preallocating Memory 367
9.1.4   Practically Speaking..368
9.1.4.1   Recording the Entire Figure....................................368
9.1.4.2   Animating a Portion of the Figure 369
9.1.5   Making an AVI Movie..371
9.2   ON-THE-FLY GRAPHICS OBJECT MANIPULATION 372
9.2.1   Simple Animation Functions 372
9.2.2   The Wrong and Right Way to Animate Graphics.....................373
9.2.3   The Need for Speed 376
9.2.4   Animating Lines 376
9.2.5   Animated Rotations...377

9.2.6   Forcing a Graphic to Leave a Trail 382

9.3   CHOOSING THE RIGHT TECHNIQUE 383

10 ELEMENTS OF GUI DESIGN   385

10.1   WHAT IS A MATLAB GRAPHICAL USER INTERFACE? 385

10.2   THE THREE PHASES OF INTERFACE DESIGN 386

10.2.1   Analysis 387

10.2.2   Design 387

10.2.2.1 User Considerations 387

10.2.2.2 The Reason for the GUI 387

10.2.2.3 Cognitive Considerations 388

10.2.2.4 Physical Considerations 389

10.2.3   Paper Prototyping 389

10.2.3.1 Appearance 389

10.2.4   Construction 390

10.3   UI CONTROL ELEMENTS 391

10.3.1   The Styles 391

10.3.1.1 Check Boxes 391

10.3.1.2 Editable Text 392

10.3.1.3 Frames 393

10.3.1.4 Pop-Up Menus 394

10.3.1.5 List Boxes 395

10.3.1.6 Push Buttons 395

10.3.1.7 Toggle Buttons 396

10.3.1.8 Radio Buttons 396

10.3.1.9 Sliders 397

10.3.1.10 Static Text 398

10.3.2   UI Control Properties 398

10.3.2.1 Uicontrol BackgroundColor 400

10.3.2.2 Uicontrol ButtonDownFcn 400

10.3.2.3 Uicontrol CData 400

10.3.2.4 Uicontrol CallBack 400

10.3.2.5 Uicontrol Enable 401

10.3.2.6 Uicontrol Extent 402

10.3.2.7 Uicontrol ForegroundColor 402

10.3.2.8 Uicontrol Font Angle, Name, Size, Units, and Weight 402

10.3.2.9 Uicontrol HorizontalAlignment 403

10.3.2.10 Uicontrol Min, Max, and Value 404

10.3.2.11 Uicontrol SliderStep 404

10.3.2.12 Uicontrol TooltipString 405

10.3.2.13 Uicontrol Position 405

10.3.2.14 Uicontrol String 406

10.3.2.15 Style 406

10.3.2.16 ListBoxTop 407

10.3.2.17 Uicontrol Units 408

10.3.2.18 Uicontrol Interruptible 408
10.3.2.19 Uicontrol Tag 408
10.3.2.20 Uicontrol UserData 408
10.3.2.21 Uicontrol Visible 409
10.3.2.22 Other UI Control Properties 409
10.3.3 Creating Uicontrol Objects 409
10.3.3.1 Uicontrol Object Layering 410
10.3.3.2 Framing Objects 411
10.3.3.3 A Stretchable GUI 412
10.3.3.4 Predefined GUIs and Dialog Boxes 414
10.4 UIMENU ELEMENTS 421
10.4.1 Uimenu Properties 422
10.4.1.1 Uimenu Accelerator 423
10.4.1.2 Uimenu CallBack 424
10.4.1.3 Uimenu Checked 424
10.4.1.4 Uimenu Children 425
10.4.1.5 Uimenu Enable 426
10.4.1.6 Uimenu ForegroundColor 426
10.4.1.7 Uimenu Label 426
10.4.1.8 Uimenu Position 427
10.4.1.9 Uimenu Separator 427
10.4.1.10 Uimenu Interruptible 428
10.4.1.11 Uimenu Tag 428
10.4.1.12 Uimenu UserData 428
10.4.1.13 Uimenu Visible 429
10.4.1.14 Other Uimenu Properties 429
10.4.2 Creating Uimenus 429
10.4.2.1 Top Level Uimenu 429
10.4.2.2 Menu Items and Submenu Titles 430
10.4.2.3 Summary 431
10.5 LOW-LEVEL MATLAB GUI PROGRAMMING TECHNIQUES 433
10.5.1 Strings of MATLAB Statements and Expressions 433
10.5.2 Programming Approaches in MATLAB 435
10.5.2.1 Creating All Graphics Elements in the Base Workspace 436
10.5.2.2 Storing Handles as Global Variables 441
10.5.2.3 Storing Handles in the UserData Properties 445
10.5.2.4 Utilizing Tags and the FINDOBJ Command 448
10.6 HIGH-LEVEL GUI DEVELOPMENT – GUIDE 450
10.6.1 The Layout Editor 451
10.6.2 The Property Inspector 452
10.6.3 The Object Browser 454
10.6.4 The Menu Editor 455
10.6.5 Saving the GUI 455
10.6.5.1 The GUIDE Created FIG-File 455
10.6.5.2 The GUIDE Created M-File 456
10.6.6 Executing a GUI 459
10.6.7 Editing a Previously Created GUI 460
10.7 COMMON PROGRAMMING DESIRES WITH UI OBJECTS 461
10.7.1 Creating Exclusive Radio Buttons 462
10.7.2 Linking Sliders and Editable Text Objects 464
10.7.3 Editable Text and Pop-Up Menu 466

10.7.4   Windowed Frame and Interruptions 468

10.7.5   Toggling Menu Labels 471

10.7.6   Customizing a Button with Graphics 472

10.8   THE MATLAB EVENT QUEUE 474

10.8.1   Event Scheduling and Execution 474

10.8.2   Execution Order of Events 475

10.8.2.1 Mouse Button Pressed Down 476

10.8.2.2 Mouse Button Released 477

10.8.2.3 Mouse Pointer Moved 477

10.8.3   Interruptible vs. Uninterruptible 478

10.8.4   Common Mouse Action Examples 479

10.8.4.1 Moving Objects with the Mouse 479

10.8.4.2 Dynamic Boxes Using the RBBOX Function 483

10.9   CREATING CUSTOM USER INTERFACE COMPONENTS 484

10.9.1   Simulating Buttons with Image Objects 485

10.9.2   Creating a Dial 489

APPENDIX : QUICK REFERENCES   493

INDEX   513

1 INTRODUCTION

IN THIS CHAPTER...
1.1 OVERVIEW 1
1.2 ORGANIZATION OF THIS BOOK 2
1.3 TERMINOLOGY AND THE MATLAB PROGRAMMING LANGUAGE 4
1.4 OTHER REFERENCES 9

1.1 Overview

As the volume and complexity of data and results continues to grow with the increasing complexity of data sources and algorithms, the need for intuitive representations of that data and results becomes increasingly critical. The graphical representation of the results is often not only the most effective means of conveying the points of the study or work which has provided the data, but is in most cases an expectation of the audience of the work. Even as computing hardware continues to increase in capability, MATLAB® continues to be one of the best applications available for providing both the computational capabilities of generating data and displaying it in a variety of graphical representations. With the advent of version 6, MATLAB has taken on a new look, a new integrated development environment (IDE), new graphics development tools, and introduces some new functions. It is in that light that we offer the "upgraded" version of this book.

Welcome to the third edition of Graphics and GUIs with MATLAB! Those of you familiar with the first and second editions will find that this third edition carries on in the same tradition of conversational style that Patrick set forth in the first two editions, as well as illustrative examples, and some details that give you a peak under the hood of MATLAB. But just as MATLAB version 6 has introduced major changes in several areas, so has this third edition. In addition to the new MATLAB specific commands and techniques, this edition offers sections on Visualization Considerations and Elements of GUI Design, which are general treatments applicable to any development software. Those familiar with the earlier editions will also be happy to find that there are now problem sets at the end of some chapters that will (hopefully) motivate the new MATLAB programmer to exercise the techniques addressed earlier in the chapter and make this book more suitable to classroom settings. But just so that you don't become too frustrated, solutions for the problems, as well as code listings for most of the examples, are available at

www.infinityassociates.com/graphics_and_guis

MATLAB is not just a computation and plotting package; it is a versatile and flexible tool which allows users with even the most elementary programming capabilities to produce sophisticated graphics and graphical user interfaces

(GUIs). The level of sophistication is only limited by one's needs, curiosity, and imagination.

As in the previous editions, it is the goal of this book to provide you with information, examples, and techniques which should give you the background you need to become a MATLAB graphics and GUI expert. If you are already conversant with the MATLAB programming language, this book will provide you a ready reference with illustrative examples. If you are new to MATLAB, you will find this book an excellent tutorial leading you to MATLAB proficiency. As in the previous editions, this book will help take you from wherever you are in your MATLAB skills, to many steps closer to where you want to be.

1.2   Organization of This Book

This book is organized into three general parts: Part 1: Information Visualization, Part 2: MATLAB Graphics Objects, and Part 3: Graphical User Interfaces. Each part is intended to provide the reader with a general introduction to the topic area before going into specific topics in MATLAB. For instance, if your main interest is in the visualization of data, the part on Information Visualization will give you a rudimentary introduction to the topic. Similarly, the part on Graphical User Interfaces will provide you with a good background useful in any programming language. Taken as a whole, the three parts will introduce you to the greater field of information visualization and GUI design in general, and with MATLAB specifically.

Part 1: Information Visualization will introduce you to visualization considerations such as when to use 2-D and 3-D techniques, the advantages and pitfalls of color, how motion can add another dimension of understanding, and how dynamic interaction with a visualization can enhance intuitive understanding. Contained in this part are the elementary aspects of plotting in two and three dimensions; MATLAB's graphics commands are discussed and applied in illustrative examples. Plot manipulation and special plots are explored, including volumetric visualization for both scalar and vector volume data. Reading, writing, and manipulating bitmap graphics is covered in this section as well as printing, exporting, and saving your MATLAB visualizations.

Part 2: MATLAB Graphics Objects thoroughly explores the concept of graphics objects by introducing the fundamentals of MATLAB's Handle Graphics™. If you consider yourself somewhat experienced with the basic plotting capabilities of MATLAB, you might well want to start with this chapter. A basic understanding of Handle Graphics needs to be achieved before you can move on to more complex and sophisticated programming of graphics and GUI applications. The first chapter in this section explores graphics objects, handles to them, properties and ways to change the values of properties. The next two chapters explore the details of powerful dimensions that can enhance the understanding of your data, specifically the properties of color, light, transparency, and animation.   Once you have grasped the concepts here you can then appreciate the power of MATLAB and will be fully equipped to comprehend the programming techniques to follow in Part 3.

Part 3: Graphical User Interfaces will bring together all you have learned by summarizing practical considerations for good GUI design. The three phases of interface design, user, and appearance considerations are covered first, followed by thorough coverage of the MATLAB Graphical User Interface Design Environment (GUIDE). Finally, user interface control elements, user actions, and the MATLAB event queue are covered so that you will be able to create GUIs that go beyond the boundaries of GUIDE.

The intent of the overall structure of this book is to lead any MATLAB programmer through a wide variety of graphics related subjects. The information, examples, and tutorials are designed to illustrate different techniques of creating graphics. These techniques can be expanded and tailored to meet your individual needs and desires.

In addition to the topic descriptions, many of the chapters contain icons in the margins to help quickly lead you to the information you need. The icons and their significance are as follows:

Speedy Solutions for those who are in a rush and don't have the time for the details.

Power Tips will especially add to your MATLAB knowledge to make you a stronger programmer. Hopefully the whole book falls under this icon, but there are some special tips that particularly enhance your abilities.

Tools describe what we especially feel are outstanding methods, techniques, and MATLAB programs that accomplish a specific job and make your life easier. The programs that get this icon are very useful and complement the standard set of programs that come with MATLAB. These include public domain M-files available from the MathWorks FTP server (ftp.mathworks.com) as well as files found at

www.infinityassociates.com/graphics_and_guis

M-file indicates that a nonstandard function is to be developed. The discussion that follows will teach you how the MATLAB code accomplishes a certain task. For your convenience, some of these M-files will be downloadable from the above website; however, we do recommend that you study the code – after all, it is there for your edification.

FAQ directs your attention to the answer to a frequently asked question about the current topic. Many of these questions come from newsgroups and classroom discussions.

Warning will call your attention to typical pitfalls.

Other visual cues will help you get around in the book. MATLAB function names that appear in the discussions will be in bold. MATLAB code examples, fragments, and listings are throughout the book and can be recognized readily by the distinctive courier font in which they are cast. For example, plotting the sine of a vector x that ranges from -2π to $+2\pi$ in 0.1 increments with the **sin** function looks like this:

```
x = [-2*pi:.1:2*pi];
y = sin(x);
plot(y)
```

This same style is used for "general forms" of MATLAB functions and commands as well; however, the general forms will use all capital letters just like in the MATLAB command line help. For example, the general form of the **view** function when using it to return the current viewing perspective is given by:

```
[AZ,EL] = VIEW
```

Finally, even after you become quite familiar with MATLAB graphics and GUIs, there are always going to be "problems" or situations that require additional thought to determine how to best accomplish a task. For just such occasions, as in the earlier editions, we have compiled a "Quick Reference" in the Appendix . These sheets provide a summarized list of helpful hints that will help ease and hopefully speed up your development process. Many of these hints have come about through our own development, consulting, and teaching experience. Included with each hint is a reference to the applicable sections of the book that provide further explanation on the topic or related topics.

1.3   Terminology and the MATLAB Programming Language

If you are new to MATLAB, it would be wise to familiarize yourself with some basic terminology and concepts. We recommend that you review the documentation included with MATLAB. A good place to start is with the "Getting Started" section of the MATLAB Help. If you have just upgraded to version 6, you will want to get familiar with the new MATLAB Desktop and the tools that make it up. When you start your MATLAB the Desktop is the first thing you see. In it you will see windows with names like "Workspace," "Command History," "Command Window," and "Launch Pad." This desktop can be configured in different ways, in essence customizing it to the way you like to work. Once you have installed and started your MATLAB, simply click on Help and then select "Full Product Family Help." From there, click on the folder entitled "MATLAB" then click on "Getting Started." The Mathworks has done a fine job of constructing a very extensive set of hyper-linked documents that allow you to get both fast answers and detailed discussions. Be sure to familiarize yourself with the MATLAB workspace, directory structure, and file types. You should understand what the MATLAB search path is and how you can add and remove directories from this search path. You should also know that in this book you will be working primarily with the MATLAB file types M-files, FIG-files, and MAT-files. The final assumption we must make is that you know what we mean by the "Command Window" and the "Figure Window." The Command Window is where you can enter commands directly to

MATLAB. The Figure Window is where you display graphics and GUIs by issuing the appropriate commands in the Command Window.

1.3.1 Getting Started

The MATLAB Desktop is what results when you invoke MATLAB on your computer and provides a convenient and easily configurable interface to the various tools that make up the development environment. Depending on how you have set preferences for your specific installation of MATLAB, it should look something like that shown in Figure1.1.

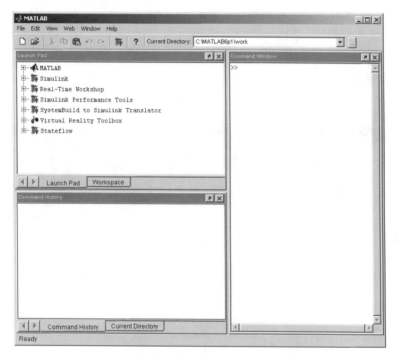

Figure1.1   The MATLAB desktop.

In order to begin, we must assume that you have already gained some familiarity with the MATLAB development environment. Of course the portion of the MATLAB desktop with which you should be most familiar is the *Command Window* as this is where you will issue commands directly to MATLAB. Specifically, you type the MATLAB statements at the Command Window prompt which is denoted by **>>** . Generally we will refer to this as the "command prompt." A few other items with which you will want to become familiar are: the *Command History* where all the commands entered in the Command Window are recorded, the MATLAB *Search Path* and how you can add and remove folders from this search path, and the three MATLAB file types that we will be mainly working with M-files, FIG-files, and MAT-files. These file types derive their names from the file extensions. We will avoid other MATLAB file types such as MEX-Files and P-Files. You will also want to become familiar with the MATLAB Figure Window as this is where you display graphics and GUIs and the MATLAB Editor/Debugger where you will create scripts and functions.

1.3.2   Getting Help

This section is intended to get you pointed in the right direction in order to familiarize you with the MATLAB environment in terms of the directory structure, the file types, and the various windows that are available to you. If you are new to MATLAB it would be wise to familiarize yourself with some basic terminology and concepts. We recommend that you review the documentation that is included on the MATLAB Documentation CD. A good place to start is with the "Getting Started" section of the MATLAB Help. You can do this easily. After starting MATLAB, just select **Help → MATLAB Help** from the pull-down menu in the Desktop. When you do, you will see something like that shown in Figure1.2.

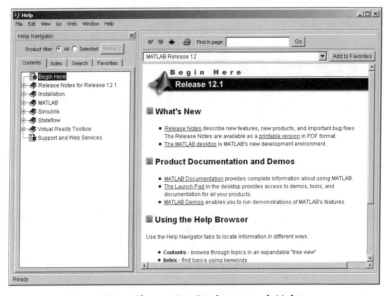

Figure1.2   Getting started: Help

From here you can dig as deeply as you wish into the many aspects of MATLAB.

The commands which are issued to MATLAB can be either from the original set of functions that came with your MATLAB package or the ones that you develop in the form of M-file scripts or functions. These are text files with a ".m" extension. Throughout this book a MATLAB "program" or "function" refers to a function M-file, that is MATLAB code that has the keyword "function" in its first line. We will use the term "script" to refer to an M-file that is simply a stored list of commands. Although the differences between M-file functions and scripts are profound, we will assume that you already understand those differences. If you don't, or would like to review M-file scripts and functions, we again refer you to the documentation that came with your MATLAB package.

One of the nice aspects about the MATLAB language is that it can be expanded by writing new functions and scripts. Moreover, any new M-file can be supplemented with on-line help. (By on-line we are not referring to the internet, but to the help available from the command prompt in the Command Window.) The on-line help feature and hypertext documentation are both useful as quick references to built-in features of MATLAB, but on-line help is something that you can provide and build into your own M-files. It practically becomes a necessity when M-files are shared among MATLAB users. A well-documented function relieves a user from the responsibility of understanding every minute detail of a function's operation, and allows the programmer to obtain desired results by following the syntax or usage of the M-file.

The on-line help feature is activated by typing **help** *filename*, where *filename* is the name of the M-file whose help contents you wish to have listed (e.g. >> help plot). Here is an example you can try in the Command Window.

```
>> help tic

TIC Start a stopwatch timer.
    The sequence of commands TIC, operation, TOC
    prints the number of seconds required for the
    operation.

    See also TOC, CLOCK, ETIME, CPUTIME.
```

Notice that what happens when you invoke help for a function is that MATLAB returns in the Command Window the first block of contiguous commented lines starting from the second line of the M-file. (Try either the Editor/Debugger or the **type** command to see the contents of TIC.M.) When creating your own functions and help comments, keep in mind that the first comment line of the M-file should be as concise and descriptive as possible, since this is the line that will appear when one executes **help** *directory-name*, where *directory-name* is the name of the directory or folder containing M-files (as an example type: help graphics).

As you continue to expand your MATLAB vocabulary, the **help** command will be a very convenient alternative to the HTML help, the documentation CD, or other printed documentation that came with your software purchase. Sometimes all that is needed is a quick bit of information or reminder of the details, and the on-line help is perfect for that.

Several other commands that are convenient in helping you enrich your command vocabulary are **more**, **type**, and **demo**. The function **more** controls the number of lines which are displayed at a given time to the Command Window. You can turn it on by issuing the command **more** *on*, so that you can read the contents of the Command Window before the next page of output is displayed. Pressing the return key displays the next line of output while the space bar displays the next page. If you wish to stop paging through the output, just press the letter "q" on the keyboard. Issuing **more** *off* turns off the paging feature.

The contents of many MATLAB commands can be viewed by either using the **type** command or by opening up the file in MATLAB's Editor/Debugger. If you want to open a file with the Editor/Debugger simply type **edit** followed by filename at the command prompt. (You can also use the pull-down menus in the desktop.) If you do this, the file will be opened in the Editor/Debugger and you can make and save changes. (Type **edit factorial** to see an example of a simple function in the Editor/Debugger.) However, if all you want is to quickly see the contents of an M-file, the command **type** *filename* allows you to list the contents of a file (if no extension is provided, MATLAB assumes and searches for an M-file) in the Command Window. We believe that viewing MATLAB programs is perhaps the quickest way to learn how to program your own MATLAB code. Using type is just the quick way of viewing the source code. However, in some cases the source code of the command is not available in a text file and cannot be typed out. For example, issuing **type line** will return "line is a built-in function." This means that this command has been built into MATLAB itself for computational efficiency and speed.

The Command History window keeps a record of all the commands you type at the command prompt. You can also select commands there and drag them into the Command Window. However, sometimes (especially when you are learning) you will want to save a log of your commands. The **diary** *filename* command can be used to keep a running record of what was typed and displayed in the Command Window. This can be useful in program development for several reasons. The first reason is that occasionally you may clear the Command History window before realizing that you have forgotten which commands you issued and what order they must be executed in to achieve specific results; but if **diary** is on you will have a file record of the commands you used. Another advantage of using the **diary** command is that you can create a script or function fairly rapidly by editing the resulting diary file and saving it as an M-file.

You can be selective as to what goes to the diary file by switching between diary states. The commands **diary on** and **diary off**, respectively allow or prevent your typed commands and MATLAB output to be sent to the file. In addition you can switch between different diary files by reissuing the **diary** *filename* command, where *filename* is the name of a different or new diary file. Diary output is always appended to the file that you specify.

Finally, we encourage you to check out the MATLAB demo packages. When you do, you will get a flavor of some of what can be accomplished with MATLAB. At the very least, you may get a jump on a solution to a problem by remembering that one of the demos did something similar to what you would like to be able to do. Once you find that demo, you can step through the code and use the ideas or techniques for your own code. The MathWorks expects and encourages you to examine their M-files so that you can learn the language quicker and with less frustration. To get a list of demos that are available to you, type **help demos** at the command prompt. Depending on which version MATLAB and which Toolboxes you have installed on your system, additional demonstrations may be provided to illustrate specific package capabilities.

1.4   Other References

In general, you might find the following reference materials of use while going through this book:

- MATLAB's Using MATLAB (version 6)

- MATLAB's Using MATLAB Graphics (version 6)

Although we try to provide some guidelines and rules of thumb concerning good visualization and graphical user interface approaches, you might consider the following texts enlightening:

- *The Visual Display of Quantitative Information* by Edward Tufte

- *Envisioning Information* by Edward Tufte

- *Scientific Visualization and Graphics Simulation* by Daniel Thalmann

- *GUI Design Essentials* by Susan Weinshenk, Pamela Jamar, and Sarah C. Yeo

2 VISUALIZATION CONSIDERATIONS

IN THIS CHAPTER...
2.1 WHY VISUALIZE? 11
2.2 CHARACTERISTICS OF GOOD DATA VISUALIZATION 12
2.3 DATA QUANTITY AND DIMENSION..13
2.4 COLOR, LIGHT, AND SHADING 14
2.5 MOTION 17
2.6 INTERACTION...17

2.1 Why Visualize?

The obvious question that is at the heart of MATLAB graphics is "Why would I ever want to visualize my data?" At its very essence, science is the quest for truth. However, some of those truths are not easily discovered, and in many cases, we don't even know how to ask the appropriate questions that will lead to the truth. Consider the fact that many natural phenomena are too fast, too slow, too large, or too small to be studied through direct observation or with traditional laboratory techniques. How can we see the *unseen* or gain enough insight into the nature of things to even know what is worthwhile to investigate? Also, consider that everything humankind has made and every fact that has been discovered were first birthed as an idea, i.e., something with form only in the mind's eye. We revere those who can see beyond the apparent and call them "visionary." A dictionary will tell us that to visualize means, "To form a mental image or vision of." Therefore, when we discuss what it means to visualize something with MATLAB or with anything else for that matter, we need to be aware of the significant role the mind plays in this discussion. However, this is a book about MATLAB, not about philosophy, but it is important that you realize that what we are really exploring in this book are ways to represent *something*, whether that something is a graphical representation of a real-world object, a hypothetical mathematical construct, or specific values of some measurable quantities. Most importantly, we want to create those representations in such a way that the human mind can understand them and then ask the right questions that lead to the discovery of new things or to a better understanding of our universe.

So why do you need visualization? Aside from making your boss look good to his superiors, the visualization of your data can help you identify and emphasize areas of interest, such as where significant events occur, or where the data exhibits a curious behavior. It can also help you to convey your thoughts, observations, or conclusions to others in a quick and intuitive way. There are probably as many applications for MATLAB as there are users of it, and every application will have its own special needs, but even amongst all that, a little understanding of some *scientific visualization* fundamentals will help you achieve the results you desire. Table 2.1.1 lists some of the reasons to visualize your data.

Table 2.1.1 Reasons to Visualize Data

Explore it	Exploit it
Emphasize some aspect of it	Analyze it
Gain new insights into it	Assess or control the quality of it
See "the invisible"	Present it
Publish it	Interact with it

One can easily argue that the need to visualize data and information has, largely, driven our technology. The entire field of computer graphics, which includes hardware and software, is devoted to furthering the science of how we represent and interact with information in effective ways.

2.2   Characteristics of Good Data Visualization

MATLAB has established itself as a preeminent computing environment. By computing environment we mean that not only does MATLAB provide the user with quick access to many data processing functions, but also allows a MATLAB programmer to create special purpose applications to be used by "domain specialists." These domain specialists are often not interested in knowing the intricacies of MATLAB programming, but are very interested in having analytical tools that are intuitive to use and in which they can have confidence. Since you are reading this book, you either have a need for visualizing some of your own data, or you are involved in developing some form of graphical user interface, either for your own analytical efforts, or to support some domain specialists who really don't want to be programmers. In Chapter 10 we will discuss the essential elements of GUI design. Here we will consider how to better represent data and results so that the salient aspects of the information contained in the data can be readily observed.

In Table 2.1.1 we listed some reasons why you would want to visualize data. Remember, the basic reason for visualization is to help you, or those you work with, solve problems. Cognitive psychologists have demonstrated that the way in which a problem is presented can determine how difficult a problem is to solve, so we "re-present" the problem in more understandable, i.e., intuitive, ways and in doing so gain insight.

Good visualizations must be meaningful; every plotted point, and each colored line needs to help with the intuitive understanding. This leads to issues of perception, and since visualization in the scope of this book is visual, we are talking about visual perception. This idea of perception has to be distinguished according to two primary areas of intent: 1) the display and communication of data, and 2) the investigation and understanding of data. The direction with which you are approaching your problem will determine largely the manner in which you visualize your data. A good visualization should distill the vast quantity of data, or the difficult-to-understand concept, into quantities and terms that are readily understandable. It is by comparison to what we know that we discover what we don't know. It is much easier to see an anomalous spike in data when it is plotted, as opposed to looking at a list of numbers on a printout. The modern scientific world is not a simple world. We have developed the scientific tools we have in order to investigate and communicate in unambiguous terms. In this communication, we must strive for clarity, precision, and efficiency. Table 2.2.1 lists some characteristics of good data visualizations.*

Table 2.2.1 Characteristics of Good Visualizations

A Good Visualization Should...	Because...
Serve a clear purpose.	We are interested in describing, exploring, or recording something.
Show the data without distorting it.	The data is what is important, or more fundamentally, the truth the data reveals.
Cause the viewer to think about the substance of the data.	Understanding will be sacrificed if graphic design, or some other "flashy" mechanism draws attention away from the content.
Present large quantities of numbers in a small space.	We are often overwhelmed by many numbers; we need to make large data sets coherent.
Take advantage of the natural tendency to make visual comparisons.	It is easy for us to see relative differences.
Reveal information at various levels of detail.	It is easier to understand the bigger picture when the details are available to support it.

2.3 Data Quantity and Dimension

Advances in technology are allowing us to gather data at an ever-increasing rate. Microphones, video cameras, telescopes, satellites, radars, etc., work round the clock gathering more data about the universe around us. X-rays, ultrasound, computed tomography, magnetic resonance images, etc., are likewise gathering more data about the universe within us. From the immensity of the universe probed with radio telescopes, to the minutiae within molecules observed with electron microscopes, we are witnessing a massive flood of

---

* If you are interested in the field of information visualization, we highly recommend the works of Edward R. Tufte, in particular his seminal text, *The Visual Display of Quantitative Information*, ©1983, published by Graphics Press.

data such as has never before been seen in human history. In addition, to compound it all, doing so at a rate well beyond the human capacity to observe or understand it. The computing capabilities that thirty years ago were the sole domain of expensive computer installations at the defense department or in university laboratories are now well exceeded on the average homeowner's desktop. Therefore, whether you are dealing with data generated by your computer from pure mathematical formulations, or measured with a physical sensor, perhaps the two most important considerations in deciding how to represent your data are likely to be the quantity of it and its dimension.

The quantity of the data might require you to consider statistical methods to show trends or occurrences of interest relative to the data set. The dimension of your data might require something more than a simple x-y plot. You might need to consider 3-D plots, slices of 2-D data, or combinations of 2-D and 3-D plots to get the emphasis you need. In any event, the old axiom, "A picture is worth a thousand words," is a mere understatement in today's world.

2.4   Color, Light, and Shading

Color is probably the most commonly used, and abused, visualization technique. For instance, bright colors can be used to indicate that a particular item should be noted in a presentation, or to quickly draw your attention to points that exceed a threshold in a plot. One should always keep in mind that the intent of any visualization is to foster the communication of some idea, whether it be overall results or stressing a specific aspect of some analysis. In our brief discussion of color, we include lighting and shading as well.

In most simple visualizations, we can effectively use color to distinguish between different data series. This is most commonly seen as multiple lines of a plot where each line is a different color. Typically, in such simple plots, the two colors need only be distinguishable to clearly define the data. Unfortunately, plots are printed and copied and often not in color, then the advantages of color are lost. In such situations, it is good to denote each data series with a distinctive marker, or line style. Figure 2.1 and Figure 2.2 show similar plots of the same data. In Figure 2.1, the lines are plotted in blue and green, poor choices for black-and-white printing and possibly confusing even in color slide presentations where lighting is poor or someone has color blindness. Although different markers were used for each data series, they are not distinct enough to really help with the problem. Figure 2.2 is the better plot. Although the color might still be a problem in black-and-white printing, the line styles have been changed so that they are easily distinguishable, and a marker is used on only one data series. Figure 2.2 will convey the data better even when copied.

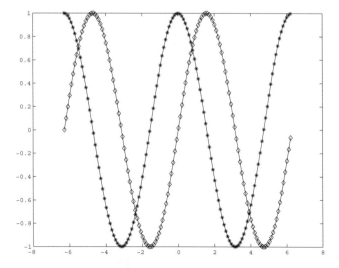

Figure 2.1 Although one trace is green and the other blue, this simple 2-line plot is difficult to read in low light or in black-and-white (as it is printed here).

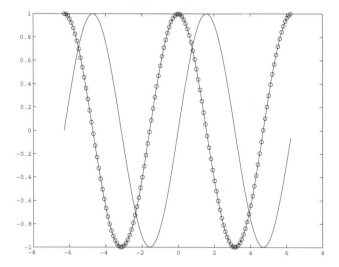

Figure 2.2   This 2-line plot is easier to read since two different line styles are used. It would look even better if it were in color.

Varying hue is good for displaying different types of objects in visualization, but in many numerical analysis cases, we are interested in ordinal, interval, or ratio data. Differences in hues do not necessarily imply differences in magnitude; it is not obvious that red has a higher value than blue. Differing brightness levels (or saturation levels of a certain color) can convey differing magnitudes, and a gradual change from one hue to another is effective in doing this. Realize, however, that not everyone perceives color in the same way. Color blindness is common, and color perception even changes with age. When you use color, be sure that the meaning of the color is unambiguous. Think of color as another dimension for representing information. Never use it to "pretty-up" a graph.

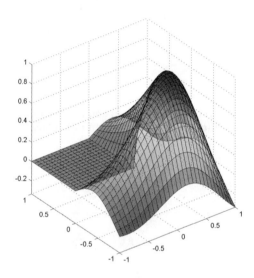

Figure 2.3   Hue and transparency in a 3-D plot .

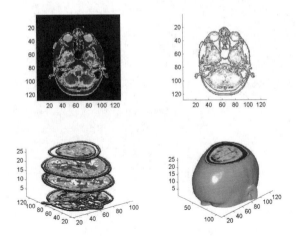

Figure 2.4 Visualizing MRI Data

2.5   Motion

Like color, motion is another representative dimension. Motion is used to represent changes over time, or to indicate sequential changes in higher dimensional data. Consider that a single observation that varies in time can be easily represented in a 2-D plot of the observation versus time. However, what happens if the measurements are 2-D themselves, such as a sequence of photographic images? This can be represented in a number of ways. Perhaps the most common is an image sequence, or frames. Like a movie, each image is a "slice in time." We will explore using motion in Chapter 9.

2.6   Interaction

The most useful data visualization methods allow the user to interact with the data by changing viewing angles, thresholding levels, applying false colors, and otherwise manipulating the presentation of the information content of the data dynamically. As you proceed through this book, you will see that MATLAB allows some simple dynamic manipulations through the Figure Window. More importantly, you will see that MATLAB provides you with a host of graphics functions that allow you to build your own custom visualizations with which you can interact to any degree you wish if you are willing to program them so. Truly, the only limit to visualization with MATLAB is your imagination!

To whet your appetite, start your MATLAB and type demo at the command prompt.

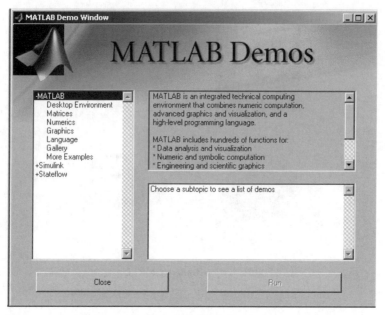

Figure 2.5   Play with the MATLAB Demos.

3 PLOTTING IN TWO DIMENSIONS

IN THIS CHAPTER...
3.1 SOURCES OF DATA 19
3.2 IMPORTING DATA..21
3.3 ELEMENTARY 2D PLOTS..25
3.4 SIMPLE 2D PLOT MANIPULATION 37
3.5 SPECIALIZED 2D PLOTTING 67
3.6 PLOT EDITING IN THE MATLAB FIGURE WINDOW......................95
3.7 ILLUSTRATIVE PROBLEMS...100

3.1   Sources of Data

What operations you perform on any given set of data as well as how you choose to visualize it are usually determined by the source of the data and by which aspects of it you wish to emphasize. In general, all the data you will ever work with will either be the result of some generating function, i.e., function data, or will be a measurement of some real-world property, i.e., measured data.

3.1.1 Function Data

Function data is data that is created by some mathematical operation. Its typical characteristics include: 1) data-uniformity, i.e., the data is not sparse or riddled with discontinuities, 2) free of corrupting noise, and 3) controllability, i.e., you can vary parameters, change algorithms, etc., and so re-create data in any form you desire. However, such ideal data rarely is representative of the real world, and in the case where generated data is intended to represent real-world phenomena a great deal of energy is expended in making generated data look like measured data. You can think of function data as any data which is the result of an algorithm, and in short you have complete control over the range, quantity, and values of the data. A simple example of function data is the mixing of two sinusoidal waves such as that described by the expression

$y(t) = \sin(20\pi t) + \sin(60\pi t)$

and shown in 0.

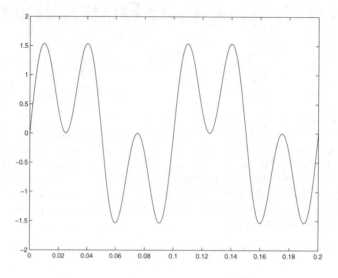

Figure 3.1 An example of function data.

3.1.2 Measured Data

Measured data results from some real-world sensing or probing. Examples of measured data include data such as daily temperature highs, g-force, velocity, etc.. The principal characteristics of measured data are: 1) measured data is only as accurate as the device making the measurement, 2) there is always some degree of uncertainty associated with the data, 3) data may take extreme excursions, and 4) measured data might be incomplete or have gaps. This last characteristic is a particularly interesting one in that it is more common than one might at first think. Consider daily temperature readings. It is common that readings do not exist for many days for a given year or in some cases, data might only be taken sporadically. (Either way, such data is called "sparse" and MATLAB provides a memory efficient means of dealing with such data.) However, it is up to you as the programmer, analyst, scientist, or engineer to determine how to deal with gaps in your data and how you choose to visualize such data is highly dependent on the intended use of it. Figure 3.2 shows a plot of average daily temperatures for the first sixty days of the years 1995 through 2000 for Anchorage, Alaska.[1] The 'o' data marker indicates the 6-year mean for that day. In this data a value of –99 indicates that no data is available, i.e., a data gap. You can see that on about the 10th day of one of the years no data was taken.

---

[1] Climatic data provided by the University of Dayton Average Daily Temperature Archive. Environmental Protection Agency Average Daily Temperature Archive, http://www.engr.udayton.edu/weather/, courtesy of J. K. Kissock.

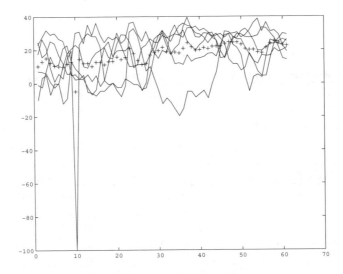

Figure 3.2 Example of measured data.

3.2   Importing Data

Whenever we are dealing with data in MATLAB, whether it is function generated, or measured, we are first faced with just how to bring that data into the MATLAB environment. Fortunately MATLAB provides a rich set of commands that support data input and output from many different standard formats. If you have a data file that was created using another application or program, the contents of that data file can be imported into the MATLAB workspace. Once you have imported the data, you can then manipulate or plot the data. However, before we consider data files from other applications, we should also understand how to import data saved during other MATLAB sessions. In many cases, you will be working with other MATLAB users and you will need to operate on their data.

3.2.1   MATLAB Data Formats

Modern MATLAB supports a broad range of standard data formats. The following tables list the data formats for which MATLAB provides built-in support and the associated import commands.

Data Formats	Command	Returns
MAT - MATLAB workspace	LOAD	Variables in file
CSV - Comma separated numbers	CSVREAD	Double array
TXT – Formatted data in a text file	TEXTREAD	Double array
DAT - Formatted text	IMPORTDATA	Double array
DLM - Delimited text	DLMREAD	Double array
TAB - Tab separated text	DLMREAD	Double array

Spreadsheet Formats	Command	Returns
XLS - Excel worksheet	XLSREAD	Double array and cell array
WK1 - Lotus 123 worksheet	WK1READ	Double array and cell array

Scientific Data Formats	Command	Returns
CDF - Common Data Format	CDFREAD	Cell array of CDF records
FITS - Flexible Image Transport System	FITSREAD	Primary or extension table data
HDF - Hierarchical Data Format	HDFREAD	HDF or HDF-EOS data set

Image Formats	Command	Returns
TIFF – Tagged image format	IMREAD	Truecolor, grayscale or indexed image(s)
PNG – Portable network graphics	IMREAD	Truecolor, grayscale or indexed image
HDF – Hierarchial data format	IMREAD	Truecolor or indexed image(s)
BMP – Windows bitmap	IMREAD	Truecolor or indexed image

Audio Formats	Command	Returns
AU – Next/Sun Sound	AUREAD	Sound data and sample rate
SND – Next/Sun Sound	AUREAD	Sound data and sample rate
WAV – Microsoft Wave Sound	WAVREAD	Sound data and sample rate

Movie Formats	Command	Returns
AVI - Movie	AVIREAD	MATLAB movie

3.2.2   Importing High-Level Data

The most straightforward method of importing data is to use the **load** command. The **load** command can read either binary files containing matrices generated by earlier MATLAB sessions (usually by use of the **save** command), or text files containing numeric data. If the data file was created in an earlier MATLAB session, simply issuing the **load** command with the filename is all that is needed. The **save** command will save the specified data in MATLAB's binary data format. The following example shows just how simple this can be.

```
>> save mydata X Y % mydata.mat created
:
:
>> load mydata % in a later session
```

The important points to remember in using **save** and **load** in this way is that MATLAB will by default attach the ".mat" extension to the data file and the file will be created or read from the current working directory.

As stated earlier, this use of the **save** and **load** commands uses the default MATLAB binary file format. Although many other applications are now being created that can read and write this format, **save** and **load** can be used to both write and read text data which can make importing and exporting data a simple matter. Either command could have been issued with the keyword –ASCII. If **save** was used with –ASCII, the data is automatically tab delimited. Otherwise, you should make sure that your data file is organized as a rectangular table of numbers, separated by blanks, with one row per line, and an equal number of elements in each row. For example, let's say that you have an ASCII data file called datafile.dat which contains three columns of data. The first column contains the integers 1 through 10. The second column lists the square root of the first column's numbers. Finally, the third column contains the square of the numbers in the first column.

datafile.dat:

1.0000 1.0000 1.0000

2.0000 1.4142 4.0000

3.0000 1.7321 9.0000

.

.

.

10.00003.1623 100.0000

The data can then be imported into the MATLAB workspace by typing:

```
>> load datafile.dat
```

You do not need to specify that the file is an ASCII format as the **load** command is smart enough to recognize that. MATLAB puts the data contained in the datafile.dat file into a matrix variable called datafile. This matrix will have 10 rows and 3 columns. New variables can be defined from the rows, columns, and elements of the datafile variable. To find out exactly how and what you can do with variables by means of their indices, take a look at the sections in the Getting Started with MATLAB manual.

3.2.3 Importing Low-Level Data

Often data files contain headers, that is, descriptive statements describing how, when, and under what circumstances the following data was collected or generated. Usually you will wish to bypass the header after you have extracted the information you need from it. Additionally, other complicating factors such as rows that have varying number of columns, or text interspersed with numerical data will inevitably be encountered. Even if your data is not in one of the standard formats, you can use the low-level file input/output (I/O) functions MATLAB provides. In such circumstances where the format of the file is known, but is not one of the standard formats, it will most likely be best to make use of the **fread** and **fscanf** commands. Both commands are used to read binary data from a file according to a specified format. Both are part of the low-level I/O commands available in MATLAB and require that certain parameters that describe the precision and location of the data in the file be specified. The general form of the **fread** command is:

```
[A,COUNT] = FREAD(FID,SIZE,PRECISION)
[A,COUNT] = FREAD(FID,SIZE,PRECISION,SKIP)
```

Here, A is the matrix returned by the **fread** command that contains the data which was read. COUNT is an optional output argument that tells you how many elements were successfully read. As you can see, **fread** expects up to four input arguments. The first argument, FID, is a required value that corresponds to the file identification number of the file to be read. This value is obtained by using the **fopen** command. The second argument, SIZE, is optional and tells the **fread** command how much data is to be read. PRECISION is a string that specifies the format of the data. Typically this consists of a data specifier such as int or float followed by an integer giving the size in bits. In general MATLAB's low-level I/O functions are based on the I/O functions of the ANSI C Library. If you are already familiar with C, then you will be familiar with these commands. The table, "MATLAB Low-Level I/O Commands" lists both the binary and ASCII low-level file I/O commands in MATLAB. The following steps are generally what is required to read and write data to data files:

1. Open the file to be read or written to using **fopen**.

2. Operate on the file:

    a. **fread** for reading binary data,

    b. **fwrite** for writing binary data,

    c. **fgets** or **fgetl** for reading text strings line by line,

    d. **fscanf** for reading formatted ASCII data,

    e. **fprintf** for writing formatted ASCII data.

3. **fclose** to close the file.

Although the following table can serve as a handy reminder, please refer to the on-line help or to the MATLAB Function Reference to learn more about MATLAB's low-level file I/O commands.

MATLAB Low-Level I/O Commands

Command	Action	Usage
FOPEN	Opens a file for reading or writing.	FID = FOPEN('FILENAME','PERMISSION')
FCLOSE	Used to close a file once reading or writing is complete.	STATUS = FCLOSE(FID)
FGETL	Reads a line from a file but discards the newline character.	TLINE = FGETL(FID)
FGETS	Reads a line from a file and keeps the newline character.	TLINE = FGETS(FID)
FREAD	Reads binary data from a file.	[A, COUNT] = FREAD(FID,SIZE,PRECISION)
FWRITE	Writes binary data to a file.	COUNT = FWRITE(FID,A,PRECISION,SKIP)
FPRINTF	Writes formatted data to a file.	COUNT = FPRINTF(FID,FORMAT,A,...)
FSCANF	Reads formatted data from a file.	[A,COUNT] = FSCANF(FID,FORMAT,SIZE)

It is not our intention to present a comprehensive discussion on the different data importing functions available in MATLAB. You can read the MATLAB helps on any of these functions as you come across a need for them. The main points to be made here is that MATLAB supports a host of data formats and provides the low-level functions to let you build a special import function if you need it.

3.3   Elementary 2-D Plots

The most basic, yet often the most useful, graph that you may wish to create is a simple line plot of numeric data. The MATLAB language provides a set of high-level commands that are used to create these simple line plots. In order to simplify the discussion and descriptions of 2-D plots, let's take a moment and list relevant graphics objects and fundamental graphics terminology. Essentially, graphics objects are the basic elements which, when assembled and drawn on your monitor's screen, generate pictures and visual information. Even the most elementary plot consists of several graphics objects. The window in which the plot appears, the lines, the axes, and the labels that make up the plot are all examples of graphics objects. The following list will help you become familiar with some of the MATLAB graphics objects referred to in this section without getting into the details which we will discuss in Chapter 7.

The following objects and terms are occasionally referred to in this section:

- figure: the window in which other graphics objects are placed

- axes: a graphics object that defines a region of the figure in which the graph is drawn

- line: a graphics object that represents the data that you have plotted

- text: a graphics object that is comprised of a string of characters and terms

- title: the text object that is located directly above an axes object

- xlabel: the text object associated with the x-axis

- ylabel: the text object associated with the y-axis

These objects and terms also happen to be the names of some of the plotting functions that can be used while creating 2-D plots.

To start, the MATLAB command **plot** will be examined in detail. Then we will look at a group of three commands (**semilogx**, **semilogy**, and **loglog**) that are variations of the **plot** command with respect to the axis scaling. After these are presented, a group of plotting commands that are more specialized in terms of their application are presented. We've placed these specialized plotting commands in the broad category of Specialized 2-D Plotting, since these are easily created with simple high-level MATLAB commands. Finally we will discuss how to edit a plot once it is created and examine the MATLAB Figure Window and its various parts as it has undergone quite a few changes in the recent releases of MATLAB.

3.3.1   A General Overview of the Plot Command

Most of the MATLAB graphics commands are straightforward and intuitive (or at least they become intuitive fairly quickly as you move along the language's learning curve). The plot command is the first one that we will explore. For example, a graph of an arbitrary set of data values assigned to the variable y can be generated using the command **plot(y)**. Let's say that your data set was the cubic of the numbers from negative five to four in step increments of one tenth. This data can be generated and plotted by typing

```
y = (-5:0.1:4).^3;
plot(y);
```

at the command prompt. You will obtain the figure shown in Figure 3.3.

Notice that the x-axis labels are not the numbers that were cubed, rather they are the subscript or index numbers of the vector y. MATLAB automatically plots your data versus the index number when only one argument is passed to the plot function. You can verify this by typing

```
length(y)
```

and seeing that

```
ans =      91
```

is returned. In the figure, you can see that the last point defining the line (in the upper right-hand corner) is at the point 91 on the x-axis and 64 = y(91) on the y-axis.

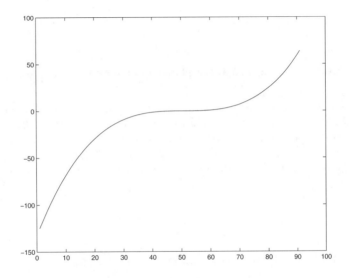

Figure 3.3 Plot of y = (-5:0.1:4).^3

Although there may be instances in which having the indices of the plotted data displayed along the x-axis is useful, in many cases it will be more informative to display the value of the input or parameter that was responsible for the data output. In order to accomplish this for our previous example, we can use

```
x = -5:0.1:4;
y = (x).^3;
plot(x,y);
```

and the Figure Window will contain the plot that appears in Figure 3.4.

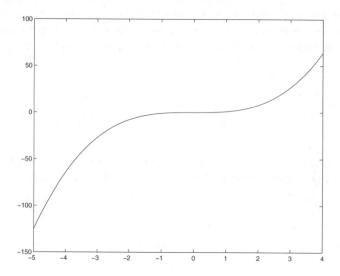

Figure 3.4 Use plot(x,y) where y = x.^3

Now we can add some labels to the x- and y-axes, a title to make the graph more informative, and a grid to assist in estimating values from the line in the graph. We will create the label "x" for the x-axis, "y" for the y-axis, and put "Graph of y = x^3" as the title to the graph. MATLAB makes adding these very simple; just type the following at the command prompt:

```
xlabel('x');
ylabel('y');
title('Graph of y = x^3');
grid;
```

Figure 3.5 shows the results of applying these commands.

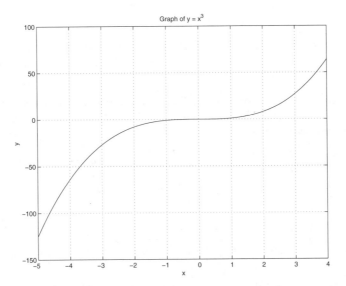

Figure 3.5 Adding labels, a title, and a grid.

The **plot** command arguments are not restricted to being vectors; the inputs may also be matrices. When passing inputs to the plotting function, there are several simple rules to keep in mind so that the appearance of the graph is what you expect. The rules can be summarized as follows:

- **plot(y)**
  - If y is a vector, you will generate a line of y versus the index numbers of y.
  - If y is a matrix, you will generate a set of lines where each line corresponds to the graph of one of the matrix columns versus the row number.

- **plot(x,y)**
  - If x and y are vectors of the same length, a graph of y versus x will be displayed.
  - If x is a vector and y is a matrix, the rows or columns of y will be plotted against x. If a column of the y matrix has the same length as vector x, then the columns will be plotted versus x. If the rows have the same length as vector x, then the rows will be plotted versus x. If the number of rows and columns of y are the same, the columns will be plotted versus x.
  - If x is a matrix and y is a vector, y will be plotted against either the rows or columns of x. If a column of the x matrix has the same length as vector y, then y will be plotted versus columns of x. If the number of rows of x is equivalent to the length as vector y, then y will be plotted versus the rows of x.

- If the number of rows and columns of x are the same, y will be plotted versus the columns of x.

- If x and y are matrices which have the same number of rows and columns, the columns of y will be plotted against the columns of x.

We have already looked at plotting a simple vector by itself or versus another vector. Let's look at a few examples which illustrate these rules. In the Command Window, type

```
x = (0:0.2:9)';
alpha = 0:5;
y = besselj(alpha,x);  % Bessel function
plot(x,y)
xlabel('x');
ylabel('y');
title('y = besselj(ALPHA,x), for alpha = 0,1,2,3,4, and
5');
```

The y variable is a 46-by-6 element matrix and x is a vector with 46 elements. The plotting results are shown in Figure 3.6. For this example it does not matter to the **plot** command if y is transposed or not; MATLAB recognizes the size of the two input variables and appropriately plots y in the orientation that matches the dimensions of x. However, the **besselj** function does require that

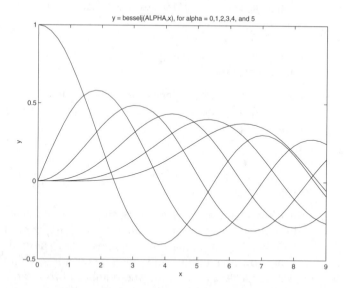

Figure 3.6 Plotting the matrix y versus the vector x.

alpha have as many columns as rows in x.

Try the same example but substitute **plot(y)** for **plot(x,y)**. Here columns of y are plotted, so you end up with essentially the same figure, but with the x-axis labels representing the row index number of y. You could also plot the rows of y by using **plot(y')**, but in this example the graph, Figure 3.7, may look interesting but doesn't provide much in the way of information.

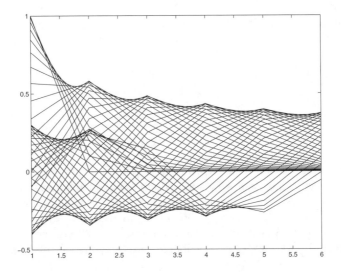

Figure 3.7 Plotting the rows of a 46 by 6 element matrix.

If you have a color monitor, you may have noticed in the previous examples that when multiple lines are plotted, they will have various colors automatically assigned to them. As you will read later on in this section, one of the ways by which the line types (e.g., solid, dashed, etc.), plot symbols (e.g., circles, stars, etc.), and line colors can be defined is by passing a string argument directly to the plot function. However, for any of the cases in which more than one line is created and where you have not defined the color in the plot statement, the color of the lines will be cycled through a specific set of colors. By default, there are six colors that MATLAB will automatically cycle the lines through. Later you will learn how to change line colors to accommodate your needs. Chapter 7 digs deeply into the objects that make up a figure and how to affect their properties such as the order in which line colors are chosen.

The number of inputs can also be extended. You can use the format **plot(x1,y1,x2,y2,...)** where the rules mentioned above apply for each x and y pair. For example, if you wanted to plot three lines representing the data sets

```
x1 = 0:.1:10;
y1 = cos(x1);
x2 = 1.5*x1;
y2 = 2*cos(x2);
x3 = 2*x1;
y3 = 3*sin(x3);
```

you could use

```
plot(x1,y1,x2,y2,x3,y3)
```

If the x1, x2, x3 and y1, y2, y3 data had been, respectively, in an X and Y matrix (in this example it is possible because the sizes of the individual vectors are the same and can be used to build a larger matrix)

```
X = [x1' x2' x3'];
Y = [y1' y2' y3'];
```

you could have also used
```
plot(X,Y)
```

Both of these plot commands would give you the exact same plot shown in Figure 3.8. Be aware of the fact that depending on the situation, there are usually many ways to achieve the same end result. If the x1, x2, x3 and y1, y2, y3 vectors could not have been used to build a larger matrix, the **plot(x1,y1,x2,y2,x3,y3)** would be more appropriate. As you continue through this book and your MATLAB vocabulary grows, you will see that there are

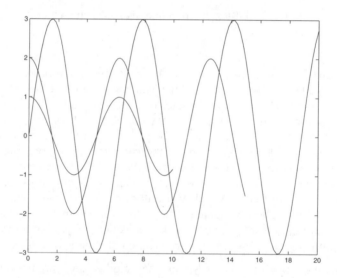

Figure 3.8 Using plot(X,Y) or plot(x1,y1,x2,y2,...).

other methods that can be used to get the same three lines on your display.

If you were able to visualize the data in your mind's eye and expected something that was close to the results in Figure 3.8, or if you were able to look at the data and associate a line with one of the data set combinations, it was most likely due to the fact that the data sets were fairly simple. In many cases this would probably not be easy to do. On your screen the lines are in color, so if you memorized the fact that by default MATLAB currently cycles through the colors blue, green, red, cyan, magenta, yellow, and black when creating multiple lines with a plot command, you would have known that the blue one corresponds to the (x1,y1) combinations, the green line corresponds to the (x2,y2) combinations, and the red line corresponds to the (x3,y3) combinations. Realistically, if you are presenting your plots to others, your audience is probably not likely to have memorized the MATLAB color cycle and unless you are some kind of savant, you probably aren't going to want to memorize it either! Of course you can always look it up in the reference guide, type **help plot**, set your own default, look at the axes ColorOrder property (an object property you'll learn about in Chapter 7), or even run a quick test. But in any event, if you print out a figure with a black and white printer, you are still going to be out of luck unless you make use of MATLAB's line types or plotting symbols. Also, since defaults may change from version to version (as the color cycle did from version 4 to version 5), if it matters what color a plot is displayed as on your monitor, it is always best to specify exactly what you want.

Fortunately, the **plot** command accepts a color-linestyle-marker string, which is a character string argument by which you can specify line types, plot symbols, and colors. For instance, if you wanted to plot a red dashed line with the vectors x and y, you could simply type the command

```
plot(x,y,'--r');        % Plot y versus x as red dashed
line.
```

The string that you create to define the characteristics of the line may use any combination of characters shown in Table 3.3.1 for the line type, symbol, or color. As you can see, your string may have from 1 to 4 characters. The order of the character sets does not matter. Later we will see how to alter the line properties using the Property Editor from the Figure Window itself. And in Chapter 7 you will learn how to directly work with figure objects.

Table 3.3.1 Line Color, Marker Style, and Line Style Strings

Line Color		Marker Style	
character	*creates*	*character*	*creates*
b or blue	blue line	.	point
g or green	green line	o	circle
r or red	red line	x	x-mark
c or cyan	cyan line	+	plus
m or magenta	magenta line	*	star
y or yellow	yellow line	s	square
k or black	black line	d	diamond
		v	triangle down
Line Style		^	triangle up
character	*creates*	<	triangle left
-	solid	>	triangle right
:	dotted	p	pentagram
-.	dashdot	h	hexagram
--	dashed		

It is important to realize that when you are using plot symbols, the symbols will appear centered on the data points. Lines, however, will interpolate (linearly) between the data points. Therefore, if you are plotting a continuous function (such as sin(x)), the relative smoothness of the line may depend on the number of samples being passed to the function and the spacing between the samples. Also bear in mind that the colors that have been listed in Table 3.3-1 are not the only colors that can be used. You will see how to create lines with other colors when we discuss the Plot Editing Mode in Section 3.6 and Handle Graphics in Chapter 7.

Let's look at an example of how a combination of line styles and symbols can be informative. First, we will create some data by squaring a range of values and adding some normally distributed random noise:

```
x1 = [-3:.2:3];
y1 = x1.^2 + randn(size(x1));
```

Given the input and noisy output, let's say we want to fit a 2nd order polynomial curve to the data. To do this we can use the MATLAB functions **polyfit** and **polyval** in the following manner:

```
p  =  polyfit(x1,y1,2);
x2 = [-3:.5:3];
y2 = polyval(p,x2);
```

Now we shall plot the data which was used as input to our curve fitting routine as green circles, and the fitted curve as a cyan dashed line.

```
plot(x1,y1,'og',x2,y2,'--c');
```

And for a flourish, add some informative labels and a grid with

```
xlabel('Input');
ylabel('Output');
title('Noisy data = "o" and Fitted Curve = "--" ');
grid
```

and voila, we have quickly created the combination plot of our results shown in Figure 3.9.

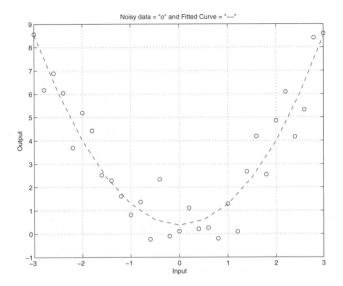

Figure 3.9 A combination plot.

3.3.2 Logarithmic Plots

Not all data lends itself to a linear scale representation. Sometimes, the range of the data to be plotted is so great that it is difficult to see just what the data is doing. MATLAB provides three forms of the **plot** function that let us view data that is better represented with a logarithmic scale, namely **semilogx**, **semilogy**, and **loglog**. Each is used just as the **plot** function, but use a logarithmic scale (base 10), for either the x-axis, y-axis, or both axes respectively.

As an example, consider the data generated by the following code.

```
x=-10:.1:10;
y=exp(x.^3);
```

The x-axis data is clearly linear but the data computed from it isn't. If we use our familiar **plot** function with

```
plot(x,y)
```

we will get the plot shown in Figure 3.10.

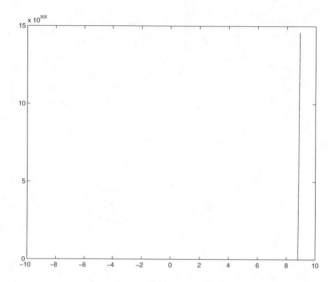

Figure 3.10 Using plot is ineffective when the data scale varies greatly.

In this case our plot looks much like a straight line and we would be hard pressed to read any values off of it. However, using **semilogy** reveals much more about the nature of the data. Simply plot the data again using,

```
semilogy(x,y)
```

and you will get the plot shown in Figure 3.11.

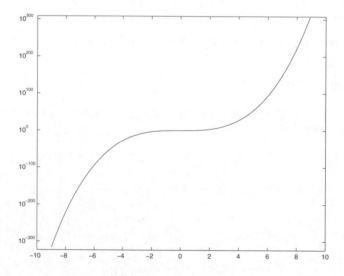

Figure 3.11 Using semilogy reveals details about the data.

If our x-axis data was better represented with a logarithmic scale, we would use **semilogx**. If both need logarithmic scales, then **loglog** could be used.

3.4   Simple 2-D Plot Manipulation

So making a simple plot is easy enough with the **plot** command. In fact, you can see how attractive MATLAB might be for making quick plots and easily adding axis labels and titles. There's lots more you can do and that is what we are going to discuss next.

3.4.1   Generating Plots with Multiple Data Sets

As you just learned, you can plot multiple sets of data with a single **plot** command. However, MATLAB does not restrict you to using a single call to the **plot** command in order to plot multiple lines in a graph. A command which you might find very useful is the **hold** command. The **hold** command allows you to add additional lines to the current graph without erasing what already exists in it. When **hold** is not used, the **plot** command will replace the contents of the current graph. The **hold** command can be used in three different ways:

hold on   tells MATLAB that you want all subsequent lines and surfaces to be added to the current graph.

hold off   is used to turn off the **hold** command, setting MATLAB back to the default mode of replacing the current graph with a new one.

hold   when used by itself will toggle the mode between the **hold on** and **hold off** state.

Here is an example where we will add three lines to a single graph using three **plot** statements to produce the graph shown in Figure 3.12.

```
x = -2:.1:2;
plot(x,sin(x),'-r');
hold on
plot(x,sin(x.^2),'--b');
plot(x,cos(x.^2),':g');
hold off
```

Both the **hold on** and **hold off** statements could have been replaced simply with the command **hold**. In fact, the **hold off** is not necessary at all. We have used it here to make sure that the graph returns to its default state. This way, if you type in subsequent examples, you will obtain results identical to those shown in the figures which are illustrated in this book. We also suggest using the on and off arguments in programs so that the hold state is not ambiguous to a person reading the M-file. If you are concerned about inadvertently plotting on an existing graph, you would be, for instance, completely safe from accidentally adding the solid red line, **plot**(x,sin(x),'-r'), to a previously existing graph, by using the clear figure command **clf** .

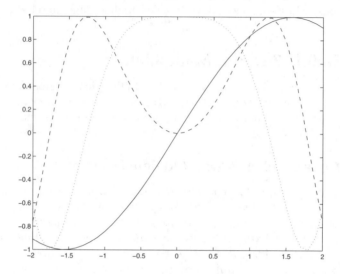

Figure 3.12 Multiple line plot using hold.

Notice that in this example we had to tell the **plot** command which color to use with each of the lines or else each line would have been plotted in blue, the default first color for MATLAB plots. This is because the **plot** command starts with the default first color each time it is called. Although **hold** is on, that fact is completely ignored by the **plot** command. The **hold** command simply keeps the current plot while subsequent **plot** commands are issued.

In some instances you will have data sets that you want to display on the same graph; however, the y-axis data values are not in the same range. MATLAB provides a useful graphics function for just such an occasion. The command **plotyy**, will help you plot these types of data sets on the same graph. This is best explained with an example.

Let's say you have created the following data sets:

```
x1 = 0:.1:20;
y1 = x1.*sin(x1);
x2 = 10:.2:25;
y2 = 50*x2;
```

If you plotted them with

```
plot(x1,y1,'-b',x2,y2,'--g');
title('y1 is the blue line, y2 is the green dashed
line');
ylabel('y');
xlabel('x');
```

As you can see in Figure 3.13 it is difficult to see the y1 values since MATLAB's auto-scaling is choosing the y-axis limits so as to display all the data points.

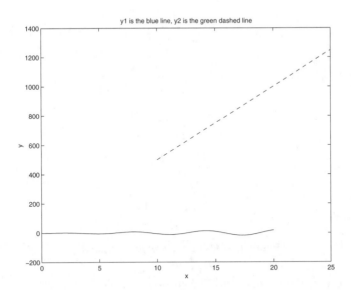

Figure 3.13 Using plot to graph data sets with a large range of y-axis values is not always acceptable.

Instead we could use **plotyy** to plot the data:

```
plotyy(x1,y1,x2,y2)
```

which would generate the plot shown in Figure 3.14 .

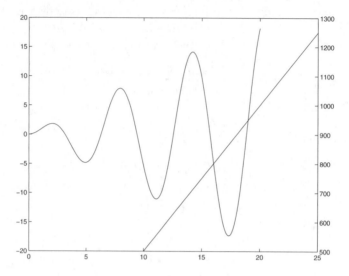

Figure 3.14 Using the plotyy command.

Observe that the y1 data took the first default color and that the y2 data took the second. Also notice how MATLAB colored the y axis appropriately. Unfortunately, the **plotyy** command does not allow us to change the color or type of line used in plotting the same way we did with the plot command. Also, using **ylabel** would only affect the left y-axis and we would not be able to label the right y-axis. However, by making use of MATLAB's handle graphics (see Chapter 7) we can do exactly what we intend, that is plotting the two sets of data with complete control over the color, line style, and even labeling each axis appropriately.

The following code will do just that:

```
[axeshandles,line1handle,line2handle]=plotyy(x1,y1,x2,y2);
set(line1handle,'linestyle','-','color','blue');
set(line2handle,'linestyle','--','color','green');
title('y1 is the blue line,y2 is the green dashed line');
axes(axeshandles(1));
ylabel('y1=x.*sin(x)');
axes(axeshandles(2));
ylabel('y2=50*x');
xlabel('x');
```

The handles (which you will learn about in Chapter 7) are used here to give us control over setting the linestyle and color attributes so that we can readily distinguish between the data sets. The **plotyy** command returns the handles to the graph's two axes in axeshandles, and the handles to objects from each plot in line1handle and line2handle. The first element in axeshandles, axeshandles(1), is the left axes and the second, axeshandles(2) is the right axis. The above code sets the appropriate properties and will give the plot shown in Figure 3.15.

Power!

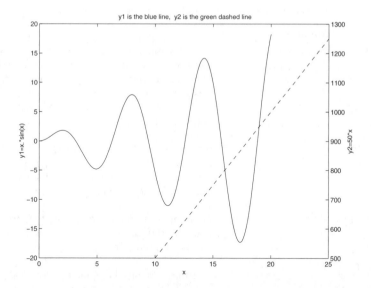

Figure 3.15 Using plotyy to show 2 data sets.

Notice how nice our graph looks with each axis labeled appropriately and with our choice of color and line styles! Don't get overly worried about the use of handles and object properties here. You will learn all about that in Chapter 7. Just keep in mind that to get the greatest control over plots in MATLAB you will need to know about Handle Graphics.

3.4.2 Using Axis to Customize Plots

You probably noticed that MATLAB automatically scales the x-axis and y-axis to encompass the data set or sets that you are plotting. In addition, the axes are automatically labeled and in a standard Cartesian coordinate system with the origin in the lower-left corner. Often you will want to display a different region of the graph than what MATLAB's default settings have provided. The **axis** command can be used to manipulate the attributes of a graph's axes. Table 3.4.1 summarizes the uses of this function with respect to 2-dimensional plots.

Table 3.4.1 The axis function summary for 2-D plots.

Speed!

Function	Action
axis([xmin xmax ymin ymax])	set the minimum and maximum x- and y-axis limits. xmax and/or ymax can be set to Inf to force MATLAB to autoscale the upper x- and y-axis limits. xmin and/or ymin can be set to -Inf to force MATLAB to scale the lower x- and y-axis limits.
axis auto	returns the axis scaling to its default, automatic mode where, for each dimension, 'nice' limits are chosen based on the extents of all lines in the graph.
axis manual	freezes the scaling at the current limits, so that if **hold** is turned on, subsequent plots will use the same limits.
axis normal	puts the axes into the default (automatic) state and restores the current axis box to full size, removing any restrictions on the scaling of the units. This undoes the effects of **axis square**, and **axis equal**.
axis square	forces the axes to have square dimensions.
axis equal	forces the unit spacing, i.e., the tic marks, on the x- and y- axis to be equal.
axis ij	puts origin of graph in upper-left corner. The x-axis scale numbering increases from left to right. The y-axis scale numbering increases from top to bottom.
axis xy	forces the axes to use a standard Cartesian coordinate system with the origin of the graph in the lower-left corner. The x-axis scale numbering increases from left to right. The y-axis scale numbering increases from bottom to top.
axis tight	forces the x- and y-axes limits to the minimum and maximum data values, i.e., the range of the data.
axis off	turns off, i.e., hides, the axes labels, tic marks, and box by making them invisible.
axis on	turns on, i.e., makes visible, the axes labels, tic marks, and box.

We haven't said anything about this before, but MATLAB commands also have a functional form. This is called *command-function duality*. The **axis** command is as good a command as any to explain this. For instance, you can use the command form by typing

```
axis square
```

at the command prompt, or you could use the function form by typing

```
axis('square')
```

MATLAB treats each method the same. The utility of the command form is that you can compound a couple of these axis manipulations at the same time, such as with

```
axis equal tight
```

which will force the unit spacing to be the same on the two axes and force the limits to the ranges provided in the plotted data sets.

Depending on the data used to create your graph, you may decide that only a specific portion of the graph is important or has relevance. You can always determine which elements are of interest and then re-plot only those elements of the data. This is inconvenient, time-consuming, and may still not give you exactly what you want. The **axis** command provides the easiest and most straightforward way to manually define the x- and y-axis limits.
For instance, if you plot the following data

```
x = -10:.1:10;
```

with

```
y = exp(x).*sin(x).*(x.^3);
plot(x,y)
xlabel('x');
ylabel('y');
```

you will obtain the results illustrated in Figure 3.16.

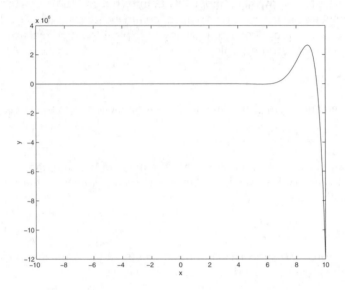

Figure 3.16 Automatic scaling can be misleading.

This graph shows all the data, but the graph seems to indicate that the expression for y is flat for x between negative 10 and 3. To see what is going on for these values of x, we can "zoom" in on this region by using the axis command. To use this function, pass a vector containing the minimum and maximum values of the x- and y-axes that you want shown (e.g., **axis**([xmin xmax ymin ymax])). Let's say we want the x-axis to run only from negative ten to three and the y-axis to run from negative six to seven. To achieve this, type

```
axis([-10 3 -6 7])
```

which will give the results shown in Figure 3.17.

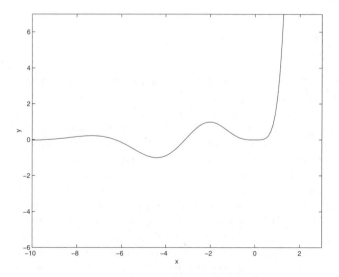

Figure 3.17 Manually defining the axis reveals details of the data.

Simply typing

```
axis auto
```

at the command prompt will plot all the data again in the graph.

When you write MATLAB M-files that create graphics there will be times when you might need to know the current graph's axis limits. This is useful for things such as determining how to appropriately redefine them based on their value, or perhaps you are interested in them for some other purpose. To get and put the current axis limits into a variable, use

```
variable_name = axis;
```

When the current graph is 2-dimensional, *variable_name* will be a row vector with 4 elements ([xmin xmax ymin ymax]). The following example illustrates the case in which you want use the minimum limits of the x- and y-axes that MATLAB determines, but want to customize the maximum values for both axes.

```
x = 0:0.1:(5*pi);
plot(x,7.5*sin(x));
axis_limits = axis
desired_max_x = 10;
desired_max_y = 15;
axis([axis_limits(1) desired_max_x ...
     axis_limits(3) desired_max_y]);
new_limits = axis
```

The axis limits before redefining them are

```
axis_limits = [0 16 -8 8]
```

and after redefining them, the limits are

```
new_limits = [0 10 -8 15]
```

Earlier we showed you how by using **hold on** you can create graphs with multiple lines by separately issuing the **plot** command for each line.   If the data for a particular line exceeds the boundaries of the x- and y-axis limits, MATLAB redefines the axis scales to include the new data. In some instances this may not be desirable. To keep the automatic scaling from occurring, you just need to define the axes limits using **axis**([xmin xmax ymin ymax]) or **axis**(**axis**). By setting the axis limits to something other than 'auto,' the axis mode is set into a manual mode instead of automatic.   Therefore, any subsequent plots that are added to the current one will not change the axes scales. The **axis**(**axis**) method of defining the limits freezes the current axis scaling limits because you are calling the axis function twice. The call that is performed within the parentheses returns a vector of the current axes limits that in turn is passed to the **axis** function. The axes limits are not changed, but since you have manually defined the axis limits, they will no longer change to accommodate the minimum and maximum values of subsequent data plots.

The **axis** command also provides a quick way to change the aspect ratio of the axes. By default, the axes will size themselves to fill up most of the Figure Window, independent of how you have sized the Figure Window. Depending on the data you have plotted, you may want the axes to be square in their physical dimensions. To illustrate this, create a Figure Window by typing

```
figure
```

Then resize this Figure Window so that its dimensions are rectangular. Now let's create a circle with a radius of two units, using

```
x = 2*cos([0:10:360]*(pi/180));
y = 2*sin([0:10:360]*(pi/180));
plot(x,y)
axis([-5 5 -5 5])
```

At this point the circle probably has a slight elliptical shape such as shown in Figure 3.18.

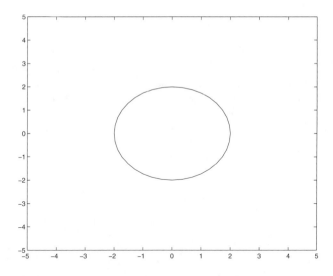

Figure 3.18 Circular data appears elliptical without customizing the axes with the axis command.

Now we can force the axes size to be that of a square instead of a rectangle with the command

```
axis('square')
```

which will result in the plot of Figure 3.19.

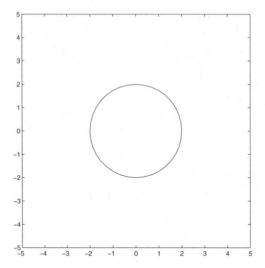

Figure 3.19 Using axis('square').

Now this looks more like a circle! But before we can be confident in our use of the **axis**('square') command, let's take our understanding a little further by looking at the plot of an ellipse.

Keep in mind that using **axis**('square') does not necessarily keep the unit spacing on the x-axis the same size as the unit spacing on the y-axis, it merely forces the size of the axes to be a square instead of the default axes size, which tries to make the most of the Figure Window real estate. To illustrate, create an ellipse by typing

```
x = 2*cos([0:10:360]*(pi/180));
y = 4*sin([0:10:360]*(pi/180));
plot(x,y)
axis('square')
```

which will produce the plot shown in Figure 3.20.

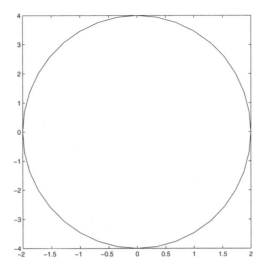

Figure 3.20 Elliptical data looks circular with axis('square').

Now you can clearly see that we have an ellipse that looks like a circle. Such a representation could lead to problems that are merely annoying or potentially devastating.

To insure that you have the correct aspect ratio, use **axis**('equal'). Typing

```
axis('equal')
```

after the **axis**('square') command of the previous example will produce the graph shown in Figure 3.21.

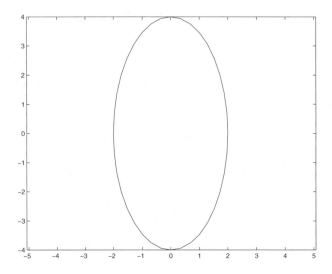

Figure 3.21 Using axis('equal') to graph the ellipse.

You can also force the range of the axis limits to adjust to the minimum and maximum data values of the ellipse by using **axis**('tight'). If you do, you will get the plot shown in Figure 3.22.

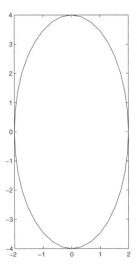

Figure 3.22 Using axis('tight') to adjust the axes to the ellipse data.

Recall that the default Cartesian axes has its origin in the lower-left corner of the plot. The x-axis lies horizontally along the bottom of the figure with the axis scale values increasing from left to right. The y-axis lies vertically along the left side of the Figure Window with the axis scale values increasing from bottom to top. You can tell MATLAB to locate the origin of the axes in the upper-left corner with the y-axis scale values increasing from top to bottom by using the command **axis**('ij'). To revert back to the default Cartesian axes use axis('xy'). In Chapter 7 you will learn how to reverse the direction of the x-axis numbering using the axes object properties.

If you think that your data is not correctly proportioned, use **axis**('equal') to make sure that the scaling of the axis is indeed equal. Also, be aware that the **axis**('auto') command only assures that the scaling can accommodate all the data in the graph; it will not necessarily undo the 'square' and 'equal' settings. To be sure you are back in the default mode, use **axis**('normal').

One last feature of the axis function that we will mention here is that you can make the axes and all labels associated with the axes invisible by typing **axis**('off'). In this mode, the graphics that were plotted in the axes will remain visible. This is useful if all you need is the data. To see the axes and labels again simply type **axis**('on').

3.4.3   Creating Supporting Text and Legends

In the last section of this chapter we will explore some of the new interactive ways MATLAB lets us edit our plots. You will see just how quick and easy it is to edit plots in the plot editing mode. Although convenient for quick edits, it is still necessary to understand how to add text to your plots using the specific text commands. You have already seen how to add text to the x-axis, y-axis, and at the top for a title using **xlabel**, **ylabel**, and **title**. Although these are sure to be the most common commands you will use when creating plots, MATLAB provides you with additional means for adding supporting text in any arbitrary position to your graph using the **text** command. We will also show you how you can place text with your mouse using **gtext**. Finally you will learn how the **legend** command can quickly add a legend to your multi-line plots.

Before we begin with the built-in functions of MATLAB, we will present a quick diversion to a handy tool called **sidetext**. The **sidetext** function provides a simple means to placing text at the right side of the axes with its orientation identical to that created with **ylabel**. The **sidetext** function uses handle graphics to manipulate the position and orientation of the string you provide. Once you have downloaded **sidetext** and placed it in your working directory, try the following example:

Tools

```
t = 0:0.02:2;  phi_0 = 45*pi/180;
y = sin(2*pi*t + phi_0);
plot(t,y);
grid on;
xlabel('t');
ylabel('y');
title('Plot of Sin(2*pi*t + phi_0)');
sidetext('phi_0 = 45 degrees');
```

which produces the graph shown in Figure 3.23.

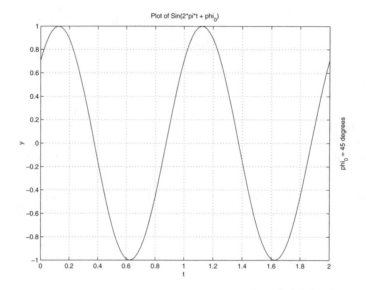

Figure 3.23 SIDETEXT places a string on the right side of the axis.

Now change the scale of the axes with

```
axis([0 3 -2 2])
```

and the graph will be that shown in Figure 3.24. The important point to note here is that changing the scale of the axes will not affect the position of the text created with **sidetext** since it positions the text relative to the axes.

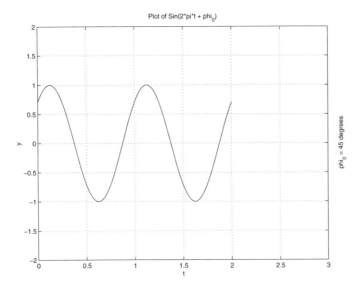

Figure 3.24 SIDETEXT is unaffected by axes scale.

You can place text interactively with the mouse by passing a string to the function **gtext**. After this command is issued, the mouse pointer will change from the standard arrow to a crosshair when the pointer is in the Figure Window. Position the crosshairs over the location in the figure where you wish to place the text and press either the mouse button or a key on your keyboard. The text string will appear left justified and vertically on top of the data point that was selected. For example, create a graph and type

```
gtext('This text was placed with gtext')
```

Now, scale the axes to something other than what is currently shown in your figure. Notice that this text string changes its location relative to the axes border, but not relative to the data point that was selected. This function is useful when you have completed a graph and want to add a few additional lines of text. Also as you will learn in the last section of this chapter, the current version of MATLAB lets you add text in the plot editing mode. However, if you are creating multiple plots, a fair number of text strings, or just want to automate this process in your MATLAB program, most likely you'll find that the **text** function is better suited for these types of tasks.

The **text** function is both a high- and low-level graphics function that can be used to add character strings to a graph. For now, we'll look at it as a high-level text placement command. After you learn about Handle Graphics in Chapter 7 we will explore the techniques and ways in which these graphics objects can be manipulated.

The most elementary way that text can be added to the current graph is with **text**(x,y,'text') where the data point (x,y) corresponds to a location in the current axes. As an example, let's have MATLAB draw a line plot and label the maximum data point as such.

```
% Create and plot the x and y data
x = -2:0.1:2;
y = 3 - (x+1).^2;
plot(x,y);
xlabel('x'); ylabel('y'); title('y = 3 - (x+1).^2');
grid on
axis([-2 2 -5 5]);
% Determine the maximum y data value
[max_y_value,max_y_index] = max(y);
corresponding_x_value = x(max_y_index);
% Put a red circle symbol at the maximum data point
hold on
plot(corresponding_x_value,max_y_value,'or');
hold off
% Create a string vector
our_string = sprintf('%g is the maximum data point',...
             max_y_value);
% Put the string into the graph at the max y value
text(corresponding_x_value,max_y_value+0.5,our_string);
```

This script will create the plot and text shown in Figure 3.25. Here we have added the 0.5 to the max_y_value variable so that the text will not overlap the line. By default, the text string will be placed left justified and vertically centered on the (x,y) data point that is provided. The addition of the 0.5 can be avoided by passing properties to the text function to keep the text vertically above the data point. To learn about all the text object's properties and how they can be used to manipulate the text's attributes, see Chapter 7.

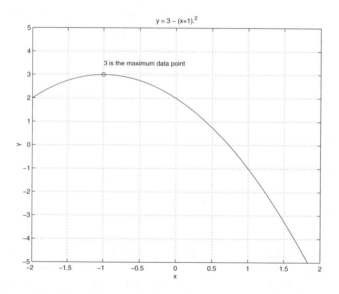

Figure 3.25: Using the text function to add text to your plot.

The **text** function can also be used in a manner very similar to the way **plot** is used. For instance, if you want to create a scatter plot of the percent change in the consumer price index versus unemployment between 1965 and 1980, you can type

```
% Source of data: Economic Report of the President, 1986.
cpi_data = [1.7 2.9 2.9 4.2 5.4 5.9 4.3 3.3 6.2 ...
            11.0 9.1 5.8 6.5 7.7 11.3 13.5];
perc_unemploy_data = [4.5 3.8 3.8 3.6 3.5 4.9 5.9 ...
            5.6 4.9 5.6 8.5 7.7 7.0 6.0 5.8 7.0];
year_strings = ['1965';'1966';'1967';'1968';'1969';...
            '1970';'1971';'1972';'1973';'1974';...
            '1975';'1976';'1977';'1978';'1979';'1980' ];
plot(perc_unemploy_data,cpi_data,'o');
% In this next text command two text properties were made
% use of so that the plot would look better.  You will
%learn how to manipulate these in Chapter 7.
text(perc_unemploy_data,cpi_data,year_strings,...
            'fontsize',10,...
            'verticalalignment','bottom');
axis([0 10 0 14]);
xlabel('Percent Unemployment');
ylabel('Percent change in CPI');
```

will create the plot shown in Figure 3.26.

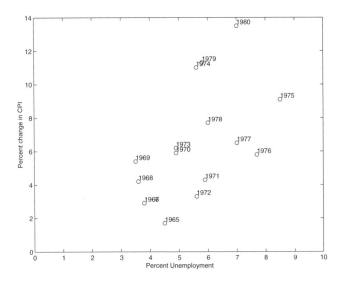

Figure 3.26 Scatter Plot with text labels.

You might have been wondering how to add a block of text, i.e., multiple lines of text to labels and titles. Perhaps it occurred to you that you could create multiple lines by repeated use of the **text** command, but then you would be faced with a kind of trial and error approach in order to get the location of your text to look right. Fortunately MATLAB provides a way to accomplish this without resorting to such manual methods. All of the built-in MATLAB text functions will accept a cell array of strings where each string contains the text for each line. This code for example,

```
string_array(1)={'This will be the first line.'};
string_array(2)={'This will be the second line.'};
string_array(3)={'And so on...'};
gtext(string_array);
```

will place the three lines of text wherever you position the mouse pointer in the figure. If you know exactly where you want to place the block of text, you could use

```
text(0.5,0.5,string_array);
```

The final text placement command we will discuss is this section is the **legend** command. This function creates a legend of the line types that you have used in the current graph and associates these line types with the text strings that you pass to it. The order in which the lines are created is the order in which they are associated with the legend strings. For example,

```
x = 0:.1:(2*pi);
sx=sin(x);
cx=cos(x);
plot(x,sx,'-r',x,cx,'--c');
axis([0 2*pi -1.5 1.5])
legend('Sin(x)','Cos(x)');
```

will produce the result shown in Figure 3.27.

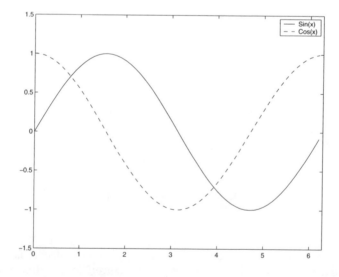

Figure 3.27 Creating a figure legend.

If you are not sure in which order the lines were created or you want only a few of the lines put into a legend you can use this form of the **legend** function:

```
legend(linetype1, string1,linetype2,string2,...)
```

This is probably the safest way to insure that when you create the legend the text string is correctly associated with the line you wanted. Implementing this for the previous example, we see that the **legend** command is replaced with

```
legend('-r','Sin(x)','--c','Cos(x)');
```

If you do not like the position that was automatically chosen by the **legend** function, you can use the mouse to click and drag the legend to a location of your choice.

3.4.4 Text Placement

There are several ways to place text in a position relative to the figure instead of the axes. For instance, you may wish to have a calendar date always located in the lower right-hand corner of the figure, even when you are displaying multiple axes, or subplots (see the following section). One method you can use is to create invisible axes that cover the entire figure space. Then place text within the invisible axes where location (0,0) is the lower left-hand corner and (1,1) is the upper right-hand corner of the figure. If you use this technique we recommend that, until you learn more about graphic objects and their handles, you create the invisible axes and the specially placed text after the rest of your plot looks the way you want it. The following code will produce the plot shown in Figure 3.28.

```
plot(0:.1:10,cos(0:.1:10))
date_string = date;
axes('position',[0 0 1 1],'visible','off');
text(1,0,date_string,'horizontalalignment','right',...
        'verticalalignment','bottom');
text(0,0,'Lower Left String',...
        'horizontalalignment','left',...
        'verticalalignment','bottom');
text(1,1,'Top Right String',...
        'horizontalalignment','right',...
        'verticalalignment','top');
text(0,1,'Top Left
String','horizontalalignment','left',...
        'verticalalignment','top');
```

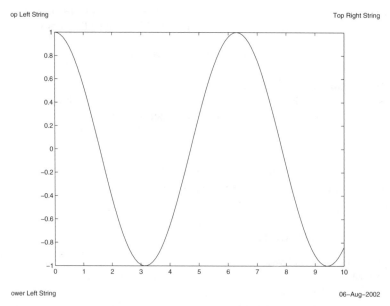

op Left String   Top Right String

ower Left String   06–Aug–2002

Figure 3.28 Placing text using invisible axes.

The problem alluded to above is that if you were to subsequently add another plot with

```
hold on
plot(0:.1:10,sin(0:.1:10))
```

you end up plotting to the invisible axes as shown in Figure 3.29. This happens since **plot** commands always apply to the most recently created axes, unless you take advantage of handle graphics (and that's not until Chapter 7).

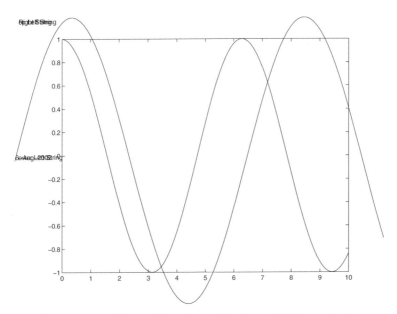

Figure 3.29 Problems arise when adding to a plot after using the invisible axes method.

This plot looks bad because the text that we placed earlier moved to new locations when the invisible axes limits automatically scaled to accommodate the limits of the new data.

A way we can overcome the restrictions of this method is to normalize the position of the current axes to the figure and then place text in normalized units. With this approach it does not matter when you generate the text as long as there is at least one plot on the screen. This means that after you have created a plot with some specially placed text, you can then, for example, add more lines to the plot without affecting the text positions or without worrying about plotting to an invisible axis. To make this process easier we can create a new function, **norm2fig**, which will return the normalized text positions.

Tools

```
function normtxtpos = norm2fig(normfigpos)
% Pass this function normalized  positions in the figure
% and it will return the positions relative to the
current
% axes.
%
%    passing a [0 0] would refer to lower left corner
%    passing a [0 1] would refer to top left corner
%    passing a [1 0] would refer to lower right corner
%    passing a [1 1] would refer to top right corner

apos = get(gca,'pos');
normtxtpos = [(normfigpos(1,1)-apos(1,1))/apos(1,3) ,...
              (normfigpos(1,2)-apos(1,2))/apos(1,4)];
```

This function will let us quickly generate the text positions we want for the previous example to get the results shown in Figure 3.30.

To duplicate this result, first plot the sine, be sure that **hold** is on, then do the following:

```
plot(0:.1:10,cos(0:.1:10))
date_string = date;
tpos = norm2fig([1 0]);
text(tpos(1,1),tpos(1,2),date_string,...
        'units','normalized',...
        'horizontalalignment','right',...
        'verticalalignment','bottom');
tpos = norm2fig([0 0]);
text(tpos(1,1),tpos(1,2),'Lower Left String',...
        'units','normalized',...
        'horizontalalignment','left',...
        'verticalalignment','bottom');
tpos = norm2fig([1 1]);
text(tpos(1,1),tpos(1,2),'Top Right String',...
        'units','normalized',...
        'horizontalalignment','right',...
        'verticalalignment','top');
tpos = norm2fig([0 1]);
text(tpos(1,1),tpos(1,2),'Top Left String', ...
        'units','normalized',...
        'horizontalalignment','left',...
        'verticalalignment','top');
```

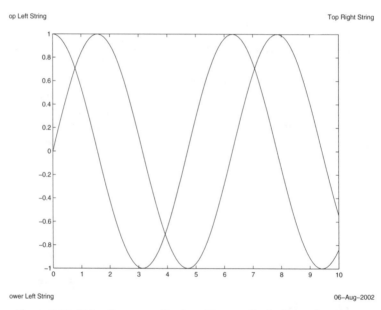

Figure 3.30 Using the normalized position method of text placement.

Short of the methods of Chapter 7, this approach works well.

3.4.5 Special Text Character Formats

You've seen that adding a string of text is relatively easy. You've even seen that you can have text in a title or on an axis label with more than one line by storing your strings in cell arrays. But in all these cases the text we provided were those characters available directly from our keyboard. What about special characters like the Greek alphabet, superscripts, subscripts, arrows, and other mathematical sysmbols?

Fortunately MATLAB has the capability of modifying text to have different styles. It does this by providing support for a subset of the TeX characters. TeX is a standard notation for special character sets. You can recognize it right away as it will have a backslash "\" prefixing a character name. For example, the Greek character Ω (big omega) is specified by "\Omega" in TeX. Table 3.4.2 lists all the TeX characters available in MATLAB.

Table 3.4.2 TeX Characters Available in MATLAB

TeX Characters	Result	TeX Characters	Result	TeX Characters	Result
\alpha	α	\upsilon	υ	\sim	~
\beta	β	\phi	φ	\leq	≤
\gamma	γ	\chi	X	\infty	∞
\delta	δ	\psi	ψ	\clubsuit	♣
\epsilon	ε	\omega	ω	\diamondsuit	♦
\zeta	ζ	\Gamma	Γ	\heartsuit	♥
\eta	η	\Delta	Δ	\spadesuit	♠
\theta	θ	\Theta	Θ	\leftrightarrow	↔
\vartheta	ϑ	\Lambda	Λ	\leftarrow	←
\iota	ι	\Xi	Ξ	\uparrow	↑
\kappa	κ	\Pi	Π	\rightarrow	→
\lambda	λ	\Sigma	Σ	\downarrow	↓
\mu	μ	\Upsilon	Υ	\circ	∘
\nu	ν	\Phi	Φ	\pm	±
\xi	ξ	\Psi	Ψ	\geq	≥
\pi	π	\Omega	Ω	\propto	∝
\rho	ρ	\forall	∀	\partial	∂
\sigma	σ	\exists	∃	\bullet	•
\varsigma	ς	\ni	∋	\div	÷
\tau	τ	\cong	≅	\neq	≠
\equiv	≡	\approx	≈	\aleph	ℵ
\Im	ℑ	\Re	ℜ	\wp	℘
\otimes	⊗	\oplus	⊕	\oslash	∅
\cap	∩	\cup	∪	\supseteq	⊇
continued on next page					

TeX Characters	Result	TeX Characters	Result	TeX Characters	Result
\supset	⊃	\subseteq	⊆	\subset	⊂
\int	∫	\in	∈	\o	
\rfloor	⌋	\lceil	⌈	\nabla	∇
\lfloor	⌊	\cdot	·	\dots	…
\perp	⊥	\neg	¬	\prime	′
\wedge	∧	\times	×	\0	∅
\rceil	⌉	\surd	√	\mid	|
\vee	∨	\varpi	ϖ	\copyright	©
\langle	⟨	\rangle	⟩		

In addition to recognizing the special characters already listed, the MATLAB TeX interpreter also recognizes the following stream modifiers that control the font used.

Table 3.4.3 TeX Stream Modifiers

TeX Stream Modifier	Description
\bf	Bold font.
\it	Italics font.
\sl	Oblique font (rarely used).
\rm	Normal font.
^	Make part of string superscript.
_	Make part of string subscript.
\fontname{fontname}	Specify the font family to use.
\fontsize{fontsize}	Specify the font size in FontUnits.

The first four modifiers are mutually exclusive so you can't use them together. However, you can use \fontname in combination with one of the other modifiers. Also, stream modifiers remain in effect until the end of the string or only within the context defined by braces { }. The following code illustrates using the TeX interpreter and produces the plot shown in Figure 3.31.

```
plot(0:.1:2*pi, sin(0:.1:2*pi))
xlabel('\tau = 0 to 2\pi','FontSize',16)
ylabel('sin(\tau)','FontSize',16)
title('\it{Value of the Sine from 0 to 2
\pi}','FontSize',16)
```

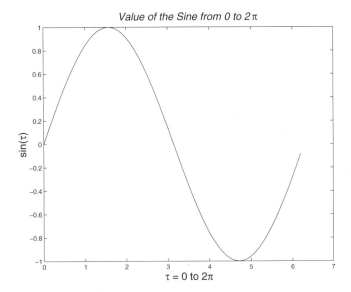

Figure 3.31 Using TeX for special characters.

What about if you want to print a "\", "{", "}", "^", or "_"? Since these have meaning in the TeX interpreter you will need to tell the interpreter to ignore the command. This is achieved by using a backslash "\" right before them.

3.4.6 Using Subplot to Create Multiple Axes

You've seen that you can have multiple plots on an axis, either by plotting multiples or by using the **hold** command and issuing another **plot** command. However, you are not limited to having one axes object in a Figure Window. The easiest way to create multiple axes in a Figure Window is to make use of the command **subplot**. This function breaks up the Figure Window's space into subregions or panes and is very useful for showing related information that is better viewed in individual plots. Calling the command with three arguments creates these subregions; the first two specify how many regions there will be in terms of rows and columns, and the third argument specifies which region you wish to plot in. For example, **subplot**(m,n,p) subdivides the Figure Window into m-by-n regions and creates axes in the pth region, where regions are numbered from left to right and top to bottom within the figure. For example, the following will break up the Figure Window into three distinct regions and create an axes object in the second one.

```
subplot(3,1,2)
```

After you have created an axes object in one of the regions, you can then use any plotting command you want. The axes created with **subplot** can be treated in the same way as the ones that are created when no subregions are specified. In fact, you can create an axes object which encompasses the entire Figure Window's space by issuing **subplot**(1,1,1) and get the same axes that you would have gotten by using plot commands without the subplot function.

If an axes already exists in the subregion identified with the **subplot** command, the existing axes becomes the current axes to which all subsequent graphics commands are issued. You can flip back and forth between these regions by re-issuing the **subplot** command to set one of the other axes or regions as the current axes.

As an example, let's break up the figure space into four panes configured in a 2-by-2 fashion. The following code will create a figure space with four subregions configured as a 2-by-2. We will plot three different shapes separately in the first three regions and then have the fourth subregion contain all three shapes superimposed on top of one another.

```
% CREATE THE X and Y SHAPE DATA.
x_square = [-3 3 3 -3 -3];
y_square = [-3 -3 3 3 -3];
x_circle = 3*cos([0:10:360]*pi/180);
y_circle = 3*sin([0:10:360]*pi/180);
x_triangle = 3*cos([90 210 330 90]*pi/180);
y_triangle = 3*sin([90 210 330 90]*pi/180);

%PLOT THE CIRCLE IN THE UPPER LEFT SUBREGION.
subplot(2,2,1)
plot(x_circle,y_circle,'--g'); axis([-4 4 -4 4]);
axis('equal');
title('Circle');

%PLOT THE SQUARE IN THE UPPER RIGHT SUBREGION.
subplot(2,2,2)
plot(x_square,y_square,'-r'); axis([-4 4 -4 4]);
axis('equal');
title('Square');

%PLOT THE TRIANGLE IN THE LOWER LEFT SUBREGION.
subplot(2,2,3)
plot(x_triangle,y_triangle,':b'); axis([-4 4 -4 4]);
axis('equal');
title('Triangle');

%PLOT THE COMBINATION PLOT IN THE LOWER RIGHT SUBREGION.
subplot(2,2,4)
plot(x_square,y_square,'-r');
hold on;
plot(x_circle,y_circle,'--g');
plot(x_triangle,y_triangle,':b');
axis([-4 4 -4 4]); axis('equal');
title('Combination Plot');
```

Figure 3.32 shows the results of this script.

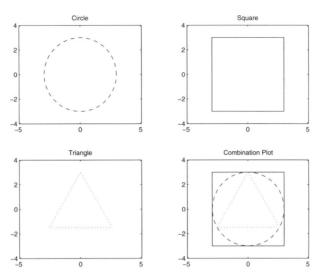

Figure 3.32 Multiple axes with subplot.

Keep in mind, that if at any time you create a subplot that breaks up the figure into a new M-by-N configuration, the set of existing axes will be deleted!

A useful (and undocumented) manipulation of the **subplot** can be used to place a different number of subplots on a row. For instance, instead of having four subplots in a two-by-two matrix, we could have two subplots on the top row and one subplot on the bottom row that spans the figure. The following code will create the plot shown in Figure 3.33.

Power!

```
subplot(2,2,1),ezplot('sin(x)')
subplot(2,2,2),ezplot('cos(x)')
subplot(2,1,2),ezplot('sin(x)^2/x^2')
```

The first two uses of **subplot** appear familiar enough; however, the final call specifies the large axis shown at the bottom of the figure. To create a similar figure with the single plot at the top use:

```
subplot(2,2,3),ezplot('sin(x)')
subplot(2,2,4),ezplot('cos(x)')
subplot(2,1,1),ezplot('sin(x)^2/x^2')
```

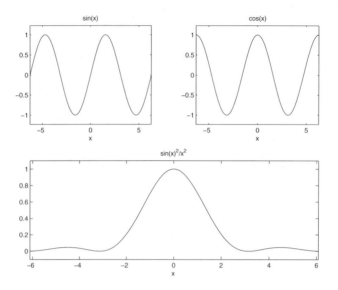

Figure 3.33 Odd axis made with subplot.

To create a subplot with one large plot axis on the left side and two small plot axes in a column on the right, use:

```
subplot(2,2,2),ezplot('sin(x)')
subplot(2,2,4),ezplot('cos(x)')
subplot(1,2,1),ezplot('sin(x)^2/x^2')
```

After you have created some subplots, you may have noticed that if you use the **title** command a title is created on top of the current axes. In some cases you may wish to have a title that is centered at the top of the Figure Window instead. Certainly you could use any of the text placement functions that were discussed previously such as **gtext** and **text**. However, if you do not want to be prompted or waited upon to place the text as with **gtext**, or you don't want to determine the desired position's relative location to the current axes as which would need to be done with **text**, you can use a function we include here called **toptitle**. This function will perform the calculations required to place a title string at the top of the figure regardless of the region to which you are currently plotting. The format for using **toptitle** is simply **toptitle**(string_vector) where string_vector is a character string containing the set of characters that you wish to have appear at the top of the figure. Don't worry about the details of the code for now, but after you have read Chapter 7 you will be able to understand it readily.

```
function toptitle(string)
% TOPTITLE
%
% Places a title over a set of subplots.
% Best results are obtained when all subplots are
% created and then toptitle is executed.
%
% Usage:
%               h = toptitle('title string')
%

% Patrick Marchand (prmarchand@nvidia.com)
% Thomas Holland (tholland@infinityassociates.com)

titlepos = [.5 1]; % normalized units.

ax = gca;
set(ax,'units','normalized');
axpos = get(ax,'position');

offset = (titlepos - axpos(1:2))./axpos(3:4);

text(offset(1),offset(2),string,'units','normalized',...

'horizontalalignment','center','verticalalignment','middl
e');

% Make the figure big enough so that when printed the
% toptitle is not cut off nor overlaps a subplot title.
h = findobj(gcf,'type','axes');
set(h,'units','points');
set(gcf,'units','points')
figpos = get(gcf,'position');
set(gcf,'position',figpos + [0 0 0 15])
set(gcf,'units','pixels');
set(h,'units','normalized');
```

M-File

3.5   Specialized 2-D Plotting

MATLAB provides several high-level plotting routines to facilitate the creation of some of the more common types of graphs and certain special or application specific graphs.   Some of these routines are similar to those typically found in plotting packages or spreadsheet applications. This section will make you aware of the types of specialized plots that are available and how they are used. We will start with the common types of graphs such as the bar graph and histogram type plots. Then we will look at plots that help show statistical distributions of data or discrete data and how to generate plots in other coordinate systems. Finally this section will touch on plotting complex data and how to generate a polygon of your own creation.

3.5.1   Bar Graphs

A bar graph can quickly be created with the **bar** command.   The **bar** function can be used to plot bars with heights specified by the variable argument, bar_height_vector, versus the index number of that variable by using

```
bar(bar_height_vector);
```

If instead of the index to the variable, you want to plot bars versus another variable, you can use **bar**(x,y), where x and y are equal length vectors, and vector x contains values which are both in ascending order and evenly spaced. If x is not evenly spaced or in ascending order, the routine will do the best it can do, but the results will most likely not be what you wanted.

If, for example, you want to create a bar graph of the percentage of widgets that passed quality tests versus the assembly line number, you can type

```
assembly_line_number = [1 2 3 4 5 6 7];
percentage_passed = [85 93 87 91 95 71 98];
bar(assembly_line_number,percentage_passed);
xlabel('Assembly Line Number')
ylabel('Percentage Passed')
```

which will produce the plot shown in Figure 3.34.

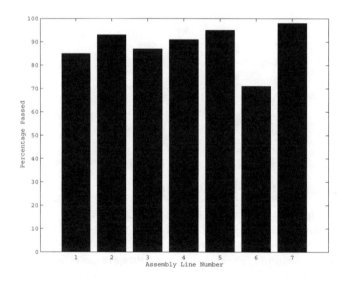

Figure 3.34 Using the bar function.

When you plot bar graphs, you may wish to have labels other than the numeric ones that automatically appear on your x-axis. In these cases, the simplest way to plot your bar graph is with the **bar**(bar_height_vector) format. Then see the section in Chapter 7 about axes properties to find out how you can manipulate the axis labels.

You also have the option of passing a string argument to define the color and line style of the bars with

```
bar(x,line_style_string);
```

or

```
bar(x,y,line_style_string);
```

The string, line_style_string, takes on the same format as the string used in the **plot** command. The **bar** function can also be used to create the data which defines the lines making up the bars. This is done by requesting that the bar function return two output variables with either

```
[x_line_data,y_line_data] = bar(x);
```

or

```
[x_line_data,y_line_data] = bar(x,y);
```

In this mode of operation, the **bar** function does not draw anything. However, these output variables, x_line_data and y_line_data, can be used with the **plot** command (e.g., plot(x_line_data,y_line_data)) to generate the bar graph.

The **bar** plotting function has the ability of clustering multiple data sets, stacking and generating horizontal bar plots. When you pass a matrix to the **bar** function, a bar will be generated for each element of the matrix. The bars associated with the elements in a specific row will be clustered together, while at the same time maintaining color properties for the bars generated from the matrix elements in a specific column. For example, if we have a 2-by-4 matrix, there will be 4 groups of 2 bars clustered around each x-axis data point associated with a row element in the matrix as shown in Figure 3.35.

```
x = [1 3 4 6];
Y = [ 3 1 ; 4 2 ; 2 3 ; 2.5 2];
bar(x,Y);
grid on;
```

If we did not provide the x-axis data points such as by using **bar**(Y), MATLAB would have used the row number and the four clusters would have been evenly spaced.

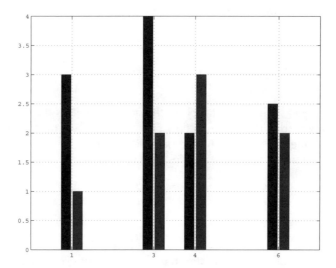

Figure 3.35 Clustered bar graph.

In the next example, several variations of the same data set (the x and Y used in the previous example) are generated in the four subplots with **bar** and **barh**. (The **barh** function is essentially identical to **bar** except that the bars are plotted horizontally.) The stacked bar plots are created by specifying the 'stack' bar style ('group' is the default). A bar's width and the relative amount of separation between bars within a clustered group can be specified by providing a scalar argument; the default value for this scalar is 0.8. A value less than 1 makes the bars thinner and separates them more, a value of 1 makes the bars in a group touch one another, and a value greater than 1 makes the bars overlap. The plots are shown in Figure 3.36.

```
subplot(221);
bar(Y,'stack');
subplot(222);
bar(x,Y,.5) % The 0.5 specifies that the grouped bars be
            % separated by more than the default of 0.8.
subplot(223);
barh(Y,'stack');
subplot(224);
barh(Y,1);  % The 1 specifies that the bars in a group
            % touch one another
```

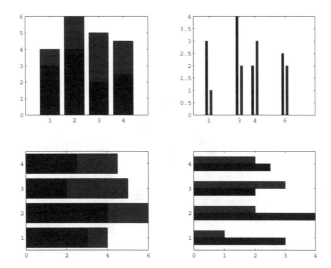

Figure 3.36 Using various combinations of the bar and barh functions.

The examples done here can be repeated using MATLAB's **bar3** and **bar3h** commands to give the plots a 3-D look. Just repeat the examples and substitute **bar3** for **bar** and **bar3h** for **barh**. The only other thing you have to know is that the default style for 2-D bar graphs is grouped while the default style for 3-D bar graphs is 'detached'. Therefore, when the style is not explicitly stated in the example, you will have to provide the 'grouped' style to get the 3-D counterpart. For example, the 3-D counterpart to the plot shown in Figure 3-32, is created with

```
bar3(x,Y,'grouped');
```

and is shown in Figure 3.37.

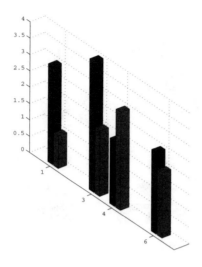

Figure 3.37 3-D version of Figure 3.35.

3.5.2   Histograms

Histograms are essentially a kind of bar graph that is created by first specifying the number bins that specify a range of values, and then counting the number of occurrences of a data set that fall within each bin. One of the most common uses of the histogram is in image processing where we are often interested in the spectrum of the color (or gray-scale) of an image. For a 16-color image, we could indicate on the x-axis 4 bins, each covering four of the colors, and plot the number of pixels that fall into each bin on the y-axis. There are many uses of the histogram and it is frequently used to give insight into the occurrence of events relative to some categories of interest.

The MATLAB **hist** function can be used to automatically create a histogram of the data you pass to it. If you use **hist**(y), the function will create a histogram with 10 equally spaced bins that cover the range of values between the minimum and maximum values of the variable y. In addition, you may specify either the number of bins or the centers of the bins by respectively passing a scalar or vector as a second argument to the **hist** function. The **hist** function makes use of the **bar** function to plot the histogram, and therefore, when you pass bin centers as a vector argument, you should pass points that are equally spaced and in ascending order. If the centers are not equally spaced or in order, you may not get the results you expect. Just as with the **bar** function, you may suppress the plotting of the histogram by having the function return two output variables. For example,

```
[n,x] = hist(y);
```

or

```
[n,x] = hist(y,number_of_bins);
```

or

```
[n,x] = hist(y,bin_centers);
```

will return two vectors. The variable n is a vector containing the number of occurrences that correspond to the bins with centers specified in the variable x.

As an example, let's create exponentially distributed data and plot the histogram as shown in Figure 3.38.

```
number_data_points = 5000;
Beta = 2;
y = -Beta*log(rand(1,number_data_points));
x = 0.2:0.4:10; % Bin Centers
hist(y,x);
ylabel('Count')
```

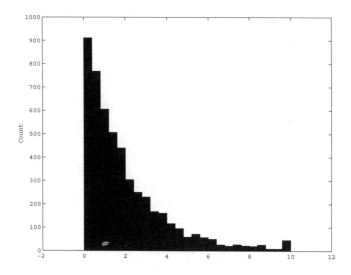

Figure 3.38 A histogram of exponentially distributed samples.

To plot the percentage of data points that fall within a particular bin on the y-axis instead of the count, we could use

```
[n,centers] = hist(y,x);
bar(centers,(n/number_data_points)*100);
ylabel('Percentage');
```

and we would get the plot shown in Figure 3.39.

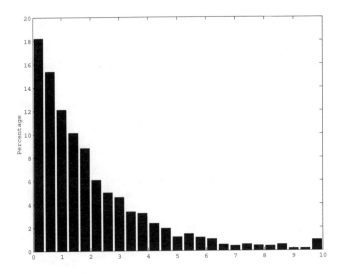

Figure 3.39 Showing percentage of occurrence with the hist function.

3.5.3 Stairstep Graphs

Instead of creating lines that directly connect your data, if you so choose you can create a plot that emphasizes the discrete nature of the data. MATLAB provides a function that will create a stairstep graph of your data. You can use **stairs**(y) or **stairs**(x,y) to draw horizontal lines at the level specified by the elements of y. This level will be held constant over the period between the values specified by the index numbers when using **stairs**(y) or the elements in x when using **stairs**(x,y). The stairstep plot is similar to a bar graph with the exception that the vertical lines are not dropped down all the way to the zero value point on the y-axis. In addition, the x values do not necessarily need to be spaced equally or in ascending order. To illustrate the use of **stairs** and to show the difference in results with respect to the **plot** function, we generate the four subplots shown in Figure 3.40 with the following code.

```
% Using unequally spaced data
x = [linspace(0,2*pi,20) linspace(2*pi,4*pi,10)];
subplot(221); stairs(x,cos(x));
title('stairs(x,cos(x))');
subplot(222); plot(x,cos(x)); title('plot(x,cos(x))');

% Using non-strictly increasing data.
x2 = [1:9 4:-1:1]; y2 = [1:9 8:-1:6 1];
subplot(223); stairs(x2,y2); title('stairs(x2,y2)');
subplot(224); plot(x2,y2); title('plot(x2,y2)');
```

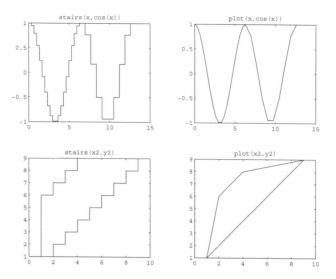

Figure 3.40 Comparing the stairs and plot functions.

As with **bar** and **hist**, you can suppress the creation of the graph by using

```
[xs,ys] = stairs(y)
```

or

```
[xs,ys] = stairs(x,y);
```

You can then use

```
plot(xs,ys)
```

to will produce a stairstep graph from the vectors xs and ys.

3.5.4   Stem Plots

Stem plots provide yet another method of visualizing discrete data sequences, such as sampled time series data. In these types of graphs, vertical lines terminating with a symbol such as a circle are drawn from the zero value point on the y-axis to the values of the elements in the vector passed along with the command, **stem**(y). If you want spacing other than that provided by the element index number, you can use **stem**(x,y), where x specifies where the line is drawn along the x-axis. Figure 3.41 is an example that can be produced with the following code.

```
x = 0:0.25:(3*pi);
stem(x,sin(x));
title('stem(x,sin(x))');
xlabel('x');
```

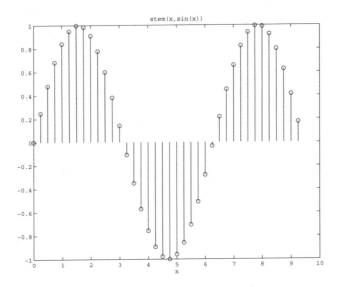

Figure 3.41 Visualizing discrete data with stem.

You can tell MATLAB to use any of the line styles and to terminate your stem plots with any of the marker types that are in Table 3.3.1. Additionally, these terminators can be either filled or unfilled. This line of code,

```
stem(x,sin(x),'-.','p','filled');
```

will generate a stem plot in which the lines are dash-dotted and the terminating symbol is a filled five-pointed star as shown in Figure 3.42.

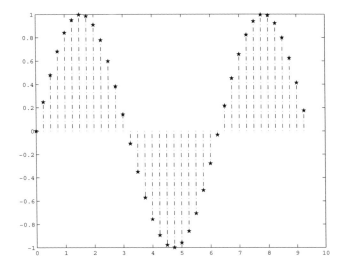

Figure 3.42 A stem plot with filled terminators and dash-dotted lines.

3.5.5 Plots with Error Bars

Error bars are used to show uncertainty in the accuracy of plotted values. With the **errorbar**(x,y,e) function, MATLAB will plot a line which passes through the set of (x,y) points with vertical lines that are called error bars centered about the (x,y) points that have lengths corresponding to twice the elements of the error vector e. When x, y, and e are same sized matrices lines with their error bars will be drawn on a per column basis. This type of plot can be useful if you are plotting data mean values, yet you wish to convey the range over which values may have fallen. If, for example, you run a simulation and want to see the effect of some input parameter, you can run the simulation many times for each value of the input parameter, so that you could determine a mean and standard deviation of the resulting output. To illustrate, the following code will generate some mean and standard deviation data and plot it with error bars that indicate the range of values that are within three standard deviations of the mean. The result is shown in Figure 3.43.

```
x_values = 1:0.5:10;
y_mean_values = 10*exp(-x_values)+3;
y_std_deviation_values = 1./x_values;
errorbar(x_values,y_mean_values,3*y_std_deviation_values)
;
xlabel('x'); ylabel('y');
title('Plot of data means, with errorbars indicating +/-3
standard deviations');
```

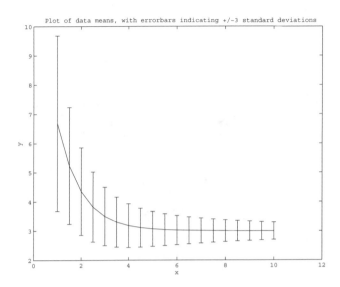

Figure 3.43 Using error bars to show deviation from a mean.

3.5.6   Pie Charts

In MATLAB pie charts display the percentage that each element in a vector or matrix contributes to the sum of all elements. They are useful when you want to show the relative proportion of data elements to one another. For example, let's say you have some data representing where government revenues come from, specifically Soc.Sec. Tax = 31%, Personal Income Tax = 36%, Borrowing = 18%, Corporate Taxes = 8%, Misc. = 7%. The **pie** function will create a pie chart of this data as shown in Figure 3.44.

```
gov_rev_percentages = [31 36 18 8 7];
h = pie(gov_rev_percentages);
```

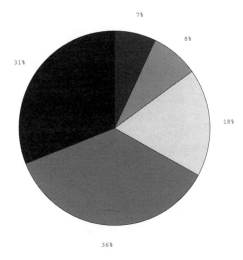

Figure 3.44 Creating a pie chart with pie.

The **pie** function will let you label each section of the pie chart, unfortunately it will replace the values that are printed. To do this, use the **pie** function passing a cell array containing the desired labels. The cell array must be the same size as the data and must only contain strings. To demonstrate this, consider the previous code with specified labels.

```
gov_rev_percentages = [31 36 18 8 7];
pie(gov_rev_percentages,{'Soc. Sec. Tax','Personal Income
Tax','Borrowing','Corporate Taxes','Misc'})
```

The result is shown in Figure 3.45. Although there are labels for each slice of the pie, we no longer have the numerical values.

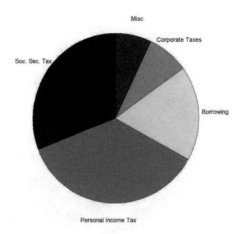

Figure 3.45 Labeling with pie omits numerical values.

We could just include the data values as text in our strings, but that is not a very elegant method and could become tedious if we had much to do. What is preferable is to have MATLAB simply add our labels to the numerical values. Unfortunately, it is not straightforward to add supporting text to pie charts; to do so requires the use of handle graphics (Chapter 7). However, if all you would like to do is quickly add labels to each pie section, we provide a handy function called **pielabel** that will do the trick. You can download **pielabel** from the web site mentioned in Chapter 1. With **pielabel** you can generate the pie chart, then add the labels you like. Here is the code that produces the desired result shown in Figure 3.46.

```
gov_rev_percentages = [31 36 18 8 7];
h = pie(gov_rev_percentages);
pielabel(h,{'Soc. Sec. Tax: ';'Personal Income Tax: ';...
        'Borrowing: ';'Corporate Taxes: ';'Misc: '});
```

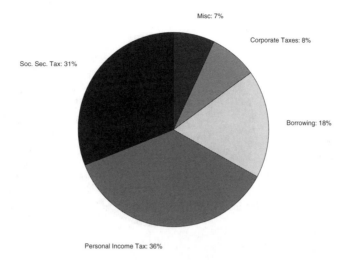

Misc: 7%

Corporate Taxes: 8%

Soc. Sec. Tax: 31%

Borrowing: 18%

Personal Income Tax: 36%

Figure 3.46 A pie chart with labels using pielabel.

As with the **pie** function, you must be sure that the cell array containing the labels is the same size as the data, otherwise **pielable** returns an error. Also be aware that since **pielabel** appends the strings to the data values, if you call it multiple times on the same pie chart you will get additional text appended to the labels.

You can emphasize a particular pie slice by "exploding" the piece out from the rest of the pie. To do this you pass one more argument to the pie function. The explode argument is a vector that is the same size as the data vector. Non-zero elements specify that the particular pie slice should be moved. As an example, the "Borrowing" pie piece could be emphasized with

```
explode = [0 0 0.25 0 0];
h = pie(gov_rev_percentages,explode);
pielabel(h,{'Soc. Sec. Tax: ';'Personal Income Tax: ';...
        'Borrowing: ';'Corporate Taxes: ';'Misc: '});
```

which will produce the pie chart shown in Figure 3.47.

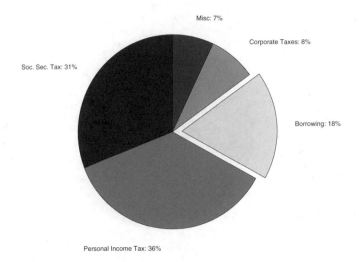

Figure 3.47 An exploded piece piechart.

Just like the 3-D looking bar chart we saw earlier, MATLAB provided a 3-D looking pie chart function called **pie3**. The **pie3** function is used in exactly the same manner as the **pie** function. If you repeat the following examples substituting **pie3** for **pie** you will get the result shown in Figure 3.48.

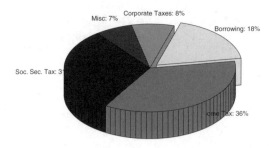

Figure 3.48 An exploded 3-D looking piechart.

3.5.7 Area Plots

The **area** function will generate a filled area plot from either a vector of data, or the columns of a matrix. When creating an area plot with a vector, the data points defined by the vector are straight line connected. Then the area between the lines and the y axis at 0 (by default) will be filled in. To change the y-axis value to which the plot is filled, you can use the form **area**(Y,ymin) or **area**(X,Y,ymin), where the ymin argument specifies the location to which the plot is filled. The area plot is generated from a patch object (discussed later in Chapter 7). Therefore, the visual attributes of the area plot can be changed using valid patch properties and property values (also discussed in Chapter 7). For example, Figure 3.49 shows a blue area plot in which the area is filled to the value 2 on the y-axis. This plot was generated using,

```
x=[0:9];
y=5*sin(x);
area(x,y,2,'facecolor','blue');
```

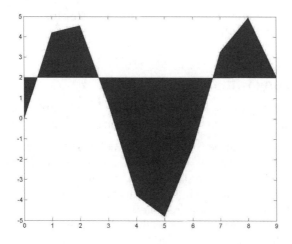

Figure 3.49 An area plot of a vector.

When using matrices, a layer in the area plot is drawn for each column in the matrix. The height of ith layer in the area plot is determined by summing the values in each row from the 1st to i^{th} column in the matrix (ie., sum(Y(:,1:i)') where i is the ith layer) . You may get some strange looking plots if your data values have negative values, but the rule used to determine the height still holds true. The colors for the area plot representing each column is automatically chosen from equally spaced intervals in the colormap. For example, we can generate the 3-layered area plot seen in Figure 3-47 with the following matrix.

```
Y=[1 2 .5; 2 1 .6; 1.5 1 .7; 3 1.5 .8; .5 1 .9; 1 1 1];
area(Y);
```

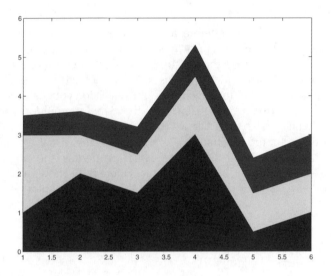

Figure 3.50 Area plot of a 3 by 6 matrix.

3.5.8 Working with Complex Data

Complex data consists of both real and imaginary components and is commonly encountered in many engineering disciplines. In MATLAB, the complex term $\sqrt{-1}$ can be represented by either of the built-in definitions **I** or **j** (depending on whether you are a mathematician or an engineer!). MATLAB knows you are giving it a complex value when the interpreter sees either one affixed to a number, that is unless you have used them as variable names and assigned another value to them. As an example, the complex number 1+3i can be entered as 1+3j; MATLAB is perfectly content with each. In fact, since both are built-in to MATLAB, 1+3*j works too. MATLAB supplies three built-in 2-D plotting functions that are especially applicable to use with complex data. These are the **plot**, **compass**, and **feather** functions. Although you have already seen the **plot** function, we will discuss its use in the case of complex data. The other two functions will follow. In most cases you can use the other MATLAB plotting functions with complex data, but the result might not be what you expect. For instance, what does it mean to plot a bar graph of complex data?

To use the **plot** command with complex data, be sure that your data is in a complex variable. The code

```
z = exp(j*(0:45:315)*(pi/180));
plot(z, '-o')
```

demonstrates this and is shown in Figure 3.51.

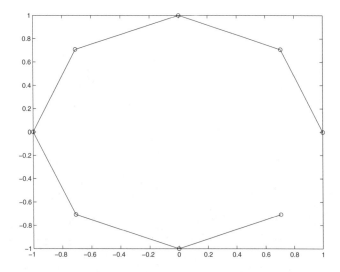

Figure 3.51 Visualizing complex data with plot.

As you can see, complex data points are placed in the axes with the assumption that the x-axis and y-axis respectively correspond to the real and imaginary components of the vector z and then connected by lines. You could also use **plot(real**(z),**imag**(z)) to generate the same results. As another example, let's say you wanted to create a plot that illustrates the complex data points, e.g., the poles and zeros of system transfer function.

```
zeros_points = [-8 -4];
poles_points = [-3+i*2 -3-i*2 -10 -9+3*i -9-3*i ];
% The next line is used since the plot command does not
% know that the zeros_points variable represents
% data in the complex plane.
plot(zeros_points,zeros(size(zeros_points)),'or');
hold on;
plot(poles_points,'xc');
hold off;
axis([-11 1 -5 5])
xlabel('Real Axis');
ylabel('Imaginary Axis');
```

The results of which are shown in Figure 3.52.

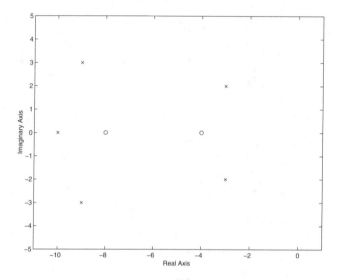

Figure 3.52 Combining complex and real data in a plot.

The **feather** function produces a plot of vectors that emanate from equally spaced points along the horizontal axis. Each arrow's length corresponds to the magnitude of a data element and its pointing direction indicates the angle of the complex data. The general form of this function is **feather**(u,v) where u contains the x-axis components and v the y-axis components, each in relative coordinates. You can call the function with **feather**(Z) where Z is complex. As with **plot** this is equivalent to **feather**(real(Z),imag(Z)). The line type can be chosen in the same way it was when using the **plot** command with the form **feather**(..., linetype_string). For an example, the following code creates some complex data by first creating some angles and corresponding magnitudes. Then it puts that in Cartesian format with **pol2cart**, and converts that result to a complex representation with **complex**. Note that this would work exactly the same with **feather**(u,v).

```
theta = (-pi/2:.15:pi/2);
r=3*cos(theta);
[u,v] = pol2cart(theta,r);
z=complex(u,v);
feather(z,'--c');
```

The result is the data plot with dashed cyan lines shown in Figure 3.53. You may have noticed that the arrowhead's size is proportional to the length (or magnitude) of the line and that they are not solid or filled in. When we explore more about object properties and Handle Graphics, you will learn how to modify the arrows to suit your needs.

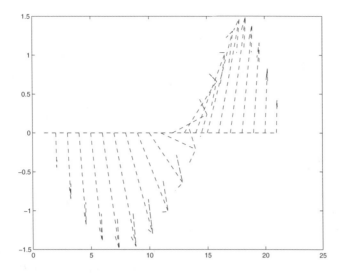

Figure 3.53 A feather plot of complex data.

3.5.9 Using the Polar Coordinate System

So far we have been discussing plotting routines that make use of the Cartesian coordinate system. MATLAB also provides functions that plot data in the polar coordinates of magnitude and angle, commonly referred to as rho and theta respectively. Here we will discuss the MATLAB commands **compass**, **polar**, **rose**, and a modified version of the function polar called **polardb**.

Although the **compass** function takes its inputs in Cartesian format, it works its way into this discussion because of its polar coordinate output. The **compass** function is similar to the **feather** function in that each arrow's length corresponds to the magnitude of a data element and its pointing direction indicates the angle of the complex data. However, whereas **feather** creates a linear plot, the **compass** function will create arrows that emanate from the origin of the axes in a polar coordinate system. As with the earlier functions, **compass** can be called with either **compass**(z) or **compass**(u,v) where the latter method is equivalent to **compass**(u+i*v). To demonstrate this function, let's create a set of arrows that increase in size from arrow to arrow in a counterclockwise manner.

```
z = [1:10].*exp(i*[1:10]*36*(pi/180));
compass(z);
```

This will produce the plot shown in Figure 3.54.

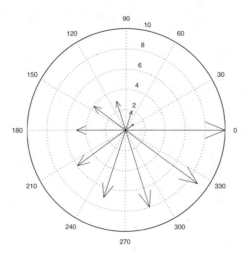

Figure 3.54 A compass plot of complex data.

As with the **feather** function, passing a character string as an additional argument to **compass** changes the style and color used for the arrows.

The **polar** function will create a polar plot from angle and magnitude data. It takes the forms **polar**(theta,rho) or **polar**(theta,rho, linetype_string) where theta corresponds to the angle (in radians) and rho corresponds to the magnitude.   As with the functions we've seen earlier, linetype_string is a character string defining the line type that is used in the plot. The variables theta and rho must be identically sized vectors or matrices.   If they are matrices the columns of theta will be plotted versus the columns of rho.   As an example, we can create the limacon with an inner loop shown in Figure 3.55 with the following code.

```
theta = 2*pi*[0:.01:1];
rho = 0.5 + cos(theta);
polar(theta,rho)
```

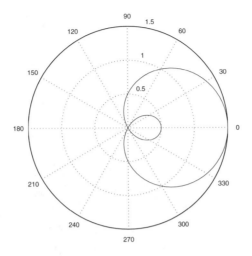

Figure 3.55 Using polar.

If the default polar axis limits are not suitable, you may need to modify the polar function so that it provides you with what you want. Changing the axis limits with the **axis** command is not exactly straightforward. For instance, you can't just use **axis**([min_theta max_theta min_angle max_angle]) as you did with **plot**, because the polar coordinate system is created with plot commands that define the concentric rings and spokes; the Cartesian x- and y-axes are hidden but can be made visible by using **axis**('on').

Often when we deal with data in a polar format, we are interested in units of relative gain or power, i.e., decibels (dB), such as with the case of an antenna gain pattern. If you need to create a polar plot with the radial units in decibels ($10\log_{10}$), you can download from the web site a version of the **polar** function that we have created called **polardb**. This function will also allow you to specify the line style by passing a string with **polardb**(theta,rho,linetype_string). With this function you can specify the rho axis limits by passing yet another 2-element vector that defines the minimum and maximum dB values (i.e., **polardb**(theta,rho,linetype_string, [min_rho_dB max_rho_dB])). The following code creates some data and illustrates using this polar plotting function. The resulting plot is shown in Figure 3.56.

```
x = -(5*2*pi):.1:(5*2*pi);
th = linspace(-pi,pi,length(x));
rho=((1+sin(x)./x));
polardb(th,rho)
```

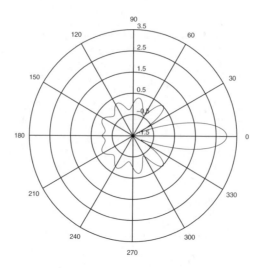

Figure 3.56 Creating a polar plot with radial units in decibels.

The function **polardb** was created by modifying MATLAB's polar function. Many MATLAB functions are available as editable M-files. As you dig deeper into MATLAB graphics, you may find it useful to look at and use code from existing MATLAB functions when developing your own specialized graphics capabilities.

The last polar plot on our list is the **rose** function. With **rose** you can create angle histograms that are drawn in polar coordinates. By using **rose**(angle_data), the function will determine how many of the angles (which are assumed to be in radians) fall within a given angular bin. By default there are 20 evenly spaced bins between 0 and 2π. The number of bins can be changed by using **rose**(angle_data_vector, number_of_bins), where the variable number_of_bins is a scalar specifying the number of bins that should be spaced between 0 and 2π. You can also specify the centers of the bins by passing a vector, bin_centers, to the rose function (i.e., **rose**(angle_data,bin_centers)). If for some reason you do not want the angle histogram to be created at the time the rose command is issued, you may specify two output arguments using any of the valid rose synopses (e.g., [t,r] = **rose**(angle_data_vector)). Then at some later time, you can create the plot by passing these two arguments to the **polar** function (e.g., **polar**(t,r)). The following code will produce a rose plot of data which is normally distributed in angle about 90°. The resulting plot is shown in Figure 3.57.

```
angle_data = angle(exp(i*randn(1,1000)))+pi/2;
rose(angle_data)
```

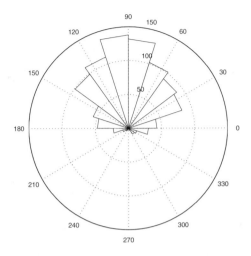

Figure 3.57 An angle histogram created with rose.

3.5.10 Plotting Functions with MATLAB

In the previous examples we have created functions to generate the data used in our plots by first defining a variable to cover the range we are interested in, e.g., x = -2*pi:.1:2*pi, and then coding the function we want, e.g., sin(x). The function **fplot** provides an alternative method of plotting functions to the method of evaluating and plotting a function at a number of defined sample points. It can be especially useful for plotting functions whose rate of change varies rapidly for certain ranges of inputs as this function adaptively determines what the required sampling rate is based on the function's rate of change.

To use **fplot** the function you pass to it must be either the name of an M-file function or a string with variable x that may be passed to **eval** function. This string can contain any combination of legal MATLAB commands or functions that you have created which resemble the form y = f(x), where f(x) is the string that you create. This function must either return a vector that is the same size as x or a matrix with columns that have as many elements as the vector x has.

The **fplot** function was designed to use adaptive step control, concentrating its evaluation in regions where the function's rate of change is the greatest. So, you use **fplot** when you don't want to determine how fine you need to sample a function. For example, y = sin(x) cos(2x), can be plotted using

```
fplot('sin(x).*cos(2*x)',[0 5*pi])
```

to generate the top graph in Figure 3.58. The plots beneath were done with **plot** where the step size was specified as indicated, demonstrating the effect of varying step size.

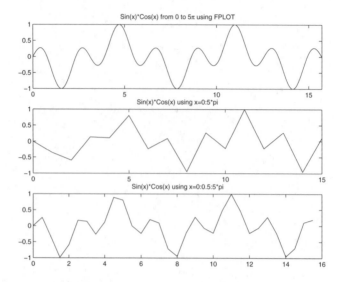

Figure 3.58 Comparing fplot to plot.

You also have the option of passing two additional arguments; one of them is a line style string and the other is a tolerance factor. The tolerance factor is by default set to $2*10\text{-}3$ and is used to determine how much sampling is required. Sampling is increased until the function and linearly interpolated value between two sampled points is less than the tolerance.

If you are really impatient and don't even want to specify a range for a function, MATLAB provides you with a convenient function called **ezplot** that will plot your function over the range -2π to $+2\pi$ and place a title above it. Here is how you would use it with function from the previous example.

```
ezplot('sin(x)*cos(2*x)')
```

The title is the string representation of the function. Also notice that we did not have to implicitly define the function, i.e., the periods weren't required before the * operators. With **ezplot** it is assumed that operations are element by element. Figure 3.59 shows the result of using **ezplot**.

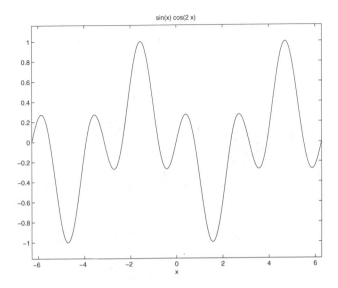

Figure 3.59 Plotting a function with ezplot.

We should point out that you could use other ranges with **ezplot** by providing a vector of the form [min, max] in which case it will perform like **fplot** but will title your plot.

3.5.11 Creating Filled Plots and Shapes

Aside from the pie charts and bar charts, our plots have been made up of lines. Although you can change the color and style of the lines, and later you will learn how to change the thickness too, MATLAB gives you a simple function called **fill** for creating 2-dimensional figures that have shapes that are filled in with a solid color.   Consider the following example, plotted in Figure 3.60, that will create several different shapes with different colored faces by specifying the coordinates of their vertices.   Then we'll fill, scale, and translate them.

```
square_x = cos([45:90:315]*pi/180);
square_y = sin([45:90:315]*pi/180);
pentagon_x = cos([36:72:360]*pi/180);
pentagon_y = sin([36:72:360]*pi/180);
octogon_x = cos([0:(360/8):360]*pi/180);
octogon_y = sin([0:(360/8):360]*pi/180);
wavy_sin_x = 0:.1:8;
wavy_sin_y = sin(wavy_sin_x);
% Create a blue square using a linetype color string.
fill(10+2*square_x,11+2*square_y,'b');
hold on;
% Create a red pentagon using the RGB color vector.
fill(1+3*pentagon_x,10+3*pentagon_y,[1 0 0]);
% Create a gray pentagon using the RGB color vector.
fill(7+2*octogon_x,7+2*octogon_y,[0.5 0.5 0.5]);
% Create a wavy shape.
fill(2*wavy_sin_x,2+2*wavy_sin_y,2+2*wavy_sin_y);
```

```
axis([0 15 0 15]);axis('equal');
```

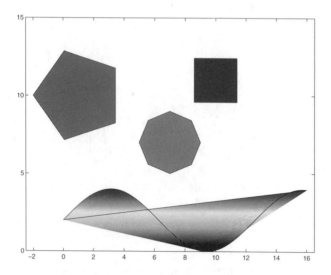

Figure 3.60 Filling shapes with the fill function.

This example illustrates several features of the **fill** function. You use **fill** by specifying **fill**(x,y,c), where the pairs of elements from the x and y variables specify the vertices of the shape, and c specifies the color. The color can be specified as a string (see the line color information in Table 3.3.1) or as a red-green-blue vector, [R G B]. Red-green-blue vectors specify the respective fractions of red, green, and blue content making up the color (e.g., [1 0 0] is equivalent to 'red', [0 0 0] = 'black', [1 1 1] = 'white'). In addition, as we see with the "wavy" shape, the color variable, c, is a vector with the same number of elements as vectors x and y. The elements of c are scaled to indices of the figure's "color map," or list of RGB combinations. If the vertices have different color map indices, the color within the shape will be bilinearly interpolated between the vertices. We will discuss these and many other details of using color in Chapter 8 when we consider color and light.

In this example each shape was created with its own **fill** function. We could have put this into one long **fill** function and eliminated the need for the **hold** on command.

```
fill(10+2*square_x,11+2*square_y,'b',...
        1+3*pentagon_x,10+3*pentagon_y,[1 0 0],...
        7+2*octogon_x,7+2*octogon_y,...
        [0.5 0.5 0.5],2*wavy_sin_x,...
        2+2*wavy_sin_y,2+2*wavy_sin_y);
```

You should know that **fill**(x,y,c) supports the use of matrices and vector-matrix combinations with the x and y variables. If both are matrices, a polygon is drawn for each column. The colors of these polygons will be a single color if c is a row vector, and interpolated if c is a matrix. If either the x or y variable is a vector and the other is a matrix, the vector will be paired up with either the columns or rows of the matrix depending on whether the length of the vector matches up with the length of the columns or rows of the matrix. If the matrix is square, the columns will be used. This might sound a little confusing at first, so an example should clear things up; the following code will create a set of colored bow ties.

```
% A row vector defining the x vertices of all the bows
x = [0 0 1 1];
% A matrix defining y vertices of the bows
y = ones(5,1)*[0 1 0 1]+[1:2:10]'*ones(size(x));
% The color argument specifies a unique color for each
bow
fill(x,y,[1:5]);
axis('equal');
```

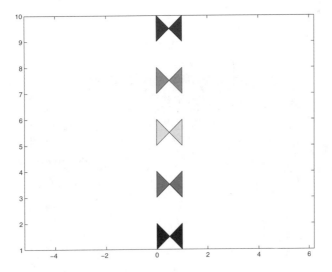

Figure 3.61 Using fill with matrices.

3.6   Plot Editing in the MATLAB Figure Window

As you have seen in the previous sections, MATLAB provides many quick ways to generate very useful plots. You've also seen that by passing various strings and vectors you can tailor certain aspects of your plot to customize it to your needs. What you have done so far is to use the specific plotting functions, and various helper functions, e.g., **title**, **legend**, etc., to annotate and adjust your plots. As you will learn now, MATLAB provides some easy high-level and low-level graphics capabilities through a, somewhat, intuitive user interface. In this section we will discuss how to change specific features of a plot using MATLAB's Plot Editing Mode, and the Property Editor. You will find these techniques very handy for one-time quick changes for your plots. This is not the way to build programs that generate plots in an automated way; that is what the various plotting functions and Handle Graphics in Chapter 7 are for. But for the one-time touch-up, the methods here can be quite useful.

It all starts with the Figure Window; the window that you have seen with the previous examples. It pops up whenever you issue a plotting function. As you might have noticed, the Figure Window has a few tool buttons and pull-downs. You've might have already discovered these, but in case you haven't we will discuss them now.

3.6.1   Plot Editing Mode

When you see the Figure Window, after you have issued a command that generates one, you will notice that there are several buttons as well as a pull-down menu. Clicking on the button that looks like an arrowhead will turn on the Plot Editing Mode. The buttons next to it let you add text and draw lines to annotate your plot. Figure 3.62 highlights some of the features available in the Plot Editing Mode.

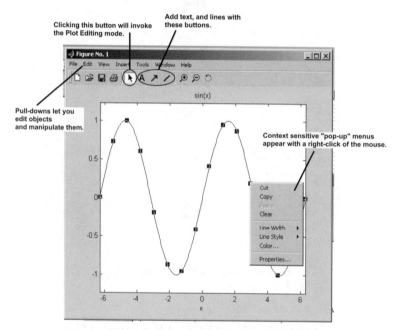

Figure 3.62 The Figure Window in Plot Editing Mode.

The plot shown in the above figure was generated with **ezplot**('sin(x)'). Once you have entered the Plot Editing Mode, when you click on an object in the plot, like the line or the axis, the object will become highlighted with distinctive markers. Clicking once on the trace above selects the line; clicking the right mouse button will reveal a pop-up menu with properties for that object. For example, by selecting the "Line Width" property the width of the line can be changed by a simple click of the mouse. You don't have to click on the object with the left mouse button and then the right to get the case-sensitive pop-up menu. If you place the cursor over the object, clicking the right mouse button will both select the object and reveal the menu.

You can also invoke the Plot Editing Mode from the command line in MATLAB with **plotedit** or **plotedit**(fig) where the former will begin the Plot Editing Mode for the current figure and the latter lets you specify a figure for editing by passing the figure number fig. In use, this is identical to selecting the Plot Editing Mode with the mouse.

3.6.2 The Property Editor

When you open the context-sensitive pop-up menu you will notice that the last menu item in the list is "Properties." Selecting this will open the Property Editor. Invoking the Property Editor on the axis in the sine plot shown in the previous figure gives the user interface shown in Figure 3.63.

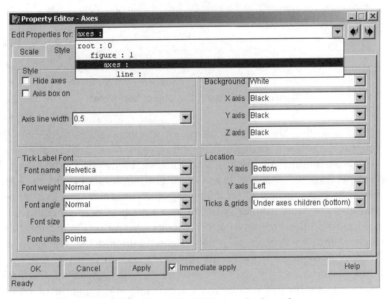

Figure 3.63 The Property Editor invoked on the axes.

As you can see in Figure 3.63, a whole host of properties associated with the axes are available to you to adjust, or mangle, as you desire. You can also start the Property Editor by double clicking on an object in the plot, such as the axis. (Note that double clicking on a text object will not start the Property Editor, but instead will give you an edit box with which you can change the text.) If you keep the Property Editor open, you can click on different objects in the Figure Window and the panels in the Property Editor will change to reveal the property selections for that object.

Another nice feature of the Property Editor is the "Edit Properties for:" selection box. If you click on the down-triangle to the right, the Property Editor will show you the hierarchy of objects. Figure 3.64 highlights this.

You can see the object
hierarchy and select other
objects in the figure here.

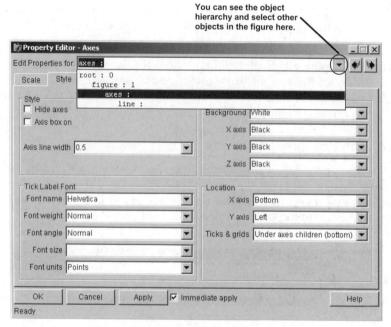

Figure 3.64 Viewing the object hierarchy.

We will leave the detailed discussion of graphics objects and the object hierarchy for Chapter 7. Unfortunately, the pages of a book make showing you the simple nature of pointing and clicking in the Property Editor a bit cumbersome. We recommend that you spend some time with an example and see what you can learn. Try changing the line color, the axis limits, the background color, and turning grids on and off on your own using the Plot Editing Mode and the Property Editor; also try adding or editing the title and axis labels. Although very useful, the Property Editor doesn't present all the properties available for an object, just the ones you are likely to use for high-level editing. Later we will explore another handy tool called the Property Inspector and see how it can help us appreciate the richness of objects in MATLAB.

3.6.3   Zooming and Rotating

To the right of the line buttons are three buttons that look like this:

With the first button here, you can zoom in on your plot by simply dragging the mouse in a box around the area at which you want to look more closely. When you do, you will see that MATLAB automatically scales the axis for the result. You can also place the cursor on a point in the image and get a 2x zoom with each click. The resulting plot will be centered at the point of the cursor. Selecting the second button, the "zoom out" will reduce the zoom by 2x each time you click it.

The third button shown here will be more useful to us after we have discussed 3-D plots. It is the rotate 3-D button and with it you can change the viewing perspective of your plot. You can use this button with a 2-D plot, but there aren't many reasons to do so.

3.6.4 Exporting, Copying, and Pasting

Once you have created a plot you probably want to save it to a file, or perhaps paste it into a word processor or presentation application. You can easily prepare your plots for this right from the Figure Window. Later we will see how we can use some MATLAB commands to produce versions of our plots to use in other applications, but for now we will focus on those available to us from the Figure Window.

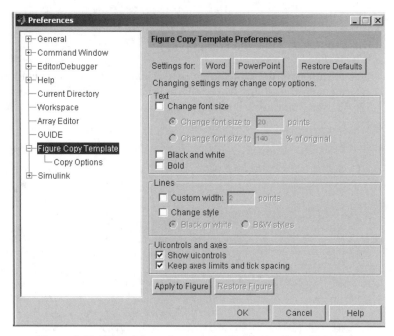

Figure 3.65 Changing preferences in the Figure Copy Template.

If you select **File→Preferences...** from the Figure Window, you will open the preferences user interface as shown in Figure 3.65. In that you can choose the "Figure Copy Template" or "Copy Options" and make changes to the way MATLAB will produce representations of your plots for uses in other applications.

3.7   Illustrative Problems

Just like a muscle, your new MATLAB muscles need exercise to get stronger. The following problems are included to help you exercise your new skills. If you feel the need, you can download the solutions from the web site mentioned in Chapter 1.

Power!

1. Use **linspace** to create a vector t that ranges from 0 to 2π and then plot the function $r = \sin(2t)\cos(2t)$ first as a x-y plot, then as a polar plot. Use **subplot** to keep them in the same figure.

2. Plot $y = \dfrac{\sin(x^2)}{x}$ over the range $-2\pi \le x \le 2\pi$. What should you do about x=0? Hint: read help on **eps**.

3. Try to duplicate the plot shown here. Hint-1: use the **sin** function itself to calculate the points where the annotation should be located. Hint-2: try using '\leftarrow' in your strings.

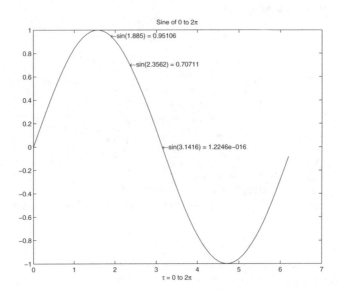

4 PLOTTING IN THREE DIMENSIONS

IN THIS CHAPTER...
4.1 ELEMENTARY 3-D PLOTTING 101
4.2 SIMPLE 3-D PLOT MANIPULATION 136
4.3 VOLUME VISUALIZATION...145
4.4 A WORD ABOUT ANNOTATING 3-D PLOTS..164
4.5 ILLUSTRATIVE PROBLEMS..165

4.1   Elementary 3-D Plotting

In Chapter 3, we discussed how matrix data could be visualized by plotting with the **plot** command. As you might recall from Chapter 2, not all data is intuitively represented with a 2-D plot. We live in a three-dimensional world and much of our information is best revealed with 3-D techniques. Fortunately, MATLAB provides you with a cornucopia of graphics functions that let you make quick 3-D plots and visualizations of your data. This chapter is intended to introduce you to these functions and lead you to a good understanding of the built-in MATLAB ability to visualize in three dimensions. We will begin by examining **plot3**, i.e., the three-dimensional counterpart to **plot**, and then examine the various surface creation techniques, followed by contour plots, and finally present MATLAB's special functions for volume visualization.

4.1.1   Using Plot3

The **plot3** function is used in almost the same way that **plot** is used, except that an additional variable, z, is used to provide the data for the third dimension. For example, let's make use of the form **plot3**(x,y,z) by typing

```
t = 0:0.1:10*pi;
x = exp(-t/20).*cos(t);
y = exp(-t/20).*sin(t);
z = t;
plot3(x,y,z);
xlabel('x');
ylabel('y');
zlabel('z');
```

to produce Figure 4.1. Notice how the axes have been labeled using **xlabel**, **ylabel**, and here we introduce a new labeling command, **zlabel,** whose form is just like that of its siblings.

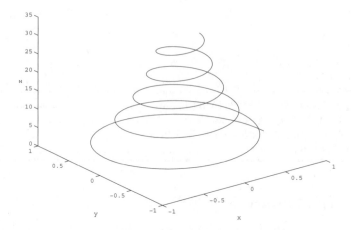

Figure 4.1   A 3-D plot using plot3.

The general form of this function is **plot3**(x, y, z, 'string'), however what it does is determined by the nature of the variables passed to it, namely:

o   If x, y, and z are vectors of the same length, a 3-D line is created by connecting the coordinates specified by the elements of vectors x, y, and z.

o   If x, y, and z are matrices which have the same number of rows and columns, several lines will be created from the columns of the matrices.

o   If some of the input variables are matrices and others are vectors, and the vectors are the same length as either the number of rows or columns in the matrices, MATLAB will "replicate" the vectors in a fashion so that multiple lines can be created. If the sizes of the vectors or matrices do not permit this, MATLAB will return an error message.

o   The variable 'string' is a 1, 2, or 3 character string made from the characters compatible with the **plot** function (see Table 3.3.1).

You can change the perspective, i.e., the viewing angle of plot by either one of two ways. First, you can select the **Rotate 3-D** tool from the Figure Window.

Doing so will let you to interactively rotate the axes of the plot by holding down the mouse button and moving the mouse about. The specific values of the azimuth and elevation will be shown in the lower left corner of the figure while you are rotating the axes.

Your second option is to use the **view** function. The general form of this function is **view**(az, el) or **view**([az,el]) and with it you can specify the exact values of azimuth and elevation by which you wish to rotate the axes. The following code will produce the different views shown in Figure 4.2.

```
subplot(2,2,1);plot3(x,y,z);
xlabel('x');
ylabel('y');
zlabel('z');
view(-10,10);
title('Default plot3');

subplot(2,2,2);plot3(x,y,z,'og');
xlabel('x');
ylabel('y');
zlabel('z');
view(-9,56);
title('Az=-10, El=10');

subplot(2,2,3);plot3(x,y,z,'xb');
xlabel('x');
ylabel('y');
zlabel('z');
view(0,90);
title('Az=0, El=90');

subplot(2,2,4);plot3(x,y,z,'dr');
xlabel('x');
ylabel('y');
zlabel('z');
view(90,0);
title('Az=90, El=0');
```

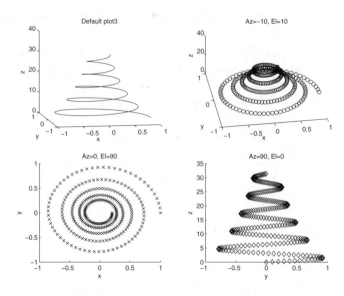

Figure 4.2 You can change your perspective by specifying az and el in the view function.

Although MATLAB's native angle unit is radians, **view** uses degrees for the units of az and el. There are a few more interesting aspects of **view** that we will save for a later discussion, but for now you need to know that the default view for 3-D plots is az = -37.5° and el = 30°. Using az = 0° and el = 90° will give the default 2-D view; you can also obtain this by using **view**(2). What if you've rotated your axes so much that you are confused and you have grown tired of trying to fix it by dragging the mouse around? You can quickly return to the default 3-D view by typing **view**(3).

4.1.2   Creating 3-D Meshes and Surfaces

As we move into more 3-D plotting methods, we are going to find that often we must deal with ordered pairs, i.e., data that is dependent on both an x and a y value. Many mathematical functions are of two variables, that is, for each pair of x and y, there is a z. You have seen this stated as z = f(x,y). One way you could compute a z for each x y pair would be to iterate through a nested loop, but one of the major advantages of MATLAB is that it can deal with matrices without resorting to looping. All you need is some way to get your data into a matrix format. If you have a vector of x values, and a vector of y values, MATLAB provides a useful function called **meshgrid** that can be used to simplify the generation of X and Y matrix arrays used in 3-D plots. It is invoked using the form [X,Y] = **meshgrid**(x,y), where x and y are vectors that help specify the region in which coordinates, defined by element pairs of the matrices X and Y, will lie. The matrix X will contain replicated rows of the vector x, while Y will contain replicated columns of vector y. This might seem a little complicated at first, but an example will help make it clear. Consider the two vectors passed to **meshgrid** here.

```
x = [-1 0 1];
y = [9 10 11 12];
[X,Y] = meshgrid(x,y)
```

MATLAB returns

```
X =

    -1       0       1
    -1       0       1
    -1       0       1
    -1       0       1


Y =

     9       9       9
    10      10      10
    11      11      11
    12      12      12
```

As you can see, X is formed by the vector x being replicated as rows for each column in y, and Y is formed by the vector y being replicated as columns for each element in x. Each element in x has been matched with each element in y. Be aware that typing **meshgrid**(x) is equivalent to **meshgrid**(x,x). The **meshgrid** function will be used in several of the examples in this section.

The first surface plotting function we will discuss is **mesh**. It creates many crisscrossed lines that look like a net draped over the surface defined by your data. To understand what the command is plotting, consider three M-by-N matrices, X, Y, and Z, that together specify coordinates of some surface in a three-dimensional space. A mesh plot of these matrices can be generated with the command **mesh**(X,Y,Z). Each $(x(i,j),y(i,j),z(i,j))$ triplet, corresponding to the element in the ith row and jth column of each of the X, Y, and Z matrices, is connected to the triplets defined by the elements in neighboring columns and rows. Vertices defined by triplets created from elements that are not in either an outer (i.e., first or last) row or column of the matrix will, therefore, be joined to four adjacent vertices. Vertices on the edge of the surface will be joined to three adjacent ones. Finally, vertices defining the corners of the surface will be joined only to the two adjacent ones. In addition to providing a visual perspective of the surface shape, this usage of **mesh** automatically chooses colors of the mesh plot to be proportional to the surface's height. Consider the following example which will produce the plot shown in Figure 4.3 .

```
[X,Y] = meshgrid(linspace(0,2*pi,50),linspace(0,pi,50));
Z = sin(X).*cos(Y);
mesh(X,Y,Z)
xlabel('x'); ylabel('y'); zlabel('z');
axis([0 2*pi 0 pi -1 1])
```

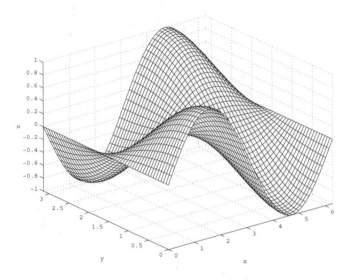

Figure 4.3 A simple mesh plot.

There are several ways to call the mesh command. We just looked at **mesh**(X,Y,Z), however, an even more general invocation of the function can be made with **mesh**(X,Y,Z,C) where the matrix C specifies the color of the mesh plot. When this C matrix is left out of the command, the function assumes that C = Z, thus providing a proportional mapping between color and surface height. For now it will suffice for you to realize that the minimum and maximum values of the matrix, C, specify the range of values that are associated with the figure's color map, i.e., a list of RGB color vectors. The minimum value of C will be associated with the first row in the color map, and the maximum value of C will be associated with the last row in the color map. All values of C that lie between the minimum and maximum shall be associated with a color in this list. For example, if an element of C corresponding to one of the vertices lies halfway between the minimum and maximum values of C, the color associated with that vertex will lie halfway between the first and last row of the color map.   We discussed general guidance for using color in Chapter 2 and an in-depth look at color maps is presented in Chapter 8.   Here is an example that demonstrates using manipulation of the color map to emphasize areas of identical slope. Consider the surface produced by,

```
[x,y] = meshgrid(-2:.1:2, -2:.1:2);
z = x .* exp(-x.^2 - y.^2);
```

We can use **mesh** to plot this surface, however **mesh** will produce colors based on the values of z. We can use the **gradient** function to examine this surface and determine where the slopes are the same according to the x-axis and the y-axis. The general form is [Cx,Cy]=**gradient**(Z) where Cx is the numerically computed solution of $\perp Z/\perp x$ and Cy is $\perp Z/\perp y$. (The actual gradient is the vector sum of Cx and Cy.) Since the derivative of function is its slope, the derivative taken at a point along the surface is the slope of the surface. By using the results of **gradient** as our color map, we can reveal those areas in the plot that have equal slope with respect to either the x, or y axes. The code that will show constant slope in the x-axis is,

```
[Cx,Cy] = gradient(z,.1,.1);
mesh(x,y,z,Cx);
```

The **gradient** function assumes an increment of 1, so we have specified it here to agree with our mesh. Figure 4.4 and Plate 1[*] shows the surface we are considering, plotted with **mesh** with its default coloring that varies according to the amplitude of z. Figure 4.5 and Plate 2 shows equal slopes with Cx from **gradient**. Figure 4.6 is the slope with respect to the y-axis dimension.

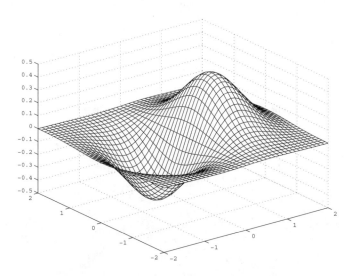

Figure 4.4 A default mesh plot with color assigned to height.

---

[*] Color plates follow page 112.

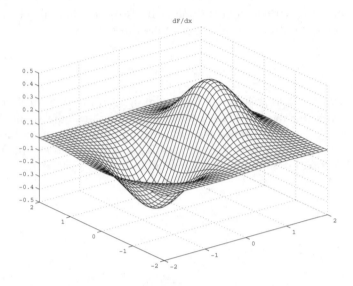

Figure 4.5 Identifying regions of slope with respect to the x-axis.

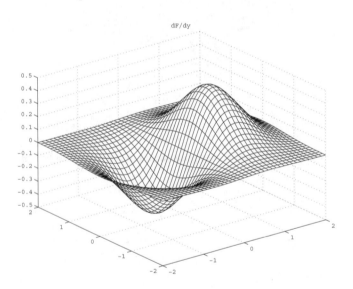

Figure 4.6 Identifying regions of slope with respect to the y-axis.

You should try this example on your computer so you can see the benefit of color better. Color, when used to actually convey information, can make a plot more informative and provides insight that may not have been achieved otherwise.

To finalize our discussion of the **mesh** function we need to mention that a mesh plot can also be created by passing two vectors, x and y, in place of the matrices, X and Y, by using either **mesh**(x,y,Z) or **mesh**(x,y,Z,C). The length vector x must be equal to the number of columns in Z, and the length of vector y must be equal to the number of rows in Z. When using this form of the command, a $(x(j),y(i),Z(i,j))$ triplet defines the vertices over the i rows and j columns of Z. If you do not provide the vectors x and y or matrices X and Y, to the function, e.g., when using mesh(Z) or mesh(Z,C), MATLAB creates a mesh plot by respectively setting the x and y vectors to the column and row number of the matrix Z.

If you want to create a mesh plot that has a "curtain" around the edge of the surface, you might want to take advantage of the function **meshz**. This function is called with the identical input argument set used with **mesh**. The curtain is created by dropping lines down from the edge of the surface to a plane parallel to the xy-plane and at a height equal to the lowest point in the surface. For example,

```
[X,Y] = meshgrid(0:.1:2*pi,-pi:.1:0);
Z = sin(X).*cos(Y);
meshz(X,Y,Z);
axis('equal');
```

will create the illustration shown in Figure 4.7.

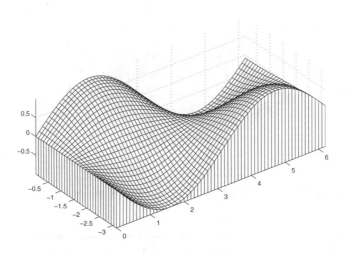

Figure 4.7 A curtain mesh plot made with meshz.

4.1.3 Waterfall Plots

Similar in appearance to the curtain mesh made with **meshz** is the function **waterfall**, which creates a mesh plot only from the row data, not from the columns. This kind of plot is often used to visualize series of data that change with each observation. It's called "waterfall" because the resulting plot looks like, well, a waterfall. The **waterfall** function takes the same form as **mesh**.

As an example, suppose a signal is received by a sensor once a second for one hundred seconds, but decays exponentially each second. The following code simulates such a scenario.

```
x=-3*pi:.25:3*pi;  %resolution of the signal
A=linspace(3,0)    %100 samples
A=exp(-A);         %exponential decay
X=sin(x).^2./(x+eps).^2;
Y=A'*X;            %the decaying signal
waterfall(Y)
```

The waterfall plot of this multiple series of data is shown in Figure 4.8 and Plate 3.

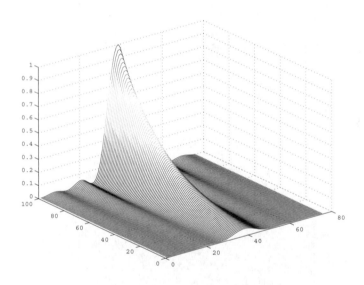

Figure 4.8 A waterfall plot of a simulated exponentially decaying signal.

Although this contrived example created its data series in row order, data analysis functions in MATLAB typically produce data in column order, that is, each data series appears as a column in a matrix. In that case, remember to transpose the matrix before calling **waterfall**.

4.1.4 3-D Plots of Non-Uniformly Sampled Data

If you are running experiments or collecting data from real world situations, you will probably encounter situations in which you do not have data points that are nicely spaced at equal increments of your input variables. Fortunately, MATLAB has a way that allows you to represent this type of data in a plot. As an example, let's pretend that you have collected samples from a process that exhibits a response similar to the function we used in the previous example, z=sin(x)cos(y). We will generate our data samples (x,y,z) with

```
x = rand(100,1)*2*pi;
y = rand(100,1)*pi;
z= sin(x).*cos(y);
```

This data, by itself, could not be viewed as a mesh or surface plot. The best you could do is generate a 3-D plot with plot3(x,y,z,'.') to see the points; however, even with plot3 it is very difficult to get a feel for what the surface defined by the data points really looks like. Therefore, we need to generate a set of evenly sampled data points that are generated by interpolating between the set of original data points. First we create uniformly sampled input variables using the meshgrid and linspace functions to create, in this example, a 40-by-40 X and Y matrix over the region defined by our data.

```
[X,Y] = meshgrid(linspace(min(x),max(x),40),...
                 linspace(min(y),max(y),40));
```

Then we let MATLAB do the work of interpolating the original data across the uniformly spaced region with the griddata function.

```
Z = griddata(x,y,z,X,Y,'cubic');
```

Finally, we can plot it with

```
mesh(X,Y,Z);   % View interpolated surface
hold on;
plot3(x,y,z,'.','markersize',10);   % View actual samples
```

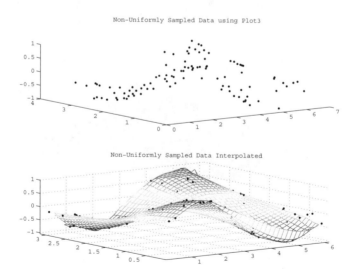

Figure 4.9 Mesh plotting helps to visualize non-uniformly sampled data.

The plot similar to that in Figure 4.9, with the exceptions being most evident around the fringes of the plot since there are not any data points outside the region with which MATLAB can estimate the surface.

4.1.5   Creating Shaded Surface Plots

Depending on the relative spacing of your data, you may want to make use of the **shading** function. You may have noticed in your mesh plots that each line segment between the mesh intersections maintains a single color attribute over the length of the segment. With some data, this may not be appropriate or may even be misleading, especially if the sampling interval is large. Previously we used color to identify surface slope and height, but the sampling interval was small. This was easy enough to do since we were dealing with function data, but with real data, we would have been forced to resample our data to get the smaller increments. The quick alternative solution to resampling data more finely is to use **shading** function with the "interp". This command will interpolate the line colors so that the color varies linearly across the length of the segment. If, after applying the interpolated shading, you determine that this is not what you want, you can always revert back to the default line colors by typing **shading** faceted or **shading** flat. We will revisit **shading** when we discuss the ways to manipulate 3-D visualizations later in this chapter. For now, we will discuss it in terms of the function **surf**.

The **surf** function is used identically to **mesh**. However, instead of the surface being represented by a screen-like grid, **surf** will produce a 3-D shaded surface. Figure 4.10 shows the example of Figure 4.4 with **surf** used in place of **mesh**.

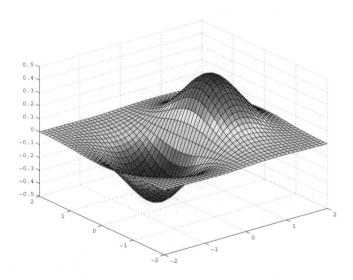

Figure 4.10 Using surf to produce a surface plot.

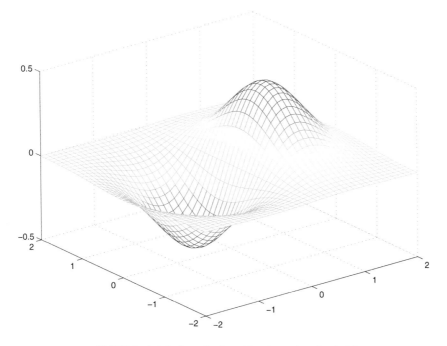

PLATE 1 A default mesh plot with color assigned to height.

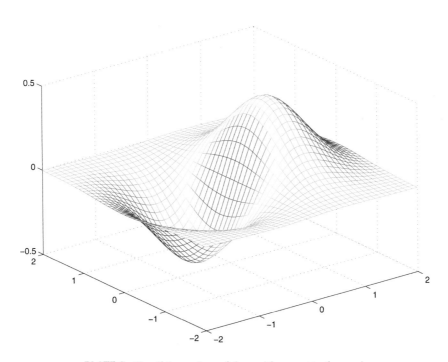

PLATE 2 Identifying regions of slope with respect to the x-axis.

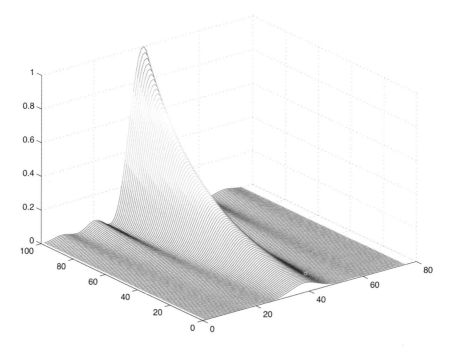

PLATE 3 A waterfall plot of a simulated exponentially decaying signal.

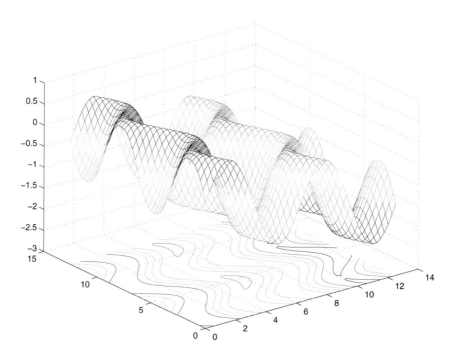

PLATE 4 Using meshc to create a mesh - contour combination plot.

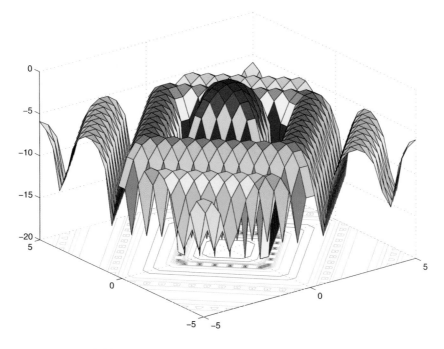

PLATE 5 A surface - contour combination plot made with surfc.

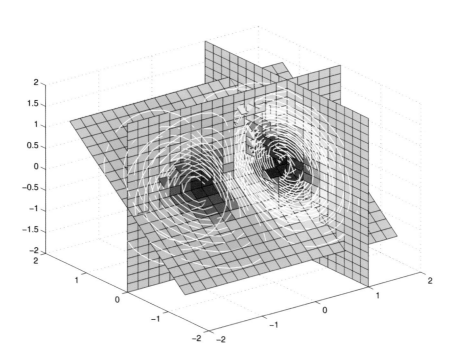

PLATE 6 Contour lines on slice planes.

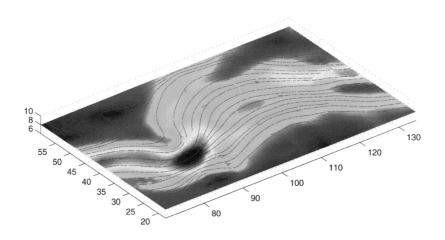

PLATE 7 Using streamslice to visualize the direction of flow.

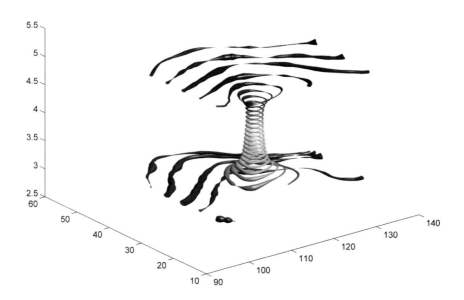

PLATE 8 Stream tubes are used to show divergence.

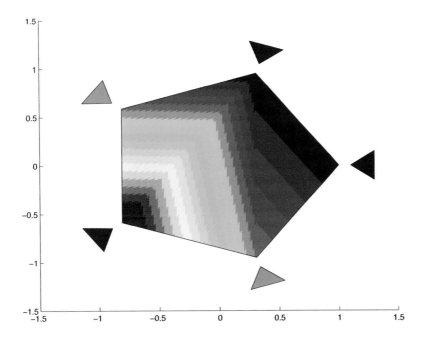

PLATE 9 Defining the color of patch vertices.

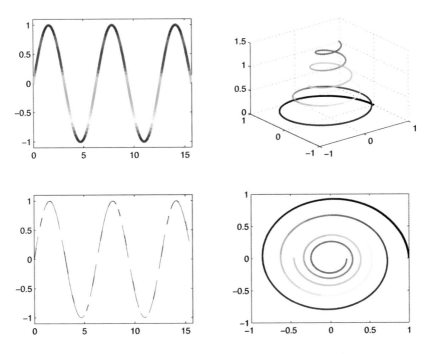

PLATE 10 Using multiple color maps in the same figure window.

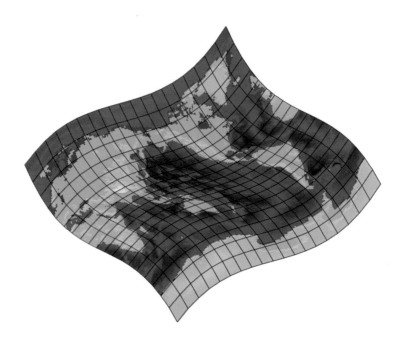

PLATE 11 A non-image data set mapped to a surface object.

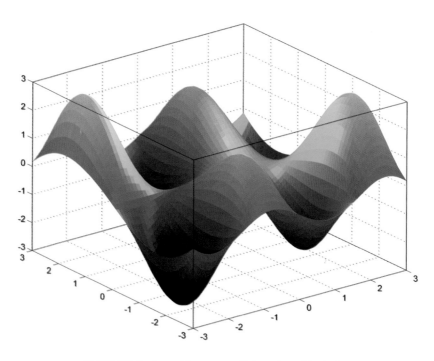

PLATE 12 Surface with one green light and one blue light.

PLATE 13 Mixing specular and diffuse reflectance models.

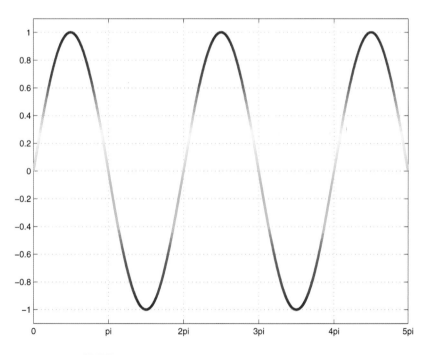

PLATE 14 Making a virtual line with surface to create a
line where the color changes as a function of the y-coordinate.

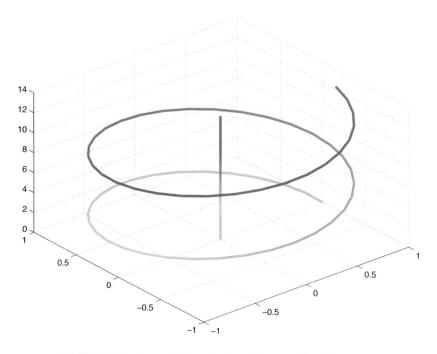

PLATE 15 Creating multiple color lines with one surface object.

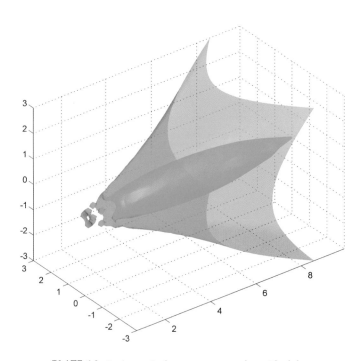

PLATE 16 Setting a single transparency value with alpha.

As you can see, Figure 4.10 has the appearance of a solid surface covered by a grid. If we did not want the grid in our plot, we can use either **shading** flat or **shading** interp. Using **shading** flat removes the grid, but the coloring is still piecewise constant, i.e., each mesh line segment has a constant color value so you see each "patch" as shown in the left plot of Figure 4.11. Using **shading** interp varies each color in a segment of the plot linearly, i.e., it interpolates the color and results in a smooth-looking surface plot as shown in the right half of Figure 4.11.

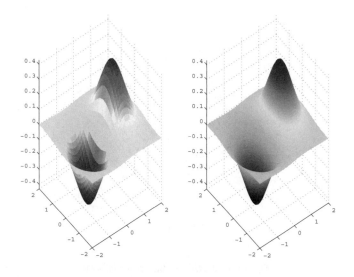

Figure 4.11 Using shading to change the appearance of surface plots.

You can reproduce the plots of Figure 4.11 with the x, y, and z that produced Figure 4.4 and applying the following commands.

```
subplot(1,2,1);surf(x,y,z); axis('tight');
shading('flat');
subplot(1,2,2);surf(x,y,z); axis('tight');
shading('interp');
```

4.1.6 Removing Hidden Lines

When you created a mesh plot, you might have noticed that the mesh lines behind the mesh surface are not visible. What you have seen with the **mesh** and **surf** functions can be likened to a solid surface made up of hills and valleys that has a multicolored net draped over it. Depending on where you are standing in this scene, you will not be able to see behind the hills and down into some of the valleys. Depending on the viewpoint, certain lines are not drawn so that the 2-D representation of the 3-D data provides a relative perspective of the surface shape and lines defining the surface. The process of eliminating some of the lines as a function of perspective is usually referred to as hidden line removal. However, in some cases, you might wish to have the hidden lines visible. The function **hidden** allows you to turn off or turn on the hidden line removal. Simply put, to hide lines use **hidden on** and to make them visible use **hidden off**. Using the function by itself will toggle between the on and off states. Figure 4.12 shows a mesh plot of the peaks function with hidden line removal on (default), and Figure 4.13 shows it with hidden line removal off. Typing **mesh**(**peaks**) will produce the plot.

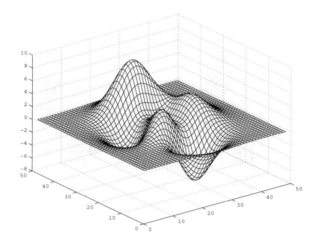

Figure 4.12 Hidden line removal on.

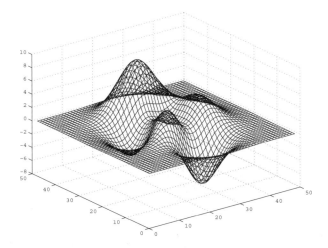

Figure 4.13 Hidden line removal off.

4.1.7   Contour Plots

Contour plots are an excellent way of visualizing some of your matrices. Contours represent the constant data values with lines called isolines. MATLAB provides both 2-D (top-down view) and 3-D (perspective view) contour plots. We cover both the 2-D and 3-D contour plots in this section since these plots are often associated with 3-D data in some way.

The simplest way to create a 2-D contour plot is to pass your matrix, Z, with **contour**(Z). MATLAB will automatically choose the number and values at which contour lines are drawn. You can also specify either the number of lines with **contour**(Z,number_of_lines) or the values at which the contour lines will be drawn with **contour**(Z,vector_of_data_levels). If you want to plot only a single contour data level, make the vector_of_data_levels a two-element vector with both elements set to the data level you want contoured.

The three methods just mentioned are plotted versus the row and column number of the matrix Z, such that the element Z(1,1) will be located in the lower left-hand corner of the figure. You also have the option of defining the x- and y-axis scaling by passing either vectors or matrices that specify the x- and y-coordinates associated with each element of the matrix Z. If these axis scaling matrices are used, they should be passed as the first two arguments to the **contour** function, i.e.,

contour(x_scale,y_scale,Z),

contour(x_scale,y_scale,Z,number_of_lines)

contour(x_scale,y_scale,Z,vector_of_data_levels)

As an example of the **contour** function, the following code will generate some data and create a contour plot as shown in Figure 4.14.

```
[x,y] = meshgrid(linspace(0,2*pi,30),...
```

```
linspace(0,pi,30));
z = sin(x).*cos(y+pi/2);
% In the next line the contour plot is created for
% data levels between -1 and 1 in 0.1 intervals
% excluding the 0 data level.
contour(x,y,z,[-1:0.1:-0.1 0.1:0.1:1])
xlabel('x');
ylabel('y');
title('Contour of z = sin(x).*cos(y+pi/2)');
```

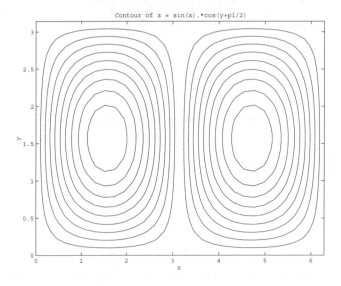

Figure 4.14 A simple 2-D contour plot.

Notice that the color of the contours is chosen in the same manner as colors are chosen when creating multiple lines with the plot command. In addition, in Figure 4.14 it is impossible to tell what value the data levels correspond to or whether there are two hills, two valleys or one hill and one valley. Fortunately, there are two options to remedy this problem; the first is to use the function **clabel**, which will attach a numeric text string to each line, the second is to create a 3-D contour plot by passing the same arguments to the function **contour3** instead of contour.

The left half of Figure 4.15 shows how to use **clabel** with the data from the previous example with:

```
c = contour(x,y,z,[-1:0.1:-0.1 0.1:0.1:1]);
clabel(c);
```

The right half shows how to specify which contour lines are labeled by passing an additional argument to **clabel** as follows:

```
c = contour(x,y,z,[-1:0.1:-0.1 0.1:0.1:1]);
clabel(c,[-1:.2:1]);
```

Note that in both cases the location of the text is randomly assigned.

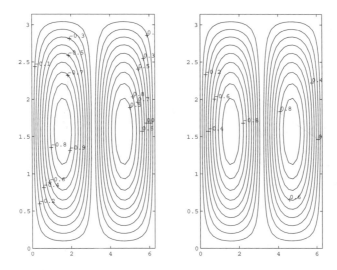

Figure 4.15 Using clabel to label contour plots.

You can also elect to manually select which contour lines are to be labeled and at the same time specify the location of the text by using **clabel**(c,'manual'). After you enter this command, a crosshair will appear instead of the normal mouse pointer arrow. Click down on the mouse button (or use the space bar) and a label will be drawn as a plus sign with a height value and attached to the contour line that is the closest to the location you clicked on. When you have labeled as many of the contour lines that you want, press the return key on your keyboard while the cursor is still in the Figure Window to indicate that you have finished.

MATLAB also provides an automatic labeling method to generate plots like that shown in Figure 4.15. To do this you must call the **contour** function and retrieve both the contour matrix and the handles of the line objects. Don't worry about the handles too much just yet as that will become clear in Chapter 7. For now, you can use the following code to produce the plot in Figure 4.16.

```
[C,h] = contour(x,y,z);
clabel(C,h);
```

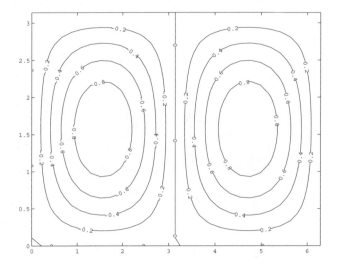

Figure 4.16 Automatic labeling of isolines.

MATLAB uses an algorithm to determine where the labels are to be placed. However, you can override this algorithm and manually place the labels with,

```
clabel(C,h,'manual');
```

As when you used **clabel** with 'manual' before, cross hairs will appear on the figure and follow the mouse pointer. The difference between this manual method and the one without the use of the handles is that this method will not produce a plus sign, but will put the value directly on the isoline.

Additionally, you can have **clabel** return the graphics handles to labels so that you can specify the properties of the labels, such as the font size or color. For instance, we can create Figure 4.17 with,

```
[x,y,z] = peaks;
% Create black dashed contours
[C,h] = contour(x,y,z,'--k');
[text_handles] = clabel(C,h);
% Modify the labels to make them bigger and blue.
set(text_handles,'fontsize',15,'color','blue');
```

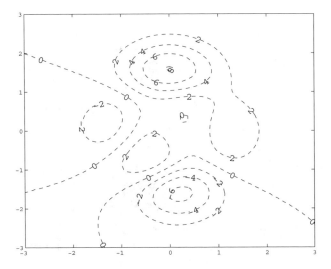

Figure 4.17 Manipulating contour label properties.

A filled contour plot displays isolines with the areas between filled with a constant color. To create a filled contour plot, use the function **contourf**. Each level of the contour is filled in with a color from the current color map. The color corresponds to the relative height of the level in the same way that color is chosen to represent the relative height of a surface plot. The following code will recreate the previous example as a filled contour plot.

```
contourf(x,y,z,[-10:10],'--k');
```

The result is shown in Figure 4.18.

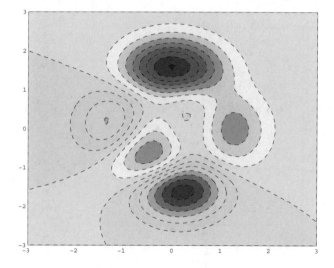

Figure 4.18 A filled contour plot.

Notice that in this code the vector [-10:10] was passed to the **contourf** function. The contour plotting functions accept a vector specifying the levels at which to plot contours.

The last contouring function we will consider is the 3-D contour plotting function **contour3**. This function allows you to see the relative heights of the isolines. As with **contour** and **contourf** you can pass a vector specifying the levels you want to plot. The following code produces the plot shown in Figure 4.19.

```
contour3(x,y,z,[-10:10],'-b')
axis tight
```

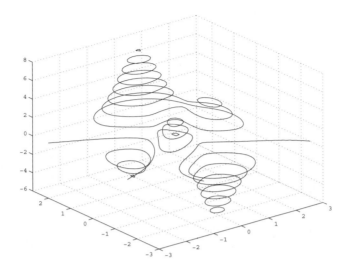

Figure 4.19 Showing the relative heights of isolines with contour3.

After you have learned more about color maps in Chapter 8, you will see just how powerful the contouring functions can be in assisting you with collecting more information about your data.

4.1.8 Quiver Plots

Quiver plots are used to visualize the gradient fields of either mathematical functions or data. For instance, you can plot arrows that point in the direction of increasing or decreasing values in a matrix and that have lengths that indicate the relative slope of the gradient at the particular locations. The graphics function that is used to create this type of plot is **quiver**. There are several different forms that can be used, but the most general is **quiver**(X,Y,PX,PY,scale,linetype_string) where the matrices X and Y define the locations of the arrows, PX and PY matrices determine the direction and magnitude of the arrows, the scale variable is used to adjust the length of all arrows by the specified factor, and the linetype_string can be used to specify the color and linestyle as was presented when the **plot** command was discussed. The partial derivatives (PX and PY) of a given surface can be obtained with the function **gradient**. To illustrate the quiver plotting function, let's look at the quiver plot of the **peaks** function shown in Figure 4.20.

```
[X,Y,Z] = peaks(20);
% Determine the spacing of X matrix elements
dy = diff(X(1,1:2));
% Determine the spacing of Y matrix elements
dx = diff(Y(1:2,1));
% Determine the partial derivatives
[PX,PY] = gradient(Z,dx,dy);
quiver(X,Y,PX,PY,1,'b');
axis([min(min(X)) max(max(X)) min(min(Y)) max(max(Y))]);
```

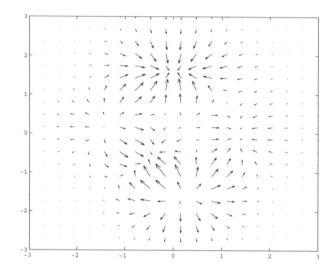

Figure 4.20 A quiver plot of the peaks function.

Figure 4.20 has arrows pointing in the direction of increasing Z. However, to change the direction of the arrows to point in the decreasing Z direction, all that is needed is to type **quiver**(X,Y,-PX,-PY). The first two arrow location defining matrices can be placed with vectors using **quiver**(x,y,PX,PY) or **quiver**(x,y,PX,PY,scale, linetype_string), where the length of x is equal to the number of columns in PX and PY and the length of y is equal to the number of rows in PX and PY. If it is not important to know the x- and y-axis locations of the arrows you can use **quiver**(PX,PY) or **quiver**(PX,PY,scale,linetype_string). The scale parameter defaults to a value of 1, indicating that MATLAB will automatically scale the arrow length. A scale value of 0 will plot the arrow length without scaling.

4.1.9 Combination Plots

Perhaps you have been wondering how you can combine different plot types in order to visually correlate the information in your data? Since 2-D and 3-D representations each tend to emphasize different aspects of the information in a plot, the combination of a surface plot with a contour plot, for example, of the same data would present a great deal of information in a compact form. There are a couple of MATLAB functions that will create useful combination plots, but it is very easy to create your own functions to produce just the combination plots you want. However, before you can design a truly custom combination plot, you will need to learn a little more about graphics objects and their properties so that you can manipulate them to your liking. Once you see how easily creating your own graphics functions can be accomplished, you will only be limited by your imagination with regard to adding new functionality in your repertoire of M-files.

If it hasn't occurred to you yet, you have already looked at some simple combination plots when we used the **hold** function to overlay line plots within the same figure. The same can be done with any of the other graphics creating functions. For example, in many cases the information provided by a flat quiver plot can be made easier to comprehend by overlaying a contour plot. Let's take the quiver plot example shown in Figure 4.20, and overlay the corresponding contour. First, create the contour with

```
[X,Y,Z] = peaks(20);
% Determine the spacing of X matrix elements
dy = diff(X(1,1:2));
% Determine the spacing of Y matrix elements
dx = diff(Y(1:2,1));
% Determine the partial derivatives
[PX,PY] = gradient(Z,dx,dy);
quiver(X,Y,PX,PY,1,'b');
axis([min(min(X)) max(max(X)) min(min(Y)) max(max(Y))]);
then type
hold on
```

and create the contour overlay with

```
[C,h] = contour(X,Y,Z,[-8:2:8]);
clabel(C,h);
```

which will produce the result shown in Figure 4.21. This plot is much more informative than the plot that either **quiver** or **contour** could have provided by themselves.

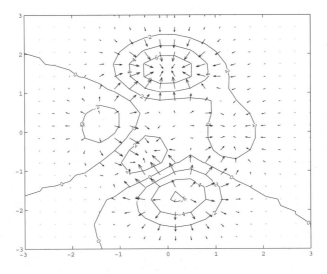

Figure 4.21 A quiver and contour combination plot.

As another example, we can create a three-dimensional quiver plot combined with a surface plot. The 3-D quiver plot can be created using MATLAB's

quiver3 function. Using the **peaks** function as in the previous example the following code will create the data and plot it as shown in Figure 4.22.

```
[X,Y,Z]=peaks(20);
% Determine the surface normals
[U,V,W] = surfnorm(X,Y,Z);
% Generate the 3D quiver plot
quiver3(X,Y,Z,U,V,W);
hold on;
% Now add the surface plot
surf(X,Y,Z);
hold off
```

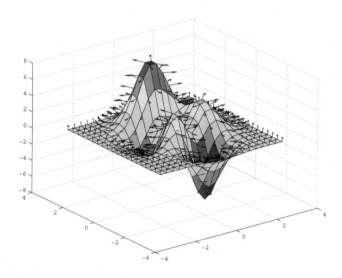

Figure 4.22 A combined 3-D quiver and surface plot.

As the previous examples show, combining different plot types can provide significant insight into data. In these cases, the plots were of the same dimension. The MATLAB plot axes are designed to allow any type of plot to be combined with any other. As such, you can readily combine 2-D and 3-D plots.

   As you have seen, the **hold** function allows different plots in the same axes. In addition to using **hold**, MATLAB provides two specific combination plots that combine a contour plot with either a mesh or surface plot. The first function **meshc** will create a mesh plot with a contour plot directly below it. The following example will help you better understand how this type of plot might be used. Consider the surface defined by the equation $z = \sin(x + \sin(y)) - x/10$. The first step is to create the surface over some values of x and y.

```
[x,y] = meshgrid(0:.25:4*pi);
z = sin(x+sin(y))-x/10;
```

The plot shown in Figure 4.23 and Plate 4 is achieved by simply plotting the surface with

```
meshc(x,y,z);
```

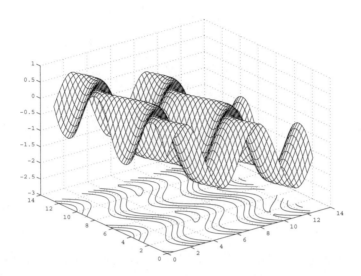

Figure 4.23 Using meshc to create a mesh – contour combination plot.

In a similar manner, a combination surface and contour plot can be created with the command **surfc**. As an example, we can use **besselj** (the bessel function) to generate some data in the following example which is plotted in Figure 4.24 (see also Plate 5).

```
[x,y] = meshgrid(-5:.4:5);
z = abs(besselj(0,abs(x)+abs(y)))+.01;
surfc(x,y,10*log10(z));
```

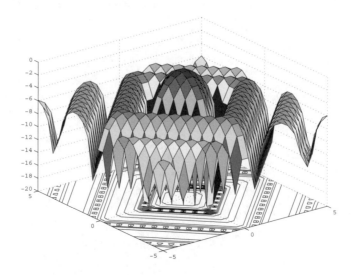

Figure 4.24 A surface – contour combination plot made with surfc.

As you experiment with the **surfc** and the **meshc** functions you will notice that the contour is always plotted at the lowest z-axis limit that appears in the figure. In many cases this is simply unacceptable since the contour can be easily obscured by the surface or mesh plot. If you are lucky enough to choose a function that shows you enough of the contour when using these two functions, then great! But if you are not so lucky, and generally speaking you won't be, it does not mean that these combination plot functions will be useless. One rather obvious work around is to simply use the view axis tool in the Figure Window, or to use the function **view**. The default perspective sets the observer at -37.5º azimuth and 30º elevation (i.e., **view**([-37.5 30])). You could just lower the elevation so as to peek under the surface a bit more, perhaps with **view**([-37.5 15]). The problem with this is that the perspective of the plot is changed, likely making it difficult to extract information from the contour lines or the surface plot, and therefore may not be desirable for some data sets. The real solution lies in using a little Handle Graphics. Although we will explore the topic rather thoroughly in Chapter 7, just as in the last example of Section 3.4.1, we will resort to a little Handle Graphics here. After you have grasped the concepts in Chapter 7, this example will seem very simple and straightforward to you. In the meantime, you can merely resort to this technique as it is, and dig under the surface of what is going on later.

Our best solution is to relocate the contour plot, i.e., offset it, to a level where the surface or mesh plots cannot obscure it. Simple enough in concept, but how is this accomplished? As we will discuss in Chapter 7, we will take advantage of one of the properties of the contour plot, that is its *Zdata*. Later you will learn that everything in MATLAB is an object and every object has properties, and you can change the value of those properties. Without further explanation, the process here requires two steps. First we must get "handles" to the part of the plot we want to affect, in this case the contour plot lines.

Second, we will use the handles to the plot lines to access the z-axis data and add an offset to it. Consider again the plot shown in Figure 4.24. Calling **surfc** as shown here will not only plot the data, but will also return the "handles" to what we want in H.

Power!

```
H=surfc(x,y,10*log10(z));
```

In this case, the first handle returned in H belongs to the surface; the remaining handles belong to the contours we want to change. We can lower the contour plane, by subtracting 5 (adding an offset of –5) from the value of each z coordinate of each contour line. Here is the code that does it.

```
H=surfc(x,y,10*log10(z));
for i = 2:length(H);
    newz = get(H(i),'Zdata') - 5;
set(H(i),'Zdata',newz)
end
```

Figure 4.25 shows the "before and after" of offsetting the contour plot from the surface plot.

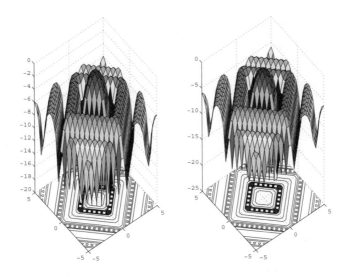

Figure 4.25 Before and after contour plot shifting.

We can't just simply subtract 5 from z since that would alter the surface portion of the plot. The solution shown here preserves the original data.

4.1.10 3-D Stem Plots

Stem plots were introduced in Chapter 3 and we discussed how they are useful for visualizing discrete data sequences such as sampled time series data. Similar to the **stem** function, **stem3** creates vertical lines terminated with a symbol but instead of emanating from the y-axis as in the case of **stem**, the lines emanate from the xy-plane. The forms of this function are:

stem3(Z) plots the discrete surface Z as stems from the xy-plane terminated with circles for the data value.

stem3(X,Y,Z) plots the surface Z at the values specified in X and Y.

Using the keyword string 'filled' will create the stem plot with filled markers just like with **stem**. Also, you can specify the style of lines and markers used just as with the **plot** function (refer to Table 3.3.1).

As an example, we can visualize the sine from 0 to 2π around a unit circle with the following code.

```
theta = 0:.2:2*pi;
x=sin(theta);
y=cos(theta);
z=sin(theta);
stem3(x,y,z);
hold on
plot3(x,y,z,'r')
plot(x,y)
title('Sine Along the Unit Circle')
zlabel('Sin(theta)')
```

This code also plots the unit circle as well as a red line through the stems as shown in Figure 4.26.

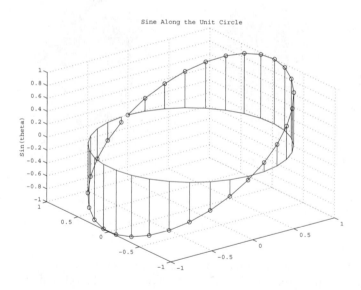

Figure 4.26 A 3-D stem plot with supporting line plots.

4.1.11 Generating Surfaces with Triangles

As you may have noticed, the **surf** and **mesh** functions use quadrilaterals as defined by neighboring vertices in your X, Y, and Z matrices to generate the 3-D mesh or surface plot. In some instances, you may have data that you want displayed by a set of triangles. The functions **trimesh** and **trisurf** can be used to generate a triangular mesh and surface plot respectively.

Both of these functions have the same synopsis and are therefore completely interchangeable. After you have learned more about object types in Chapter 7, you should revisit these two functions and notice that the two functions create the same object with only minor changes in the attributes of that object.

To help you understand how these functions work, we will look at a simple example. Let's say we have the data points as described in the following code and shown in Figure 4.27.

```
x = [0 1 1 0 0.5 0.5]
y = [0 0 1 1 0.5 0.5]
z = [0 0 0 0 1 -1];
plot3(x,y,z,'o','markersize',4,...
    'markerfacecolor','black');
axis equal;
grid;
for i=1:length(x)
    text(x(i),y(i),z(i),num2str(i),...
        'verticalalignment','bottom');
end
```

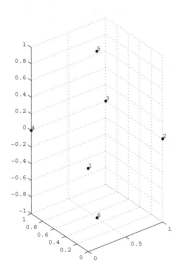

Figure 4.27 Data points for a triangular plot.

We can then create a set of eight triangles: one face that connects data points 1, 2, and 5, another for data points 2, 3, and 5, another for data points 3, 4, and 5, another for 4, 1, and 5, another for 1, 2, and 6, another for 2, 3, and 6, another for 3, 4, and 6, and a final one for 4, 1, and 6. This is done by creating an mx3 matrix, where each of the m rows represents a triangle by identifying the three indices in the x, y, and z vectors that make up the three vertices of the triangle. Continuing with the x, y, and z data we've just created, the following code will create this matrix and produce the plot shown in Figure 4.28.

```
%specify the triangles
tri=[1 2 5;
     2 3 5;
     3 4 5;
     4 1 5;
     1 2 6;
     2 3 6;
     3 4 6;
     4 1 6];
% generate the triangular mesh plot
hold on;
trimesh(tri,x,y,z,'edgecolor','black');
```

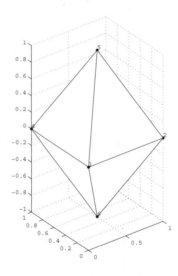

Figure 4.28 Triangular meshplot of the three data points.

Each row of the matrix tri specifies the points that constitute each face of the object.

Using the **peaks** function that we saw in the earlier surface plots, we can see that the **trisurf** function can also be used as a way to get a look at a surface from a set of non-uniformly sampled data points. Consider the following code that will generate the surface shown in Figure 4.29.

```
x = 6*rand(1,500)-3;
y = 6*rand(1,500)-2;
z = x .* exp(-x.^2 - y.^2);
tri = delaunay(x,y);
trisurf(tri,x,y,z);
grid on;
```

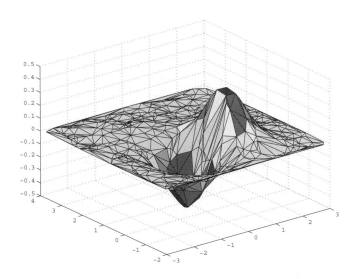

Figure 4.29 Visualizing non-uniformly sampled data points using trisurf.

The **delaunay** function creates a triangular grid for scattered data points by returning a set of triangles such that no data points are contained in any triangle's circumcircle. Put in simpler terms, each point is matched with its natural neighbors (as determined by the underlying algorithm) to produce a triangle, a circle about which will cover no other data points. This will assure that there are the required three data points to define a triangle. Try playing around with this code by running it multiple times and so producing a new data set with **rand**, and by changing the number of data points affecting the number of triangles.

4.1.12 Polygons in a 3-D Space

In Chapter 3 we saw that 2-dimensional polygons could readily be created with the MATLAB **fill** function. Just as **plot3** was the 3-dimensional counterpart to **plot**, **fill3** is the 3-dimensional counterpart to **fill**. The command **fill3** is used in the same way as **fill** but with an additional vector or matrix used to define the z-axis coordinates of the polygon. So, for example, where you used the form **fill**(x,y,color_string), you could now use **fill3**(x,y,z,color_string). When we discuss handle graphics in Chapter 7 we will consider the **patch** function, which enables you to create any sort of polygon and mix and match them. We save this for the discussion on handle graphics since you will need to have a firm grasp (pun intended) on the concepts of objects and properties.

4.1.13 Built-In Surface Functions

You have already seen that MATLAB provides a built-in surface function called **peaks**. Although useful for demonstration purposes, **peaks** isn't all that practical. Of course, in theory anyway, you can always create your own functions for any surface you desire. Fortunately, MATLAB includes three very useful surface generating functions in its base set of graphics commands. You can generate spheres, ellipsoids, and cylinders without determining what the coordinates of the surface vertices should be.

The command **sphere**(n) will generate a plot of the unit sphere. The sphere will be defined with (n+1)2 points. If you do not supply a number to this graphics function, n will default to 20. You also have the option of having the function pass the (x,y,z) coordinates of the sphere by using output arguments with the **sphere** command. When the function is used in this manner the plot will be suppressed. This allows you to alter the coordinates of the sphere and then plot it with the **mesh** or **surf** commands. For instance, we could scale a sphere and translate it in the 3-dimensional space. The following code will plot both a translated version of the unit sphere, which is centered on something other than the point (0,0,0), and a scaled version of the unit sphere. Figure 4.30 shows the result.

```
[x,y,z] = sphere(25);
surf(x-3,y-2,z);        %translated
hold on
surf(x*2,y*2,z*2);      %scaled
```

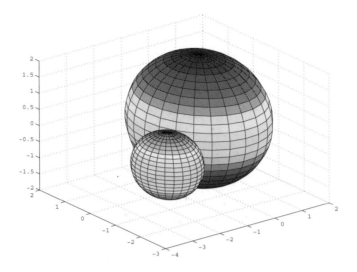

Figure 4.30 An example using the sphere function.

The **ellipsoid** function is actually based on the **sphere** function and produces x, y, z coordinates for the ellipsoid described by the equation,

$$\left(\frac{x-x_c}{r_x}\right)^2 + \left(\frac{y-y_c}{r_y}\right)^2 + \left(\frac{z-z_c}{r_z}\right)^2 = 1$$

Where x_c, y_c, and z_c are the centers of the radii and r_x, r_y, and r_z are the radii in the corresponding axis. The general form of the **ellipsoid** function is [x,y,z]=ellipsoid(xc,yc,zc,rx,ry,rz,n). As with the **sphere** function, n relates to the number of data points computed and is assumed to be 20 if it is not otherwise specified. As an example, the plot shown in Figure 4.31 depicts an ellipsoid centered at x=2, y=0, and z=2, with x-radius = 2, y-radius = 1, and z-radius = 1. (Figure 4.31 is actually a combination plot; we've included the contour in order to better visualize the elliptical shape.) The following code will produce Figure 4.31.

```
[x,y,z]=ellipsoid(2,0,2,2,1,1);
surf(x,y,z);
axis([0 4 -2 2 0 4]);
hold on
contour(x,y,z);
```

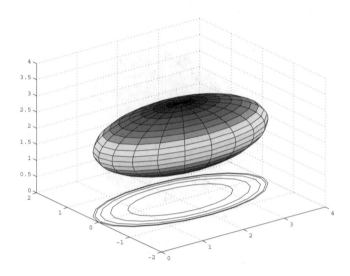

**Figure 4.31 An example using data created with the ellipsoid function
(contour included for clarity).**

Note that if you specify an ellipsoid with all radii equal to 1, you will create
the unit sphere.

The final built-in surface function MATLAB includes is **cylinder**. When
cylinder is called without any input or output arguments it creates a 3-
dimensional perspective of a unit cylinder, i.e., radius of one and height of
one, standing upright. Calling the function with output arguments will return
matrices that specify the coordinates, (x,y,z), of the vertices that define the
cylinder in the 3-dimensional space. This data is then useable by **surf** or **mesh**
to create a plot of the surface. There are two optional input arguments that
can be used in which case the function takes the form **cylinder**(R,N). The first
input argument, R, is a radius vector that defines the radius of the cylinder at
equally spaced points along the cylinder's height, i.e., the z-axis direction. A
mathematical function can be used to generate R and so create a cylinder with
radial profile described by that function. By default, the vector defining the
radius is set to [1 1]. A cone, for instance, would be created using

```
cylinder([0 1])
```

The second input argument, N, is an integer that specifies how many points
will be used to define the circumference of the cylinder. As with its
counterpart in **sphere** and **ellipsoid**, the default value is 20. The height of the
cylinder is always scaled to run between 0 and 1; but you can scale the height
by calling the function with output arguments, then manipulating the matrix
defining the z-coordinates of the vertices, and use **surf** or **mesh** to create the
surface.

We can easily make regular cylinders with **cylinder**, but it is much more
interesting to use a function to create a radial profile and then create a
cylinder with that. To illustrate what the function **cylinder** can do, let's work
with the mathematical expression

$$r = \cos(x) \cdot \sin(x/2) \cdot e^{x^2/200}$$

for x between -3π and 3π. The following code generates a plot of the radial profile of the cylinder that we are about to create as shown in the left panel of Figure 4.32.

```
% Define the x data range
x = linspace(-3*pi,3*pi,50);
% Evaluate the function
r = cos(x).* sin(0.5*x)*exp((x.^2)/200);
% Force the minimum radius to zero.
r = r - min(r);
plot(r,linspace(0,1,length(r)));
title('Radial Profile');
ylabel('z')
```

Try to imagine spinning this radial profile about the z-axis in a manner that pushes the profile into and out of the page. The elements of the radial vector, r, do not need to be all positive quantities. For example, in the previous set of MATLAB instructions,

```
r = r-min(r);
```

forced the minimum radius to equal zero. Now we can use the **cylinder** function to visualize the radial profile as shown in the right panel of Figure 4.32.

```
cylinder(r);
```

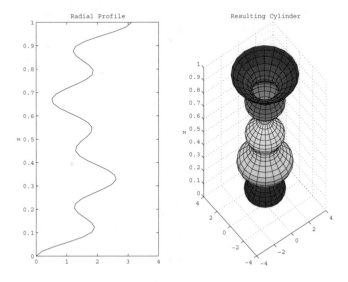

Figure 4.32 A function-described radial profile and its corresponding cylinder.

The central axis of any shape created with **cylinder** is defined by a line that is perpendicular to the xy-plane and passes through the coordinate (0,0) in this plane. If you need to redefine the central axis location or scaling in any of the coordinate directions, first obtain the vertex coordinates with

```
[X,Y,Z] = cylinder(r);
```

then scale by multiplying one or more of these matrices by some factor, or translate by adding a constant to one or more of the matrices. Finally, generate the surface with one of **mesh** or **surf**.

4.2   Simple 3-D Plot Manipulation

In Chapter 3 we presented plot editing using the tools available in the Figure Window. These tools are just as applicable in the 3-D case as they are for 2-D plots. The Insert Text, Insert Arrow, etc., all work just as in the 2-D case; however be aware that if you annotate your plot before rotating it, the annotations will not move with the plot. This can lead to confusion and frustration so the rule of thumb is to set your view before you begin annotations. Simply select the cursor icon in the Figure Window toolbar to enter the plot edit mode, or type **plotedit** at the command prompt in the Command Window. Then, just as in the 2-D case, you can access each object in the figure and edit their properties by a simple point-and-click interface.

4.2.1   The Camera Toolbar

We did not cover the camera tools in the discussion of the Figure Window tools in Chapter 3 as this is much more meaningful when dealing with 3-D plots. Although using the camera tools might seem like animation, and in a way it is, we reserve Chapter 9 for a detailed discussion of "proper" animation. Here we will only deal with "simple" uses of the camera, namely those available from the Figure Window toolbar. To facilitate this discussion, open the Figure Window by creating a surface plot of the **peaks** function.

```
surf(peaks(30))
```

Now select **View** → **Camera Toolbar**. When you do, the Camera Toolbar will appear in the Figure Window looking like Figure 4.33.

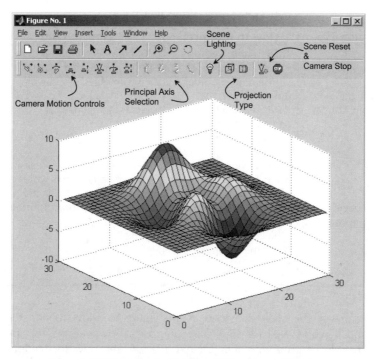

Figure 4.33 The Camera Toolbar in the Figure Window.

The Camera Motion Controls let you select different camera motion controls such as those that will orbit the camera or scene lighting, pan or tilt the camera, move the camera in and out, etc. The Principal Axis Selection tools provide a choice of axis about which some of the camera controls will operate. The Scene Lighting tool is a toggle that switches a light on or off. It can be useful in emphasizing the elevations and valleys in a surface. The Projection Type tool lets you choose between orthographic and perspective projections. Table 4.2.1 summarizes the two projection types, their consequences, and their use. The last two tool buttons, Reset Camera and Scene Light, and Stop Camera/Light Motion, let you reset the scene to the standard 3-D view and stop the camera from moving. In the next section, we will discuss the axis in general as it relates to 3-D graphics. In Chapter 7 we will explore the camera and how to program its properties using Handle Graphics.

Table 4.2.1 Projection Types

Projection Type	How to Interpret	How to Use
Orthographic Projection	Think of the "viewing volume" as a box whose opposite sides are parallel, so the distance from the camera does not affect the size of surfaces in the plot	Used to maintain the actual size of objects and the angle between objects. This works well for data plots. Real-world objects look unnatural.
Perspective Projection	The "viewing volume" is the projection of a pyramid where the apex has been cut off parallel to the base. Objects further from the camera appear smaller.	Used to create more "realistic" views of objects. This works best for real-world objects. Data plots may look distorted.

You can use the tools provided in the Figure Window to fine tune your plots, and it works well for single-use purposes, however the real power of MATLAB is in its programmability. Next, we will see how to manipulate the axis in code, as well as high-level color and shading manipulation.

4.2.2   Generalizing the Axis for 3 Dimensions

The **axis** function we used in Chapter 3 (**axis**([xmin xmax ymin ymax])) is fully generalized as **axis**([xmin xmax ymin ymax zmin zmax cmin cmax]), where xmin, ymin, and zmin are respectively the minimum x-, y-, and z-axis values, xmax, ymax, and zmax are the respective maximum x-, y-, and z-axis values, and cmin and cmax are color scaling limits. Uses such as **axis**('equal'), **axis**('ij'), and **axis**('xy') also manipulate the 3-D plot but only the x- and y-axis of the current plot are affected as discussed in Chapter 3. Table 4.2.2 summarizes the **axis** function syntax and its affect on a plot.

Table 4.2.2a Summary of the Axis Function

Syntax	Affect
axis([xmin xmax ymin ymax])	Sets the x- and y-axis limits .
axis([xmin xmax ymin ymax zmin zmax cmin cmax])	Sets the x-, y-, and z-axis limits and the color scaling limits.
v = axis	Returns a row vector containing the x-, y-, and z-axis limits, i.e., scaling factors for the x-, y-, and z-axis.
axis auto	Computes the current axes' limits automatically, based on the minimum and maximum values of x, y, and z data.
axis 'auto x' " "'auto y' " "'auto x' " "'auto xz' " "'auto yz' " "'auto xy'	Computes the indicated axis limit automatically.
axis manual	Freezes scaling of the current limits. Used with **hold** forces subsequent plots to use the same limits.
axis tight or axis fill	Sets the axis limits to the range of the data.
axis ij	Sets the origin of the coordinate system to the upper left corner. The i-axis is vertical, increasing from top to bottom. The j-axis is horizontal, increasing from left to right.
axis xy	This is the default coordinate system with the origin at the lower left corner. The x-axis is horizontal increasing from left to right, and the y-axis is vertical increasing from bottom to top.

Table 4.2.2b Summary of the Axis Function

Syntax	Affect
axis equal	Sets the aspect ratio of the x-, y-, and z-axis automatically according to the range of data units in the x, y, and z directions so that the data units are the same in every direction. This makes a sphere look like a sphere instead of an ellipsoid.
axis image	The same as **axis**('equal') but also makes the plot box fit tightly around the data.
axis square	Adjusts the x-, y-, and z-axis so that they have equal lengths. This makes the axes region of 2-D plots square and of 3-D plots cubed.
axis vis3d	Freezes the aspect ratio so that rotation of 3-D objects will not "stretch-to-fill" the axes.
axis normal	Automatically adjusts the aspect ratio of the axes and data units on the axes to fill the plot.
axis off or axis on	Turns off or on all axis lines, tick marks, and labels.
[mode,visibility,direction] = axis('state')	Returns the strings indicating the current axes settings: mode = 'auto' or 'manual' visibility = 'on' or 'off' direction = 'xy' or 'ij'

4.2.3   3-D Plot Rotation

As you recall from Chapter 3, the Figure Window provides some specific tools for modifying the appearance of your plot. Recall the zooming and rotating buttons; these are still very much functional, and even more useful, with a 3-D plot. Figure 4.34 depicts the **peaks** function plotted differently in four subplots. Each subplot has been altered using either the zoom or rotate buttons. To zoom or rotate a subplot using the buttons, simply click on the button you wish to apply, then start clicking in the subplot. Notice that zooming changes the size of the axes by a factor of two in the subplot and can quickly overwhelm the other subplots.

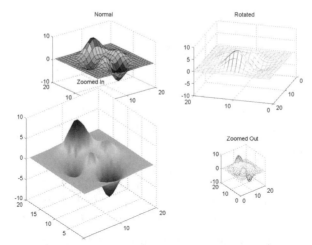

Figure 4.34 The results of using the zooming and rotation tools from the Figure Window.

As you click on a plot to rotate it, if you hold the mouse button down you will see that MATLAB creates a reference box around the plot. You will also notice that the azimuth and elevation specifying the rotation is displayed in the Figure Window, but only as long as you keep the mouse button depressed. Figure 4.35 shows what you can expect to see.

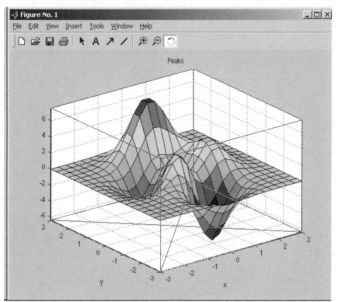

Figure 4.35 The rotation box is visible while the mouse button is depressed.

The rotate tool can also be activated from the command line or M-File with **rotate3d**. In the next section we will discuss how to exercise greater control over our point-of-view of a 3-D plot.

4.2.4   Using the View Command

In the previous section, we saw how to use the rotation button to change the aspect, i.e., our point of view, of a 3-D plot. You noticed that as you kept the mouse button depressed, the Figure Window would indicate the aspect in terms of azimuth (Az) and elevation (El). You can achieve the same results from the command line or in your M-Files but with greater control by using the **view** function. The function **view** is used to specify the aspect you want to use to view a 3-D plot. You use it by calling it explicitly with two input arguments specifying the value of azimuth and elevation, or with a single input, being a vector with the values as its elements. When called with a two-element vector as an output argument, **view** will return the aspect currently in use.

In its simplest form, the function is used by passing an azimuth (Az) and elevation (El) angles in degrees as input arguments with

```
view(Az,El)
```

or with a single vector variable with two elements,

```
view([Az El])
```

The angles are defined with respect to the axis origin, where the azimuth angle, Az, is in the xy-plane and the elevation angle, El, is relative to the x-y plane. Figure 4.36 depicts how to interpret the azimuth and elevation angles relative to the plot coordinate system.

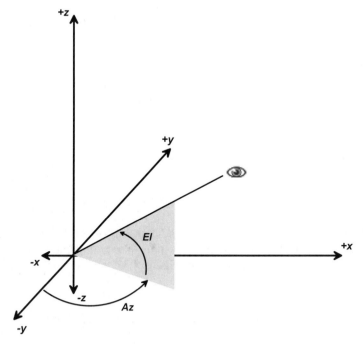

Figure 4.36 The point-of-view in a 3-D plot.

If you call **view** after creating a 3-D plot, it will return the current azimuth and elevation of the plot. If you have not previously changed these values, then this will return the default values of az = -37.5° and el = 30°. Consider again the surface plot of the **peaks** function.

```
surf(peaks(20))
```

The code,

```
[az el]=view
```

will return

```
az =

  -37.5000


el =

   30
```

which are the default values for the azimuth and elevation.

As you have seen already, you can use the rotate tool from the Figure Window and change the aspect of the view of your plot. Let's say that you have been merrily rotating away at your plot with the rotate tool, and now you have discovered that you can't tell up from down in the figure. In such a situation, which happens more often than you might expect, **view** can come to your rescue. One way to use **view** is to issue the function with the default azimuth and elevation values.

```
view(-37.5, 30)
```

However, even more convenient, the **view** function has two very simple forms that can help you when you get in such a bind. The forms of the function

```
view(3)
```

and

```
view(2)
```

will restore the current plot to the default 3-D or 2-D views respectively. Again we visit the function **peaks**, this time presenting multiple views of it using the **view** function, as shown in Figure 4.37, created with the following code.

```
azrange=-60:20:0;
elrange=0:30:90;
spr=length(azrange);
spc=length(elrange);
pane=0;
for az=azrange
    for el=elrange
        pane=1+pane;
        subplot(spr,spc,pane);
        [x,y,z]=peaks(20);
        mesh(x,y,z);
        view(az,el);
        tstring=['Az=',num2str(az),...
          ' El=',num2str(el)];
        title(tstring)
        axis off
    end
end
```

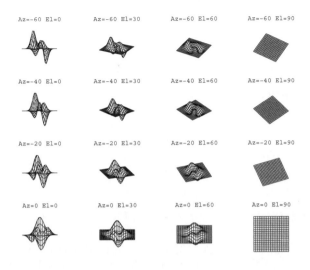

Figure 4.37 Multiple views of the peaks function.

4.3   Volume Visualization

In the 3-D visualization methods discussed so far, we have been concerned with surfaces. Volume visualization is concerned with representing a three-dimensional matrix of points, i.e., a volume, in which each point can be either a scalar (magnitude only) or vector (magnitude and direction). Scalar data is a single value for each point, while vector data for a point is either two or three values. Knowing the difference between scalar and vector volume data determines which techniques are better suited for your visualization. In short, since scalar data presents amplitude at a point within a volume, they are best visualized with isosurfaces, slice planes, and contour slices. On the other hand, vector data represents both magnitude and direction at a point in a volume so, consequently, techniques such as particle, ribbon, tube, cone, and arrow plots are more appropriate. Keep in mind, just as we have seen already, that when it comes to data visualization, a combination of techniques is often the most effective at conveying the salient information in any instance of visualization – even more so with volume visualization. The document that came with your MATLAB software (if you don't have it in printed form, it will be included in your document disk), *Using MATLAB Graphics*, presents an excellent treatment of volume visualization and includes some striking examples. We will touch on the highpoints of volume visualization in this section and use the example data that comes with MATLAB in examples here.

4.3.1   Scalar Volume Data

MATLAB includes a host of functions specifically designed for scalar volume data visualization. In general, X, Y, and Z are arrays that specify the points on the x-, y-, and z-axis at which volume data, V, is provided. Table 4.3.1 lists those functions, but be sure to read the command prompt help for each function to see all the capabilities. The functions that produce plots return handles to the graphics objects they create.

Table 4.3.1 Scalar Volume Computation Functions

Function	Action
FVC = isocaps(X,Y,Z,V,ISOVALUE)	Computes an isosurface end cap geometry for data V at isosurface value ISOVALUE and returns a structure containing the faces, vertices, and colors of the end cap which can be passed directly to the **patch** function.
NC = isocolors(X,Y,Z,C,VERTICES)	Computes the colors of isosurface vertices VERTICES using color values C and returning them in the array NC.
N = isonormals(X,Y,Z,V,VERTICES)	Computes the normals (N) of isosurface vertices VERTICES by using the gradient of the data in V.
FV = isosurface(X,Y,Z,V,ISOVALUE)	Extracts an isosurface at ISOVALUE in the volume V, returning the structure FV containing the faces and vertices of the isosurface, suitable for use with the **patch** function.
NFV = reducepatch(P,R)	Reduces the number of faces in a patch P by a fraction R of the original faces. It returns the structure NFV containing the new faces and vertices.
[NX, NY, NZ, NV] = reducevolume(X,Y,Z,V,[Rx Ry Rz])	Reduces the number of elements in a volume by only keeping every Rx, Ry, Rz element in the corresponding x, y, or z direction.
NFV = shrinkfaces(P,SF)	Reduces the size of patch P by shrink factor SF, returning a structure NFV containing the new faces and vertices.
W = smooth3(V,'gaussian', SIZE) W = smooth3(V,'box', SIZE)	Smooths the data in V according to the convolution kernel of size SIZE specified by the given string.
FVC = surf2patch(S)	Converts a surface object S into a patch object. FVC is a structure containing the faces, vertices, and colors of the new patch.
[NX, NY, NZ, NV] = subvolume(X,Y,Z,V,LIMITS)	Extracts a subset of volume data from V using limits LIMITS = [xmin xmax ymin ymax zmin zmax].
contourslice(X,Y,Z,V,Sx,Sy,Sz)	Draws contours in a volume slice plane at the points in the vectors Sx, Sy, and Sz.
patch(x,y,z,C)	Creates a patch in the 3-D space of color defined by C.
slice(X,Y,Z,V,Sx,Sy,Sz)	Draws a slice plane described by the vectors Sx, Sy, Sv, through the volume V.

4.3.1.1   Slice Planes

When 3-D surface plots or contours are not sufficient for visualizations, an example of which might be determining the heat transfer or density characteristics of a solid object, you will most likely have a need for the **slice** function. Table 4.3.1b presents the general form of the **slice** function, however it can also take on a number of other forms, based on the input provided.

The **slice** function will plot "slices" of the volumetric data, V, along planes which are perpendicular to either the yz-, xz-, or xy-axis planes at locations Sx, Sy, or Sz on the respective x-, y-, or z-axis. These can be multiple slices on each axis. This is best explained by example; consider the scalar volume bounded by

```
[x,y,z] = meshgrid(-2:.2:2, -2:.2:2, -2:.2:2);
```

and defined by

```
v = x .* exp(-x.^2 - y.^2 - z.^2);
```

We can use the function **slice** to visualize slices through the volume, in this case at planes at x = 1, y = 0, and z = 0, as shown in Figure 4.38.

```
slice(x, y, z, v,1,0,0)
axis tight
```

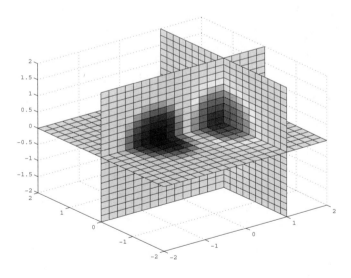

Figure 4.38 Slicing through a scalar volume.

MATLAB doesn't limit you only to slices parallel to one of the axis planes. However, creating slice planes at arbitrary angles does require just a little bit of handle graphics. Here are the steps to slice with a plane at an arbitrary angle:

1. Create the surface to slice with by defining a surface and rotating it. You will need the handle to that surface in the following steps, so get it too. Here we use the bounds of the original volume as the bounds of our slice plane; notice how the z-axis is zeroed.

```
Hslice = surf(-2:.2:2, -2:.2:2, zeros(length(z)));
```

2. Rotate the slicing surface to the desired angle using the **rotate** function. Here we rotate only about the x-axis.

```
rotate(Hslice,[-1 0 0],-45);
```

3. Use the **get** function to retrieve the data that defines the rotated slice plane.

```
xs = get(Hslice,'XData');
ys = get(Hslice,'Ydata');
zs = get(Hslice,'Zdata');
```

4. Use **slice** to plot the new slice plane.

```
slice(x,y,z,v,1,0,Inf)
hold on
slice(x,y,z,v,xs,ys,zs)
```

The result is shown in Figure 4.39.

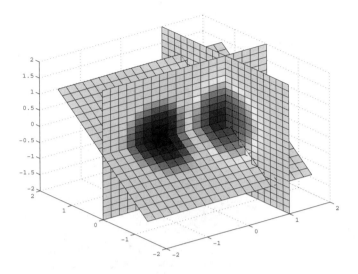

Figure 4.39 A slice at −45 degrees.

4.3.1.2   Contour Slices

Contour lines can be added to slices readily with the function **contourslice**. This function is shown in its general form in Table 4.3.1b and as with other volume visualization functions, it takes the arrays defining the volume space (X, Y, and Z), and the value for the volume (V), but it also requires the slice plane specification (Sx, Sy, Sz). By default, **contourslice** will automatically assign contour line colors based on the value of the volume, but usually when we are combining slices with contours, we want our contour lines to be a single easy to see color and let the slice provide the color indicating the value of the volume. However, to do so requires the application of a little Handle Graphics, so without apology we present here the solution, as in the previous example. Let's say we want to add white contour lines to the plot in Figure 4.39. This might seem a little challenging at first since our view includes both a vertical slice and then an intersecting slice at 45 degrees. Continuing with the previous example, here's how to do it:

```
Hcs=contourslice(x,y,z,v,1,0,Inf,20);
set(Hcs,'EdgeColor','white','LineWidth', 1.0);
```

The first line plots 20 contour lines on the x = 1 plane and returns the handles to them. The **set** function, which you will learn more about in Chapter 7, is then used to set the color of the contour lines to white and the width of the lines to 1 (which is wider than the default hairline width).

For the 45-degree plane, we must be sure to use the slice plane data that defines that slice, i.e., xs, ys, and zs in this example.

```
Hcs=contourslice(x, y, z, v,xs,ys,zs,20);
```

```
set(Hcs,'EdgeColor','white','LineWidth', 1.0);
```

Figure 4.40 (see Plate 6) shows the result. Note that you do not have to have **hold on** with **contourslice** since it will hold the current plot itself.

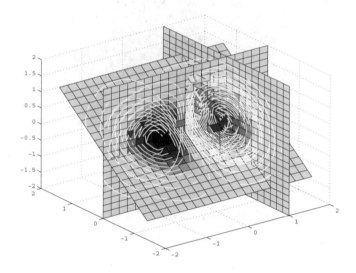

Figure 4.40 Contour lines on slice planes.

As a final note about slice planes, they don't have to be planes at all! MATLAB allows you to use any surface you care to create in defining, what is more properly stated as, the slicing surface. To illustrate this, let's continue with this same data, and slice it with the surface of a unit sphere.

First, put a slice in the original volume at x = 1. Be sure that **hold on** is activated.

```
slice(x, y, z, v,1,0,0)
hold on
```

Then get the surface definition for a sphere using MATLAB's convenient **sphere** function. Recall that **sphere** will create a unit sphere centered at zero.

```
[xss,yss,zss]=sphere;
```

Now slice the volume with the sphere surface, and adjust the perspective with **view** for a better look.

```
slice(x,y,z,v,xss,yss,zss);
view([-29,12]);
axis tight
```

Your plot should look something like the one shown in Figure 4.41.

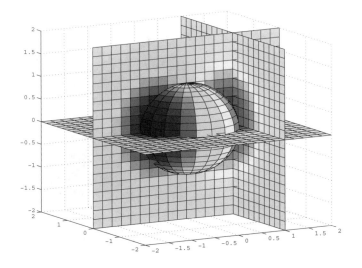

Figure 4.41 You can use any surface to "slice" a volume.

4.3.1.3   Isosurfaces and Isocaps

Another interesting and insightful method of volume visualization is to identify a surface throughout the space where the value of the volume is a constant. Just like contour lines connect values of z that are equal in a 2-D space where z=f(x,y), the function **isosurface** will outline in a volume where v=f(x,y,z) is a constant. To illustrate this, we will use a demonstration function included with MATLAB called **flow**. This is a function in three variables, and represents the speed profile of a submerged jet in an infinite tank. We like it because it produces an image with changing contours that readily illustrates interesting features of volume visualization. Let's say we want to look at the **flow** data where it is equal to -1.5.

```
[x y z v] = flow;
isosurface(x, y, z, v, -1.5);
```

The resulting plot is shown in Figure 4.42.

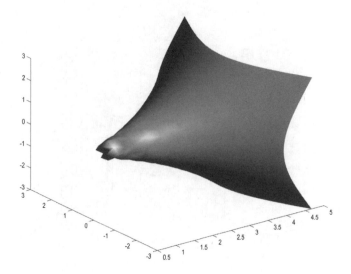

Figure 4.42 An isosurface plot of the fluid function data at a value of - 1.5.

Whereas **isosurface** outlines where a volume is of constant value, **isocaps** can be used to show what is inside the volume. Technically stated, **isocaps** computes an isosurface end-cap geometry for a given isovalue. Again, let's consider the data generated by the **flow** function, but this time use **isocaps**.

```
isocaps(x,y,z,v, -1.5);
view(3);
```

The resulting visualization is shown in Figure 4.43.

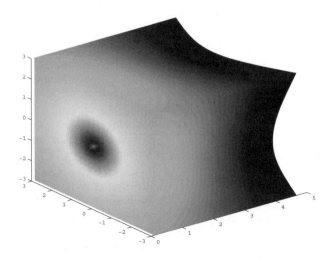

Figure 4.43 Isocaps shows what is inside a volume.

4.3.2 Vector Volume Data

A 3-D vector field has both magnitude and direction for every point in the volume. Just as with the scalar volume data, vector volume data requires coordinates for x-, y-, and z-axis, but for vector volume, each point has either a 2- or 3-element vector that describes both magnitude and direction. Table 4.3.2 summarizes the visualization functions that deal with vector volume data. In these functions, the arrays X, Y, and Z define the coordinates for velocity vector data U, V, and W, i.e., the 3-D vector field. The drawing functions can return handles to the surface objects in the plot. Be aware that these functions can take varied inputs, depending on usage, and we only show the most basic form of the function here for brevity. Please use the **help** command with the function name to get complete details.

Table 4.3.2 Vector Volume Computation Functions

Function	Action
[CURLX, CURLY, CURLZ, CAV] = curl(X,Y,Z,U,V,W)	Computes the curl and angular velocity (CAV) perpendicular to the flow of the 3-D vector field .
DIV = divergence(X,Y,Z,U,V,W)	Computes the divergence of the vector field.
VERTSOUT = interpstreamspeed(X,Y,Z,U,V,W,VERTICES)	Computes the streamline vertices (returning a cell array of vertex arrays) from vector field data U, V, and W, magnitudes (speed) by interpolation at vertices (such as those produced by **stream2** or **stream3**) specified by the cell array VERTICES.
continued on next page...	

Table 4.3.2 Vector Volume Plotting Functions

Function	Action
XY = stream2(X,Y,U,V,STARTX,STARTY)	Computes 2-D stream line data, returning a cell array, XY, of vertex arrays suitable for use with plotting functions like **streamline**.
XYZ = stream3(X,Y,Z,U,V,W,STARTX,STARTY,STARTZ)	Computes 3-D stream line data returning a cell array, XYZ, of vertex arrays suitable for use with plotting functions like **streamline**.
LIMS = volumebounds(X,Y,Z,U,V,W)	Returns the x-, y-, and z- axis coordinates and color limits for a volume as a vector LIMS = [xmin xmax ymin ymax zmin zmax cmin cmax].
[NX, NY, NZ, NV] = subvolume(X,Y,Z,V,LIMITS)	Extracts a subset of the volume data. The extent of the subset is specified in the vector LIMITS = [xmin, xmax, ymin, ymax, zmin, zmax], which contains coordinate values.
coneplot(X,Y,Z,U,V,W,Cx,Cy,Cz)	Plots velocity vectors as cones at the points Cx, Cy, and Cz in the vector field.
streamline(X,Y,Z,U,V,W,STARTX,STARTY,STARTZ)	Draws stream lines from either 2-D or 3-D vector data. STARTX, et al., define the starting positions of the stream lines.
streamparticles(VERTICES, N)	Draws stream particles using the vertices (such as those produced by **stream2** or **stream3**) in the cell array VERTICES. N is the number of stream particles drawn, or the fraction of the total if less than 1. If not specified, the default is used, N=1, or 100% of the vertices.
streamribbon(X,Y,Z,U,V,W,STARTX,STARTY,STARTZ)	Draws stream ribbons from vector data U, V, and W. STARTX, et al., define the starting positions of the stream lines at the center of the ribbons. The twist of the ribbons is proportional to the curl of the vector field.
streamslice(X,Y,Z,U,V,W,Sx,Sy,Sz)	Draws stream lines with direction arrows using the vector data U, V, and W, aligned in an x, y, z plane defined by Sx, Sy, Sz.
streamtube(X,Y,Z,U,V,W,STARTX,STARTY,STARTZ)	Draws stream tubes from vector data U, V, and W. STARTX, et al., define the starting positions of the stream lines at the center of the tubes.

Throughout this section on vector volume data, we will use an example data set included with MATLAB, *wind.mat*, that represents the air currents over North America. This data is made up of wind speed and direction vectors within a volume. You can access this data with the **load** command. Typing,

```
load wind
```

at the command prompt will load the data into the MATLAB workspace. (If you have been doing the examples as you read, you might want to first clear the workspace by issuing the **clear** command.) Once you load the wind data, you will have the volume arrays x, y, and z , and the volume vector arrays u, v, and w. We will use this data to illustrate the topics in this section.

First, we must point out that although this section is concerned with volume data that has both magnitude and direction, you can still use scalar volume techniques with vector volume data; all you have to do is convert the vectors to scalars by computing the magnitude of the vectors. In the case of the wind data,

```
wind_vel = sqrt(u.^2 + v.^2 + w.^2);
slice(x,y,z,wind_vel,[80,90,100,110,120],Inf,Inf)
axis equal
shading interp
```

produces the plot shown in Figure 4.44.

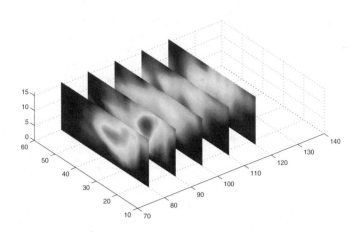

Figure 4.44 Vector data plotted as scalar data.

However, Figure 4.44 gives us no information about the direction of the wind. In fact, we have lost that information completely in the computation of wind_vel. Now we shall turn our attention to *stream plots*, and how they are useful in visualizing the direction of flow within a volume.

4.3.2.1   Stream Plots

Since vector volume data can be thought of as particles *flowing* through a volume, it is desirable to have some visualization methods that indicate the direction of flow. Stream plots do just that. The differences amongst the stream plots provided in MATLAB have to do with the way the streams are visualized. Stream plots are typically combined with other visualization techniques, such as slices, in order to provide richer information content. Table 4.3.3 summarizes the five stream plot techniques available in MATLAB and their use.

Table 4.3.3 MATLAB Stream Plot Techniques

Stream Plot Technique	Function	Use
Lines	**steamslice** **streamline** **stream2** **stream3**	Traces the path that a particle in the vector field follows.
Particles	**streamparticles**	Markers that trace stream lines. Useful for creating stream line animations. Usually used in combination with stream lines.
Ribbons	**streamribbon**	Similar to stream lines, but the width of the ribbon allows it to show *twist*, i.e., curl angular velocity.
Tubes	**streamtube**	Again, similar to stream lines, but the width of the tube can be varied. Tubes are useful for showing the divergence of a vector field.
Cones	**coneplot**	Each particle in the volume vector field is represented by a conical arrowhead or arrow, indicating both magnitude and direction.

We will now look at each of these techniques, applying them to the *wind* data as an example. Since each of these techniques is used to represent direction of flow, in many cases we will have to define starting points for the streams. This will become obvious as we explore the examples. Additionally, some of the techniques are better used with Handle Graphics and those will be deferred to later chapters.

4.3.2.2   Stream Lines

In many cases, you will want to explore your data before you dive into a specific volume visualization technique. The more you know about your data the more effective you can make your visualization. Continuing with the example we started in the previous section, we would be wise to explore the extents of our data, and then proceed with the appropriate volume visualization techniques. Earlier, we examined the *wind* data using slices, but only after we had converted the data to scalar data. This gave us a feel for the magnitude of the data, but in doing so, the direction information was ignored. In fact, our choice of slice planes was somewhat arbitrary. Now, we will try to do better.

The function **streamslice** is designed to show particle flow in a slice through the volume. Consider again the plot in Figure 4.44. Here we see that the data extends in the z-direction from about 0 to somewhere around 15. Let's say we are interested in the wind velocity and direction at a slice exactly midway of the z-data. Rather than estimating from our previous plot, we can determine exactly the midway plane by examining the extents of the volume in the z-axis and so do better than guessing. The following code generalizes the approach.

```
minz=min(z(:));
maxz=max(z(:));
midz=(maxz-minz)/2;
```

Now we can create a slice midway in the z-plane using **slice**. Then we will use **streamslice** to visualize the direction of the flow in the plane.

```
slice(x,y,z,wind_vel,[],[],[midz])
streamslice(x,y,z,u,v,w,[],[],[midz]);
axis equal
shading interp
```

Figure 4.45 (see Plate 7) shows the result of combining these two plots.

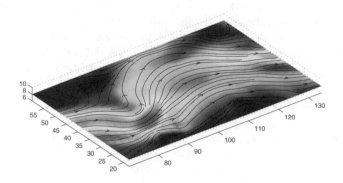

Figure 4.45 Using streamslice to visualize the direction of flow.

As you can see, the general direction of the flow is in the positive x direction. It is also easy to see the zones of high and low velocity by using the scalar data, i.e., the magnitude. This plot gives us a very intuitive sense of the data.

The other stream plotting function you should be familiar with is **stream3**. Refer to Table 4.3.2, for the form of the function. The *wind* data is almost ready for use by **stream3**, but we have to also provide starting points for the stream lines. The following example produces a plot much like Figure 4.44 but, rather than guessing, the limits of the volume are found using **volumebounds**. Five slice planes are located linearly spaced along the x-axis.

```
lims=volumebounds(x,y,z,u,v,w);
slice(x,y,z,wind_vel,...
    [linspace(lims(1),lims(2),5)],[],[]);
```

Next, **meshgrid** is used to define the start for the stream lines. The x-axis value is set to the lower limit of the volume, that will make all the stream lines start from the lower x –boundary. The y- and z-axis are incremented through their extents. (Yes, **linspace** could have been used here like it was with the **slice** function.)

```
[sx sy sz] = meshgrid(lims(1),...
    lims(3):5:lims(4), lims(5):5:lims(6));
```

Finally we use **streamline** to plot the result returned by the **stream3** function.

```
streamline(stream3(x,y,z,u,v,w,sx,sy,sz));
shading interp;
axis equal;
```

The result is shown in Figure 4.46.

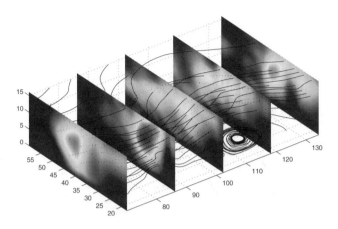

Figure 4.46 Stream lines with volume slices.

4.3.2.3   Stream Particles

Stream particles are used to put markers in the vector field. They can be used to show both position and velocity of the stream line. They are particularly useful in animation of stream lines, a topic we will visit when we discuss animation in Chapter 9. Here, we will deal with stream particles in the static case.

The function **streamparticles** relies on a variable called *vertices* which is a cell array that contains the vertices of the volume. This is the type of cell array that is typically returned by the **stream3** function, and in fact, is what was used by **streamline** in the previous example. You can simply replace **streamline** in the previous example with **streamparticles** and MATLAB will reward you with success, but what you will see will look a bit messy. However, we can demonstrate using particles by noting that there is an interesting phenomenon between about 110 and 120 on the x-axis and between 15 and 40 on the y-axis. To investigate this area closer, we shall first redefine where we start our stream.

```
[sx sy sz] = meshgrid(100, 15:5:40, 3:2:6);
```

Then plot using **streamline** enhanced with **streamparticles**.

```
streamline(stream3(x,y,z,u,v,w,sx,sy,sz));
streamparticles(stream3(x,y,z,u,v,w,sx,sy,sz),...
'markers',2)
```

```
view(3)
axis tight
```

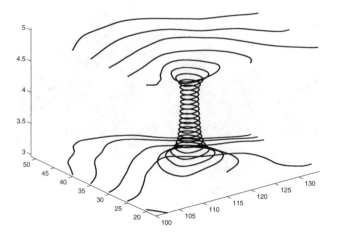

Figure 4.47 Stream particles combined with stream lines.

The string 'markers' is a keyword to **streamparticles** for "marker size" so that it will know that the number that follows specifies the size of the marker to use. Unfortunately, in this plot, the particles don't convey any more information than the stream lines do. We will see this and other properties when we revisit stream particles and put them in motion in Chapter 9.

4.3.2.4   Stream Ribbons

Stream ribbons are used to represent flow, just like a stream line, but unlike a line, a ribbon depicts the data direction and rotation about the axis of flow. This rotation is mathematically known as "curl" and we will not delve into the mathematical formulation for that here. If you know what curl is, then its use is obvious to you. If the concept of curl is new to you, you will find it most useful for vector field data such as that used in the example here. In this kind of visualization, the curl looks like a twist in the ribbon. In its basic form, the function **streamribbon** allows you to specify an angle for the twist for each vertex in the ribbon. However, **streamribbon** will determine the curl if you do not specify it. Note that the function **curl** is available for this type of computation.

The simplest way to use **streamribbon** is shown in the following code. Here, no twist is specified, so **streamribbon** will compute a twist proportional to the curl of the vector field. The width is constant and determined automatically. Figure 4.48 is the resulting plot.

```
load wind
```

```
[sx sy sz] = meshgrid(100, 15:5:40, 3:2:6);
streamribbon(x,y,z,u,v,w,sx,sy,sz);
shading interp;
view(3);
%some camera and lighting controls we will
%discuss later
camlight; lighting gouraud
```

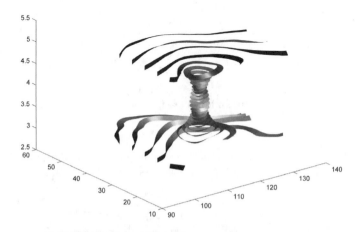

Figure 4.48 Stream ribbons convey "twist" information.

As with other such plotting functions, **streamribbon** can return handles to the objects it creates. You are encouraged to read the helps on this function once you have read about Handle Graphics and Color and Light in Chapters 7 and 8.

4.3.2.5   Stream Tubes

Since curl represents the "twist" about a vertex in vector fields, you might wonder what mathematical method is used to represent direction at each point, relative to the flow direction. This is called "divergence" and is most readily visualized with stream tubes that vary in diameter based on magnitude. The function **streamtube** is used much like **streamribbon**, and replacing the **streamribbon** function in the previous example with

```
streamtube(x,y,z,u,v,w,sx,sy,sz)
```

produces the plot shown in Figure 4.49 (see Plate 8).

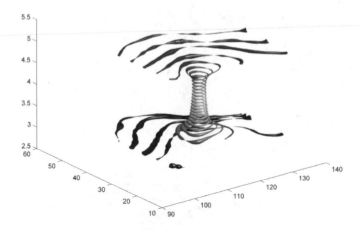

Figure 4.49 Stream tubes are used to show divergence.

To better see the detail, you might zoom in on an interesting part of the plot as shown in Figure 4.50, or select a smaller subset of the volume.

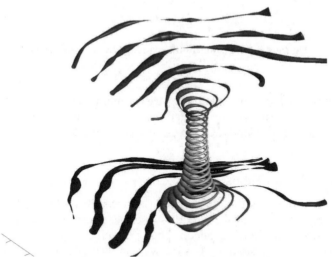

Figure 4.50 A closer view of stream tubes reveals divergence.

4.3.2.6   Cone Plots

Cone plots represent the data in a vector field as a cone having direction and length proportional to the velocity at that point in the field. The function **coneplot** can be used to produce such a plot. As with the previous vector volume plots, you need to specify both the volume and volume vectors, but instead of just indicating where to start a stream, with **coneplot** you must specify the position of the cones within the volume.

Cone plots are very effective when combined with other volume visualization techniques. Using the wind data provided with MATLAB we will set about demonstrating using **coneplot** and introduce some new functions too. Consider the following code.

```
load wind
%extract a portion of the volume
[x y z u v w] =...
subvolume(x,y,z,u,v,w,[105,120,nan,30,2,6]);

%compute the magnitude of the wind
wind_vel = sqrt(u.^2 + v.^2 + w.^2);

%slice at the extremities
lims=volumebounds(x,y,z,u,v,w);
slice(x,y,z,wind_vel,...
[lims(1),lims(2)],[lims(4)],[lims(5)])

%specify where to put cones
xrange = linspace(lims(1),lims(2),8);
yrange = linspace(lims(3),lims(4),8);
zrange = linspace(lims(5),lims(6),6);
[cx cy cz] = meshgrid(xrange,yrange,zrange);
coneplot(x,y,z,u,v,w,cx,cy,cz,wind_vel,1);

%pretty it up a bit
shading interp
axis equal
```

Here we use the function **subvolume** to extract only the portion of the volume we want to consider. Notice that the minimum y axis data is specified as NaN (Not-a-Number) which tells the **subvolume** function to start with the beginning of the data on that axis. As with our examples demonstrating stream lines, we find the wind velocity and this time make some slices at the boundaries of our data; the function **volumebounds** makes doing this convenient as it returns the extents of the volume in vector form. Now we specify the vertices where we want cones and pass those values to **coneplot**. Here we use **coneplot** in the form where we specify a matrix for color, in this case wind_vel. We also specify a comfortable size for the cones. The result is the plot shown in Figure 4.51 where the direction of each cone is the direction of the wind and both the length and color represent the velocity. We will revisit this plot in Chapter 8 where we will improve on the presentation of it by manipulating the lighting and color.

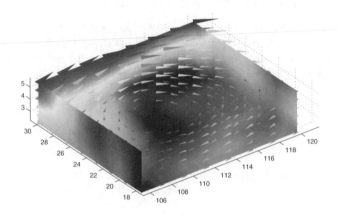

Figure 4.51 Cones visualize magnitude and direction in a vector volume field.

4.4   A Word About Annotating 3-D Plots

The approach you take to adding text to 3-dimensional plots is very similar to doing so for 2-dimensional plots (see Chapter 3, Section 4). The only real difference is that you now have a z-axis that you can label. You have already seen this in action in this chapter such as with the example of Figure 4.26 where we used the function **zlabel**.

In addition to the function **zlabel**, the function **text** can be specified with x-, y-, and z-coordinates, allowing you to place text anywhere in the 3-D space. The example of Figure 4.27 demonstrates the use of specifying a z-coordinate with **text**. Bear in mind, that placing text with either **text** or **gtext** simply puts the text where you specify; if you change the plot by re-plotting or changing the axis, the text will not likely be where you want it. As far as labeling axes, use **x-**, **y-**, or **zlabel**.

4.5 Illustrative Problems

1. Create a 3-dimensional pyramid using the fill3 function. Can you create each face individually with vectors defining the x-, y-, and z-coordinates? What about creating it with one fill3 command and a set of x-, y-, and z-coordinate matrices?

2. Load the MRI data mri.mat (provided with MATLAB). What variables were loaded? Read about the function **squeeze** in the MATLAB helps. Then try the following code

```
D = squeeze(D);
image_num = 4;
image(D(:,:,image_num))
axis image
colormap(map)
```

You can treat this MRI data as a volume because it is a collection of slices taken progressively through the 3-D object. Try using contourslice to display a contour plot of a slice of the volume.

5 IMAGE GRAPHICS

IN THIS CHAPTER...
5.1 IMAGE FILES AND FORMATS...167
5.2 IMAGE I/O 170
5.3 IMAGE TYPES AND PROPERTIES 176

5.1   Image Files and Formats

With digital cameras and scanners available at ridiculously low prices, practically everyone is familiar with images on their computer. Images can convey a great deal of information in a very intuitive form. Images don't have to be of the real world, they can certainly be representations of mathematical phenomena; in fact, that's exactly what we have been exploring in the pages of this book. But whatever the case, at some point in your use of MATLAB, you will be faced with either one of two image file related issues; either you will want to read in a scanned (or similarly digitized) picture and operate on it, or you will want to generate an image file from a plot or some graphic you have created in MATLAB and now wish to share with the outside world. In fact, bitmap images are a natural "data type" for MATLAB since images can be simply thought of as 2-D arrays. Consequently, all the array manipulation and operations you are familiar with are applicable. The only issue is how to get the images into or out of your computer. Of course, you could write your own low-level I/O functions, assuming you are suitably familiar with the image file format you are dealing with, but fortunately, MATLAB provides some convenient and powerful ways of getting images in and out, and also of viewing them by means of a robust set of image file specific functions. Table 5.1.1 summarizes the image specific functions that come with MATLAB. If you have the image processing toolbox, then you will have many more image functions.

Table 5.1.1 Image Graphics Specific Functions

Function	Action
[X,Map] = imread(fname, fmt)	Reads an image from a graphics file.
image(C)	Displays the matrix C as an image.
colormap(Map)	Sets the current image's colormap to Map.
imagesc(C)	Same as image, except that the data is scaled to use the entire colormap.
continued on next page...	

Function	Action
A = imfinfo(fname,fmt)	Gets information about the fmt formatted image file fname and returns it in the structure A.
imwrite(X, fname, fmt)	Writes matrix X to fname in fmt format.
newmap =brighten(beta), brighten(fig,beta), brighten(map,beta)	Lightens (or darkens) the current color map, the current figure, or a specified color map.

Even if we are not conversant with the technical details of the file formats, most of us are at least familiar with the names of a few of the common image data formats. You have probably heard of "bitmaps," "tiffs," "jpegs," and "gifs," but you might not know why there are different formats or what the differences are between them. We will not go into an extended discussion of the different formats in this book, but we will try to give you enough information to help you with your image file work in MATLAB.

5.1.1   Common Image File Types

Image data files fall into three general formats. The first is *non-compressed* bitmaps, the second is *compressed* bitmaps, and the third is *vector* graphics.

Non-compressed bitmaps, like early versions of the BMP (bitmap) and TIFF (tagged image file format) formats, are image data files that simply store each pixel of image data in an array. Compressed bitmaps do the same thing, except some mathematical function is applied to the data to attempt to reduce the size of the file. Compressed bitmaps can be further broken down into *lossy* and *lossless* formats as well. Lossy formats actually lose some of the original data during compression and it is usually a judgment call as to how much of the data can be lost before it is no longer suitable for its intended use. The JPEG format is lossy to varying degrees depending on the compression ratio. You will always get the image file size reduced with JPEG, but you might do so with severe degradation of the data. The simple images shown in Figure 5.1 illustrate the possible effects of compression. The top image is in an uncompressed TIFF format; the bottom is JPEG with only moderate compression. In this case, compression introduces artifacts because of the strong contrast edges in the image; a natural trait of the JPEG compression process.

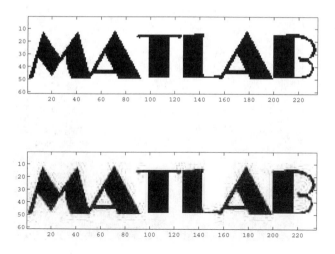

Figure 5.1   Compression introduces artifacts.

Figure 5.2 and Figure 5.3 demonstrate the appropriate use of compression. The image in Figure 5.2 is uncompressed and has a file size of 3.6M Bytes, while the compressed version in Figure 5.3 only takes 65K Bytes in a file. The differences between the two are revealed only under close inspection.

Figure 5.2   An uncompressed image stored in a 3.6MB file.

Figure 5.3 The same image after JPEG compression uses only 65KB.

Once you compress an image with a lossy method, the data is lost and cannot be recovered. Lossless formats like PNGs, GIFs, and compressed TIFs offer varying degrees of compression based on the data, but do not compress to the point of losing data. This is achieved through methods such as Run Length Encoding (RLE) in which recurring pixels are stored as a single pixel and a count value.

The third format, vector graphics, relies on a kind of descriptive computer language to tell either your computer screen or your printer how to draw the graphic. Postscript is a kind of vector graphics format and we will usually talk about images that are "encapsulated postscript." The "encapsulated" part of EPS is that usually a TIFF or JPEG image is contained within the Postscript file. An EPS file will generate beautiful images of plots, line art, and letters, and accommodates images where subtle changes in shading occur by encapsulating a bitmap image format. Keep in mind that Postscript is dependent on the device (usually a printer) that the file is output to, i.e., an EPS file cannot be readily printed on a non-Postscript printer (there are software converters that will allow this).

5.2   Image I/O

If you are generating an image file for use by others, you will likely choose a format that is commonly viewable with most image editing software. If your image contains only a few colors or is made up of mostly constant color areas, then the color-mapped formats, e.g., PNG and GIF, will do well; however, if you are sharing your images on the Internet, you will probably choose a format that will compress your data and reduce file size. If your images contain many shades of color without distinct boundaries, e.g., using **shading interp** or photographic images, then you will probably pick a true-color image format. Finally, if you are reading an image provided by someone else you need to be able to read whatever format is provided. Fortunately, MATLAB gives you a great deal of choice in all cases without having to know a lot about the details of a particular image file format. Table 5.1.2 lists the image file formats that MATLAB can read and write.

Table 5.1.2 MATLAB Readable Image File Types

Extension	Type	Read Write	Use
bmp	bitmap or device independent bitmap	R/W	Native format for Microsoft Windows. Can support up to 24-bit color. Originally uncompressed, run-length encoding (lossless) compression is now supported.
jpg	jpeg – joint photographic experts group	R/W	24-bit (true color) support. Created to support the photographic industry. Compression can result in noticeable loss of image quality in some images or annoying "artifacts." Compression ratios of 25 or 30 to 1 with photographic images producing good results are not uncommon, but the more you compress, the poorer the picture quality. No transparency support.
tiff	tagged image file format	R/W	Originally created in the 1980s to support data output from scanners (raster scan). Limited to 4GB of data. Can contain information about colorimetry calibration, gamut tables, etc., such as occurs with remote sensing and multi-spectral applications. Can support various compression algorithms in compressed modes.
gif	graphics interchange format	R	Very common and used extensively on the Internet. Works well for illustrations or clip-arts that have large areas of flat colors. Does not work so well with photographic images or images with continuous tones. Limited to 256 colors that are "dithered" to look like more colors. Supports animation (GIF89a standard). Use for logos, bullets, or clip-arts where few colors are used. Typically 5 to 1 compression ratio.

continued on next page...

Extension	Type	Read Write	Use
png	portable network graphics	R/W	Similar to GIF, very efficient lossless compression, supporting variable transparencies (alpha channels), and gamma correction, but not animations.
pcx		R/W	Similar to bmp, up to 24-bit color and lossless compression.
hdf	hierarchical data format	R/W	A data interchange format championed by the National Center for Supercomputing Applications.
xwd	X-Windows Dump	R/W	Used on Unix workstations.
ico	Windows icon format	R	Used by Windows for icon graphics. 32 x 32 bits by default. Can have multiple images in one file (animations). No compression.
cur	Windows cursor format	R	Used by Windows for cursor graphics. Can contain animations. No compression.

The three principal image I/O functions in MATLAB are **imread** (for reading graphics files), **imwrite**(for writing data to a graphics file format), and **imfinfo** (for retrieving information about a specific graphics file).

5.2.1   Reading a Graphics Image

MATLAB includes a JPEG image of the complex planetary nebula NGC6543A, a.k.a. the "Cat's Eye Nebula," in the file ngc6543a.jpg. This should have been automatically placed in a folder on your MATLAB path when you installed MATLAB. You can use **imfinfo** to retrieve information about the image file. Its general form is

```
imfinfo(filename, fmt)
```

where both input variables are strings, the first being the name of the file and the second being the image file format. To retrieve the file data about the Cat's Eye Nebula, you could type,

```
imfinfo('ngc6543a','jpg')
```

which will return

```
ans =

            Filename:
'C:\MATLAB6p1\toolbox\matlab\demos\ngc6543a.jpg'
         FileModDate: '02-Oct-1996 23:19:16'
            FileSize: 27387
              Format: 'jpg'
```

```
   FormatVersion: ''
          Width: 600
         Height: 650
       BitDepth: 24
      ColorType: 'truecolor'
FormatSignature: ''
```

From this we can see that ngc6543a.jpg is a 600x650 truecolor image. Note that you don't have to specify a file extension in *filename* provided the extension is the same as *fmt*. The function **imfinfo** returns a structure, so we could have used,

```
info = imfinfo('ngc6543a','jpg');
```

and then accessed specific data from the fields, such as

```
info.Format
```

which would return the string 'jpg'.

The **imread** function has the general form

```
[X,C]=imread(filename,fmt)
```

where *filename* and *fmt* are strings specifying the name of the file and its format, just as in **imfinfo**. X is the returned image data, which can be MxN for indexed images, or MxNx3 for true color images, and C is the colormap if the image is indexed. For example, you can read the Cat's Eye Nebula image with the following code.

```
[x,c]=imread('ngc6543a.jpg','jpg');
```

Since this is a truecolor image, this will return a 650x600x3 uint8 array in x and an empty array for c since there is no colormap with JPEG images. Knowing this is a JPEG format, we know there would not be a colormap so we could have used the form,

```
x=imread('ngc6543a','jpg');
```

As with **imfinfo**, we don't have to specify an extension with the file name if the extension correctly corresponds to the specified file format *fmt*.

5.2.2 Displaying a Graphics Image

Displaying the image you have just read is achieved by simply typing

```
image(x)
```

This will open a Figure Window and plot the image on the axis. If you have read the JPEG image above, **image** will readily accept the 3-D array and display the RGB images. Notice that the same plotting considerations of relative axis scale arise here as well; you might wish to use **axis equal** to correct the perspective. If the image you read is an indexed image, i.e., a 2-D index array with a corresponding Nx3 colormap, you will need to use the **colormap** function to get the correct coloring of the plot. At the website you can download the indexed image usflag.dib (Windows device-independent-bitmap). Once you have that file you can read it, then view it with,

```
[x,cm]=imread('usflag.dib','bmp');
image(x)
```

You should see the image shown in Figure 5.4.

Figure 5.4 An indexed image without its colormap.

Notice that the colors are not as you would expect. You might have something very strange indeed if you have been using other colormaps since MATLAB will apply the last colormap used in the current Figure Window. To get the expected patriotic colors, you will next need to load the appropriate colormap, in this case cm.

```
colormap(cm)
```

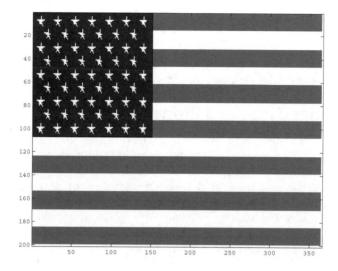

Figure 5.5   An indexed image with the appropriate colormap.

In Section 5.3 we examine indexed images in more detail.

5.2.3 Writing a Graphics Image

Writing the contents of a Figure Window to an image file is just as simple as reading one. The **imwrite** function provides a means to create image files of the formats indicated in Table 5.1.2. The general form of **imwrite** is,

```
imwrite(A,filename,fmt)
```

where *A* is the image, either grayscale if NxM, or truecolor if NxMx3, *filename* is a string containing the name of the file to be created, and *fmt* is a string indicating one of the write formats indicated in Table 5.1.2. For the case of an indexed image, i.e., one containing an image and colormap, **imwrite** takes the following form.

```
imwrite(X,C,filename,fmt)
```

In this case, X is an NxM array of indexes into colormap C. Using **imwrite** can be easily demonstrated by loading one of the color image data files that is distributed with MATLAB and writing to one of the image file formats. In the following example, we will use the **load** function to load an image of a clown. Typing

```
load clown
```

at the command prompt will load *X*, a 200x300 double array, and *map*, an 81x3 double array, into the workspace. Notice that this data is an index array and colormap. We can then create, for instance, a PNG format image by typing,

```
imwrite(X,map,'clown.png','png')
```

which will create the file clown.png in the default working folder.

5.3   Image Types and Properties

Whenever we deal with images we need to be aware that there are fundamentally three types of image data formats. You have already seen some of this in the previous examples. The first image data format is *indexed* images where two arrays are used to describe the image. The second image data format is *intensity level* images, comprised of a single array where each element indicates the relative intensity of a pixel. The third data format is called *truecolor* and uses three intensity level arrays, where each is the relative intensity of red, green, and blue primary colors.

5.3.1   Indexed Images

By indexed we mean that the image is created from information in two arrays: the first is an array of indexes into the second, which is a three-column array containing the red, green, and blue contributions for each pixel. The following code will load an image of a clown and display it in a Figure Window, looking like that shown in Figure 5.6.

```
load clown
image(X)
colormap(map)
```

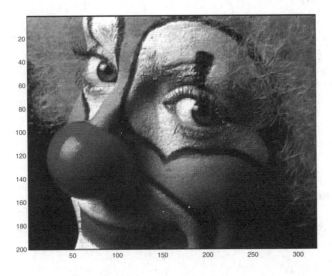

Figure 5.6   Who is that clown?

When we load clown.mat the arrays *X* and *map* were placed in the MATLAB workspace. The array *X* is an array of indices; row indexes actually, specifying which row from *map* (an 81x3 array) the pixel color is to be taken from. Looking at a few pixels from the upper left corner, we see

```
X(1:5,1:5)

ans =

    2     2     2     2     2
   61    69    69    69    69
   69    61    69    61    69
   61    69    61    61    56
   69    55    61    44    61
```

Each element is a number corresponding to a row in the color map. We can find the color components for each of these pixels by

```
map(unique(X(1:5,1:5)),:)

ans =

   0.1250        0        0
   0.8672   0.4141   0.1250
   0.8672   0.5781        0
   0.8672   0.5781   0.1250
   0.9961   0.5781   0.1250
   0.9961   0.7031   0.1250
```

The **unique** function returns only one instance of an element, and orders the results, so the first row here corresponds to an index value of 2, the second 44, the third 55, etc.. We can see these first few pixels, shown in Figure 5.7, with the following code.

```
image(X(1:5,1:5))
colormap(map)
```

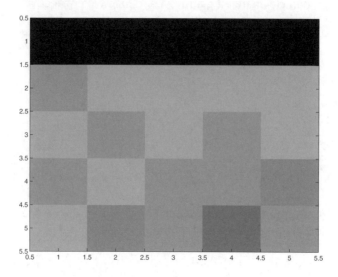

Figure 5.7   The upper left 5x5 of the clown image.

Try this.

```
image(X(20:60,50:100))
colormap(map)
```

Use unique again to see how many different colors are used in this portion of the image.

```
map(unique(X(20:60,50:100)),:)
```

5.3.2   Intensity Level Images

An intensity image file does not provide a color map. Instead, the array describes the image by relative pixel amplitude. An example of such an image is the "raw" image from an imaging device such as a CCD (charge-coupled device) imager. These devices, being digital, assign a value to the pixel based on the intensity of the light falling on it . Most are either 8-bit or 16-bit devices, meaning that for any pixel the intensity range is either 2^8 (0-255) or 2^{16} (0-65,535) where 0 is no light and the maximum is fully illuminated. When an image from such a device is viewed, we typically do so in shades of gray. However, since MATLAB uses color maps when it plots an image, intensity image or not, we need to tell MATLAB what kind of color map should be used. The default data type in MATLAB is 64-bit floating-point numbers, i.e., double precision. However most image data formats are designed to use no more file space than is necessary. It is indeed wasteful to use an 8-bit imaging device and store its output with 64-bit numbers. The MATLAB image functions typically will deal with images in their native format as either 8-bit (uint8) or 16-bit (uint16) unsigned integers.

When plotting intensity level images, you will find that MATLAB's built-in color maps don't always agree with the scale of your image.   With the following code, a 256-level grayscale image is viewed using **image** along with the color map *gray*.

```
image(X); colormap(gray);
```

Since the color maps in MATLAB are 64 levels, in this case, since there actually are 256 shades of gray, shades will be lost when using a built in color map as shown in Figure 5.8. Shades are lost since all indexes above 64 are mapped to the highest level in the map, in this case white.

Figure 5.8   A 256-level gray scale image loses quality when used with a 64-level color map.

One simple solution is to create a 256-shades-of-gray color map in this manner:

```
gray256 = linspace(0,1,256)';
gray256 = (repmat(gray256,1,3))';
```

This will create a 256-shade color map and produce much better results as shown in Figure 5.9.

Figure 5.9 Same image using gray256.

Another solution to the color map-scaling problem is the function **imagsc**, which serves the same purpose as **image**, but scales the image data to use the full extent of the color map. The following code will take an 8-bit intensity image X and scale it to the built-in MATLAB color map *gray*.

```
imagesc(double(X),[0 255]); colormap(gray);
```

Note that **imagesc** requires the image data to be of type *double*. Although upon close inspection once can discern loss of shades, since the color map has been used across the full extent of the image's intensities, the results are quite good as shown in Figure 5.10.

Figure 5.10 The result of mapping a 256-gray-level image to 64 gray levels using imagesc.

5.3.3 Truecolor Images

Truecolor images are in essence a set of three intensity image arrays where one is red filtered, another green filtered, and the last one blue filtered. For any given pixel in the image, the corresponding element from each of the arrays contributes a proportionate amount of red, green, or blue.

The following function can be used to translate an indexed image into RGB format. Not only useful in seeing the relative contributions of red, green, and blue, but later we will see the need for converting indexed bitmaps to RGB in MATLAB when we present CData.

Tools

```
function rgbimage = makergb(bitmap,colormap)
%RGBIMAGE = MAKERGB(BITMAP,COLORMAP)
%where BITMAP is a NxM array, and COLORMAP is
% a Cx3 double array
%RGBIMAGE will be a NxMx3 double array.
%Makes an RGB image from an array of indexes
% (BITMAP)into
%a color map (COLORMAP).
%MAKERGB will determine if the index array needs
%to be 1-shifted.

bitmap=double(bitmap);
if min(bitmap(:))==0 %is it 0 indexed?
    offset=1
else
    offset=0
```

```
end

[rows,cols]=size(bitmap);

for L=1:3
    layer=colormap(bitmap(:,:)+offset,L);
    layer=reshape(layer,rows,cols);
    rgbimage(:,:,L)=layer;

end
```

We can see the relative contribution of red, green, and blue using **makergb** along with,

```
load clown
rgbclown=makergb(X,map);
subplot(1,3,1),imagesc(rgbclown(:,:,1)),axis square,
title('Red');
subplot(1,3,2),imagesc(rgbclown(:,:,2)),axis square,
title('Green');
subplot(1,3,3),imagesc(rgbclown(:,:,3)),axis square,
title('Blue');
colormap gray
```

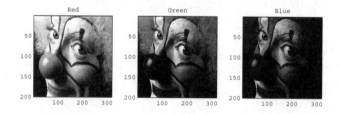

Figure 5.11 The red, green, and blue components of the clown image shown in relative intensity.

6 GENERATING OUTPUT

IN THIS CHAPTER...
6.1 THE QUICKEST WAY TO PAPER 183
6.2 PRINTING COLORED LINES TO BLACK & WHITE PRINTERS 185
6.3 ELECTRONIC OUTPUT 186
6.4 USING THE PRINT COMMAND 187

In Chapter 5 we saw how to read and write image files of the standard image file formats. You also need to know how to print your output to printers and how to create graphic output that can be used in other computer applications such as word processors. There are several ways to do this depending on the intended use of the image and the format of image output you need.

6.1   The Quickest Way to Paper

The first output goal we will explore is how to get the graphics of your Figure Window to paper. The simplest way to do this is to use the **File → Print** pull-down from the Figure Window as shown in Figure 6.1.

When you select this option, your plot will be sent to your printer via your operating system's print manager. Before you go off sending everything to your printer, notice the other printing commands available in the pull-down window; these are **Page Setup**, **Print Setup**, and **Print Preview**. The **Page Setup** item will open a window with tabs that allow you to change various properties of the page to be printed such as the size and position of the plot on the page, portrait or landscape orientations, color or black and white lines and text, and also some axis properties such as whether or not MATLAB should recalculate the axis tic marks based on the printed size, or to keep the tic marks as seen on your computer display. The **Print Setup** window will open the Windows printer setup panel. The options you have there depend on the printer or printers you have installed. The **Print Preview** window can be used to see a representation of what your printed page will look like. If you don't like what you see, a button is available to take you to the **Page Setup** window. When you are satisfied, clicking on the **Print** button will print your plot.

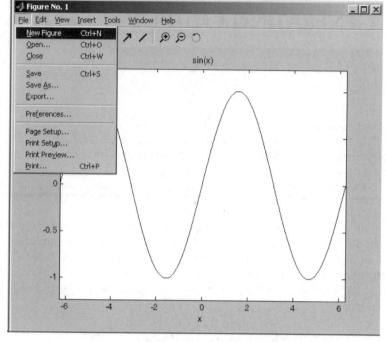

Figure 6.1 Printing from the Figure Window.

6.1.1 Page Setup

When you print in MATLAB, the contents of the Figure Window are what is printed. This includes the axes, labels, titles, annotations, and any other objects that you see. By default, MATLAB assumes a print area centered on 8.5-inch by 11-inch paper that is 8 inches by 6 inches with no window frame. The default figure and axis background color is white. Tic marks and axis limits are automatically calculated to accommodate the printed size, so it very likely will be different from what you see on the computer's display. However, you can alter many of the figure attributes that affect the appearance of your output from the **Page Setup** window. By selecting **File → Page Setup** you will open a window comprised of four tabbed sections. These are listed in Table 6.1.1.

Table 6.1.1 Page Setup Tabs

Tabbed Section	Actions
Size and Position	Choose to automatically print the figure at the screen size, centered on the page, or manually set the size and position of the figure on the page.
Paper	Set paper size and orientation.
Lines and Text	Select to print lines and text in their colors or in black and white.
Axes and Figure	Choose to use the tic marks and axis limits you see on the screen, or have MATLAB recalculate them based on the printed size of the figure. Here you can also choose to print the figure with its color background, or force it to white background.

If you are not printing to an 8.5-inch-wide by 11-inch-high piece of paper, you will want to go to the Paper tab to choose one of the other twenty-two paper types, or to create your own Custom paper type. The Lines and Text tab lets you choose to print in black and white for lines and text. Notice that in each tab, there is a graphic depicting the placement of your figure on the page. You can hold the mouse button down on this representation and move it to different places on the page, or change the size of the figure on the page.

6.2   Printing Colored Lines to Black & White Printers

Unless you have changed the default line style and color properties of the root object, MATLAB will by default plot lines in the "solid line" or '-' setting. If you have several lines up at once, they may or may not be different colors, depending on how you plotted them. When you print these to a black and white printer, the lines might be difficult to distinguish from one another.

For one-time printing, you can use the MATLAB Property Editor as discussed in Chapter 3 and change the style of the lines so that you can easily distinguish them when printed.

Another approach that will let you set the line style and color once for a session is with the following code.

```
% Make default color black
set(0,'defaultaxescolororder',[0 0 0]);
% Change the line style order to anything you want,
% for example, (solid,dashed,dash-dotted,dotted,and
% circles.
set(0,'defaultaxeslinestyleorder','-|--|-.|:|o')
```

This uses a little Handle Graphics to change the default axes color to black only, and sets a line style order to continuous, dashed, dash-dotted, dotted, and circles. With this code, the styles are changed at the "root level," meaning that the properties are altered for all figures during the session. To return to the original MATLAB defaults, use

```
set(0,'defaultaxeslinestyleorder','factory')
set(0,'defaultaxescolororder','factory');
```

If you want to make a different line style order, or axes color order the default every time you use your MATLAB, all you need to do is include the **set** code shown above in a file called *startup.m*; MATLAB looks for this file each time you start a MATLAB session. There is also a master MATLAB startup file called *matlabrc.m* and you can add or modify code there, but it is not advisable. If you want to change your MATLAB startup options, you should read the on-line helps on startup and matlabrc.

6.3   Electronic Output

In addition to printing to paper, MATLAB supports various electronic formats for graphics output. You have already learned how to read and write image files using **imread** and **imwrite**. Now you will learn how to use the pull-downs from the Figure Window and the **print** function from the Command Line to create image files of your plots in standard image file formats that can be easily imported into various applications.

6.3.1   Using File Export

Selecting **File → Export** from the Figure Window will open a dialog window that will allow you to name and save your plot in any of the image file formats listed in Table 6.4.1.

Table 6.4.1 Export Plot Image Formats

File Format	Extension
Enhanced Meta File	EMF
Windows Bitmap	BMP
Encapsulated Postscript (4-types)	EPS
Adobe Illustrator	AI
JPEG	JPG
TIFF (compressed or non-compressed)	TIF
Portable Network Graphics	PNG
Paintbrush (24-bit)	PCX
Portable Bitmap	PBM
Portable Graymap	PGM
Portable Pixmap	PPM

Simply select the folder where you want to save the file, give it a name and specify the format in which you wish to save the plot. Although exporting is convenient, aside from selecting among four types of encapsulated postscript and compressed or non-compressed TIFF, you have very little control over the specific image formats. For instance, the default quality level of 75 is used in the JPEG format.

6.3.2   Using the Windows Clipboard

If you have MATLAB installed on a Windows system, you can copy a figure to the clipboard by using **Edit → Copy Figure** and then paste the figure into another Windows application. When you do paste the image, you have the choice either of pasting it as a picture, or as a Windows Metafile. The advantage of pasting it as a Windows Metafile is that you can edit various parts of the image. For instance, if you paste your figure into Microsoft Word, you can select the image, ungroup it, then edit the axis lines, text, plot lines, etc.

Before you use the **Copy Figure** pulldown, you should select **Edit → Copy Options** which will open the Preferences Dialog Window to a section called Figure Copy Options. Here you can alter the specifications of various properties of the figure before it is copied to the Clipboard. Here you have the choice of making the plot keep the axis tic marks that you see on the screen or use the settings established by **File → Page Setup**. Notice that this is a subset to the Figure Copy Template. If you select Figure Copy Template, you can use built-in settings that will optimize the size of lines and fonts in your figure for Microsoft Word or PowerPoint as well as make some limited specific changes to the Clipboard copy.

6.4   Using the Print Command

Certainly the most powerful way to print your Figure Window, either to a printer or to an image file, is by using the **print** function. In its simplest use, **print** will send the contents of the current Figure Window to the computer's default printer. The general form of **print** is

```
PRINT -DEVICETYPE -OPTIONS
```

Since both the *devicetype* and *options* are optional, if you do not supply them, MATLAB will determine the default values for your system. The printer command and device type that **print** uses by default can be determined by typing

```
[printcommand,devicetype] = printopt
```

The function **printopt** will return the current printer type and output destination. It can also be edited to change your default printer type and destination.

6.4.1   Creating Hardcopy with PRINT

Every time you issue the **print** command, it determines the default printer settings by issuing a call to the **printopt** function. The default device type on Windows systems (-dwin) is to print through the Windows Print Manager. The default type is a Level 2 black and white PostScript (-dps2) for Unix and Macintosh systems. As we mentioned in the last section, you can alter the device type and print command by editing the *printopt.m* file, but we suggest that you consult the Full Product Family Help or the MATLAB Reference Guide for more information about *printopt.m* before you do. Typing **help print** in the Command Window will list all the print options, device drivers, etc., available in MATLAB.

6.4.2   Creating Graphics Files Using Print

Essentially, the only thing you need to do to generate electronic copies of your figure is to add a file name to the **print** statement you used when generating paper copies. The full synopsis for the print command when generating files is

```
print -devicetype -options filename
```

For example, the following code segment will generate a file called graphics1.eps in level 1 black and white encapsulated postscript form.

```
print -deps graphics1.epsf
```

If you specify that MATLAB generate an encapsulated PostScript file, it will not be sent to the printer and MATLAB will generate a file. If you do not give it a file name, MATLAB will warn you that it cannot print encapsulated postscript to a printer and will create a file in the current working directory.

6.4.3   Adding Additional Figures to a File

To keep the number of electronic files to a minimum, you may find it useful to append figures to the same file. Normally, if you type

```
print -deps filename
```

and the file filename already exists, it will be overwritten without warning. However, by adding on an argument as follows

```
print -deps -append filename
```

the figure is added to the end of the file as a new page. Later, the entire set of figures can be conveniently viewed or printed at once.

6.4.4   Publishing Using 4-Color Separation

When generating PostScript and Encapsulated PostScript files, you have the option of generating figures using cyan, magenta, yellow, and black (CMYK) color values instead of red, green, and blue (RGB). This is often needed when sending files to particular color printers that can take advantage of matching figure colors more precisely with CMYK color values. As an example, you might type

```
print -depsc -cmyk filename
```

to generate an Encapsulated color PostScript file.

6.4.5   EPS with a Preview Image

If you are creating encapsulated postscript images to be included in a document, it is useful to have a preview image instead of the postscript message box when you include a figure in your word processor. This makes it easier to remember what the figure contains when you are editing a document. Most modern word processors allow the use of encapsulated postscript with a preview image.   If you have a Figure Window open and you type,

```
print -depsc -tiff -r300 figure1
```

MATLAB will save the entire contents of the Figure Window at a resolution of 300 dpi in the file figure1.eps and generate a TIFF bit map of the entire Figure Window, always at 72 dpi, that can be used by other applications as the preview image.

6.4.6   Rendering Method with -zbuffer or -painters

There are many methods by which an image can be rendered, i.e., interpreted for printing. When using the **print** function, by default MATLAB will determine which rendering method to use when printing a figure, however, there are some situations in which you will want to have control over which method gets used. In general, if you are plotting lines and less complex figures, it will be to your advantage to use the *painters* algorithim (-painters). If your plots are more complex, say you are plotting surfaces or using lights, it will be better to use the *Z-buffer* method (-zbuffer). *OpenGL* is another rendering method that is available on many computer systems. This method is generally faster than *painters* or *zbuffer* and in some cases enables MATLAB to access special graphics hardware that is available on some systems.

If you want control over the printing resolution of a figure use the -zbuffer option and specify the resolution with the print command's -rnumber option (where number represents the number of dots per inch, dpi). If you are only going to display the figure on your screen or plan on using it in a web page, changing the resolution to a number higher than the default will not provide you with better image quality, but it will increase your file size. The same can be said when sending the file to a printer; only choose a number that is no greater than what your printer can support. Since the Z-buffer method uses raster graphics to draw the figure, the size of the file and memory needed to print will depend on 3 factors: the resolution, the size of the graphic, and whether or not you are using a color or grayscale driver. The OpenGL method can take advantage of compatible video hardware, if it is available, and significantly reduce the time to render an image. Although OpenGL has the potential of being very fast, there are some caveats to consider: 1) OpenGL does not do color map interpolation which means that plots created using **surface** that use index color with interpolated faces or edges will not be as you expect; 2) Similarly, the *phong* value for face lighting and edge lighting of surfaces is not supported.

6.4.7   Indicating Which Figure Window to Print

When you have multiple Figure Windows open, you can specify the Figure Window number that you want to print from the command line using the form,

```
print -fhandle
```

where *handle* is the figure number. By default, MATLAB will print the current figure (usually the last one created or the last one that you clicked your mouse in). Later you will be able to appreciate the ability to specify the figure to be printed. Soon you will learn how to create GUIs in MATLAB. In a GUI, for instance, you could provide a user with a GUI window that lets the user create figures in a different window. You might want to have a print button in your GUI, then the callback, which you will learn about later, of that button would need to use the form of the **print** function in which you pass the figure's handle so that you can be sure that the figure you intended to print gets printed.

6.4.8   Saving Figures for Future Use

You will notice that from the Figure Window, you can select **File → Save**. If you do so, MATLAB will give you the option to save your Figure Window as a FIG-file with a *fig* extension to the file name. FIG-files are binary files that contain a complete description of the Figure Window and all that is in it. In previous versions of MATLAB (prior to version 6), figures were saved by saving an M-File and corresponding MAT-File. The FIG-file format did not exist in earlier versions of MATLAB. In those earlier versions, figures were saved as an M-file that generated the figure from data that was stored in a MAT-file. You ran the M-file and it generated your figure. The **print** command still supports this older approach. The option *dmfile* tells the **print** command to create the M-file and if need be the MAT-file to reproduce the figure. Please be aware however, that you will not find reference to it in the helps for **print**.

```
print -dmfile filename
```

To regenerate the figure, just execute the M-file by typing the name of the file you saved.

MATLAB version 6 was a radical departure in many ways from earlier MATLAB versions. There were changes in appearance to some extent, but many of the changes occurred "under the skin" of MATLAB and the FIG-file format was created to accommodate those changes that the earlier figure saving method could not. For example, the earlier method could not keep figure annotations. With the FIG-file, annotations and other changes you make with the plot editing tools are saved. In Chapter 10 we will see how being able to easily save and create FIG-files can lead to rapid GUI development.

7 HANDLE GRAPHICS

IN THIS CHAPTER...
7.1 GRAPHICS OBJECTS 191
7.2 GRAPHICS OBJECTS HIERARCHY 193
7.3 GRAPHICS OBJECTS HANDLES 198
7.4 PROPERTIES 202
7.5 OBJECT SPECIFIC PROPERTIES 214
7.6 SETTING DEFAULT PROPERTIES..295
7.7 UNDOCUMENTED PROPERTIES...296
7.8 USING FINDOBJ..297
7.9 ILLUSTRATIVE PROBLEMS...300

7.1   Graphics Objects

By this time, you have already explored many of MATLAB's graphics capabilities. What you've done so far can be thought of as "high-level" graphics, i.e., it didn't require you to get very deep into what was really going on in MATLAB. Nevertheless, we have already had to "slip" a little Handle Graphics in a few places in order to get the results we wanted. This chapter however, marks your departure from the high-level use of MATLAB graphics, and begins your journey into a deeper understanding of the basic mechanism behind everything that happens in MATLAB. Here you will learn how to use "low-level" functions to manipulate every aspect of graphics objects. You will learn more about graphics objects and how to affect their properties, which will give you the knowledge you need to become a master MATLAB programmer.

All of the high-level graphics functions that have been discussed so far either create or manipulate graphics objects. The term "graphics object" may conjure up mental images of computer generated spheres or cubes, or it might bring to mind some of the objects that you have already created either with examples in this book or on your own, such as lines (using **plot**, **plot3**, etc.), surfaces (created with **surf**, **mesh**, etc.), and text (using **text**, **xlabel**, etc.). However, you have already seen many other objects and probably didn't even realize it. The computer screen, the individual Figure Windows, the axes, and images are all MATLAB graphics objects. As you proceed in this book, you will learn more about these objects, as well as some more graphics objects such as user interface controls and user interface menus. You should think of these graphics objects as drawing primitives, i.e., the elementary building blocks that are used to generate plots that are more intricate in the Handle Graphics system.

In addition to the computer monitor's screen, which is considered the *root object*, there are 12 other graphics object types.

The 12 low-level graphics functions that create graphics objects, as well as the root object which is present at the time you invoke MATLAB, are either the name of the object or an abbreviation of the name and are listed here in the following table.

Graphics Object	Low-Level Creation Function	Description
Figure	figure or figure(H)	A window to show other graphics objects.
Axes	axes, axes(H), or axes('position',RECT)	The axes for showing graphs in a figure.
UIcontrol	Uicontrol	The user interface control is used to execute a function in response to the user.
UImenu	Uimenu	User defined menus in the figure.
UIcontextmenu	uicontextmenu('PropertyName1',value1,...)	A pop-up menu that appears when a user right-clicks on a graphics object.
Image	image(C) or image(x,y,C)	A 2-D bitmap.
Light	light('PropertyName','PropertyValue',...)	Light sources that affect the coloring of patch and surface objects.
Line	line(x,y) or line(x,y,z)	A line in 2-D or 3-D plots.
Patch	patch(x,y,c) or patch(x,y,z,c)	A polygon that is filled with some color or texture and has *edges*.
Rectangle	rectangle, rectangle('Position',[x,y,w,h]), or rectangle('Curvature',[x,y],...)	A 2-D shape; can be rectangle or oval created within an axes object.
Surface	surface(X,Y,Z,C), surface(X,Y,Z), surface(Z,C), surface(Z)	3-D representation of data plotted as heights above the x-y plane.
Text	text(x,y,text_string) or text(x,y,z,text_string)	Character strings used in a figure.

The x, y, and z variables that define the coordinates of the line object are the arguments to the **line** graphics function. These variables can be either vectors or matrices. If the variables are matrices, an individual line object will be created for each column of data.

The x, y, and z variables that are passed to the patch function specify the vertex coordinates of the patch object. If the variables are matrices, the patch command will draw a polygon for each column of the matrices. The c variable is used to specify the color of the patch objects. For now, consider this variable to be defined as a RGB triplet or as a string to provide a uniform color across the polygon. Later we will present other forms of specifying the color.

Low-level surface object creation may use X, Y, and Z matrix variables to specify the corner points of the surface's quadrilaterals. X and Y may also be vectors, in which case the length of X must be equal to the number of columns in Z and the length of Y must be equal to the number of rows in matrix Z. If provided, the matrix C defines the color of the surface object and must either be the same size as Z (**size**(C) = **size**(Z)) to allow for interpolated shading or have one less row and column (**size**(C) = **size**(Z)-1) for flat shading. We will see more about how to use the C matrix in a later chapter.

Text object locations can be defined with the variables x, y, and z. By default, the string will be placed so that the first character is left justified and vertically centered about the point specified by the coordinate (x,y) or (x,y,z).

You cannot see light objects, but you can see their effects on patch and surface objects. They essentially specify light sources to which you can control style, color, and location. The properties of light objects will be discussed in this chapter and in the next chapter we will embark on an in-depth exploration of light, color, and transparency. The other objects in the table that deal with user interfaces will be encountered in Chapter 10.

There are two main advantages of using low-level graphics functions. The first is that they never clear the axes or alter any of the current attributes of the existing graphics. Recall that subsequent calls to the **plot** function would clear the current Figure Window (unless **hold** was set to on, or another Figure Window was open.) The second advantage is that you can pass property name/value pairs as additional arguments to these functions to control various aspects of the graphics objects at the time of creation. In Section 7.4 we will examine the different properties for each of the graphics objects, but first we shall explain how all these objects relate to each other in the graphics environment.

7.2   Graphics Objects Hierarchy

The MATLAB graphics system is an effective and powerful *object oriented* approach based on the simple paradigm of *parent-child* relationships between some of the objects. Each object has its own identity and characteristics as defined by its attributes or *properties*. In the previous section, some examples of the typical types of graphics objects that fall under each object category were presented; now we shall discuss how these objects relate to one another. In the next section, we will look at the properties of all the objects.

The parent-child relationship in MATLAB graphics is straightforward. Essentially, a child object cannot exist without the existence of its parent object. For example, before a surface object can be drawn, both a figure and an axes object must be present. Fortunately, you are not required to specifically create a parent object by typing the low-level graphic function in the Command Window (or by writing them in an M-file) before generating children objects. If the parent figure and axes objects are not present, MATLAB will automatically create them and then draw the surface object. Although you could very well create these objects yourself with,

```
figure;
axes;
plot(1:10)
```

you could have just typed

```
plot(1:10)
```

and MATLAB would create the figure and axes (assuming that another Figure Window was not already present). If you then close the Figure Window with the **close** command, neither the axes nor surface object will remain on the screen, since they cannot exist without their respective parents. Later we will learn about the **delete** command, which allows you to delete specific graphics objects from the set of objects contained within the current graphics environment. The primary point here is to understand that when a parent is deleted, so are its children.

So you might be wondering which graphics objects can be parents and which ones can only be children? The screen object is the most basic of all and is the foundation on which all other objects must rest. As previously mentioned, the screen object is referred to as the *root* object because the organization of graphics objects can be cast into a tree-like hierarchy, where, without the roots, the rest of the tree's components are not able to survive or exist. Figure 7.1 depicts where the various graphics objects are in the hierarchical tree.

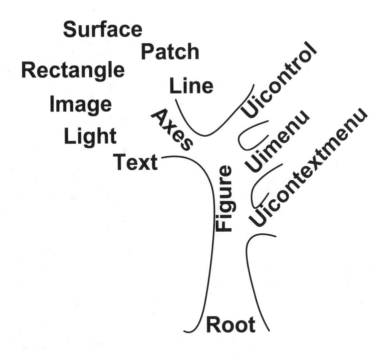

Figure 7.1   **The graphics objects hierarchy tree.**

Figure Objects are the windows in which all other graphics objects are displayed. High-level graphics functions, e.g., **plot**, will automatically create a Figure Window if one is not already present. They can also be invoked with the command **figure**. In fact you have already been relying on Handle Graphics for, as you have already seen, you must use this command to create multiple Figure Windows. The Figure Window is a child of the root object, i.e., the screen object. The root object can be the parent to as many Figure Windows as you want, provided your computer system has enough memory. When there are multiple figure objects displayed, subsequent calls to the **plot** function will create plots in the *current figure*, i.e., the last Figure Window on which an action was made. The simplest way to make a given Figure Window the current one is to explicitly select it with

```
figure(figure_number)
```

where *figure_number* is the integer displayed in the border at the top of the Figure Window. The current Figure Window is the one which subsequent graphics commands will affect.

Figure objects have four different types of children. The children can be either a type of user interface object, specifically user interface control (UIcontrol), user interface menu (UImenu), and user context menu (UIcontextmenu), or axes objects. A figure can have multiple children, and not all its children need be of the same type.

The UIcontrol objects are used to generate graphical user interface controls. Their positions can be over any region of the Figure Window. There are various styles of controls that can be defined and each UIcontrol style allows a user to provide MATLAB with input data or a stimulus for initiating the execution of a predetermined set of actions. The UImenu objects are used to generate graphical user interface menus. UImenus appear at the top of the Figure Window when using MATLAB on X-Windows and MS-Windows systems. On a Macintosh system, the menu objects that are children of the current figure will appear at the top of the screen (by default, the menus provided by MATLAB and the system software will also appear). However, no matter what system you are using, these menus are children of a specific figure. The UIcontextmenu objects are used to generate menus that appear when a user right-clicks on a graphics object. Chapter 10 will examine these objects in detail and show how easily they can be used to create sophisticated interfaces.

An axes object specifies a region of the Figure Window that will contain any collection of the seven axes children. In the previous chapter, we saw that the **subplot** command could be used to designate multiple regions in the Figure Window for displaying multiple plots in the same Figure Window. Essentially, this high-level command creates an axes object in a location which is dependent on the input arguments. The axes object can be the parent to line, patch, surface, rectangle, image, light, and text objects. Instead of having high-level commands create an axes object, they can also be created explicitly with the **axes** function.

Line objects are the basic drawing primitives used to create 2-D and 3-D plots and contours. Specifically, they are used by the **plot**, **plot3**, **contour**, **contour3**, **ezplot**, **fplot**, and other specialized high-level commands. These objects do not have any children, but can have many siblings, which do not necessarily need to be other line objects.

Patch objects can also be used in both 2-D and 3-D contexts (unlike images they can be viewed from any perspective). These objects are usually thought of as filled polygons whose edge and face colors can be independently defined. Their colors can be defined with either a solid or an interpolated color or even with no color (making them transparent). These objects have no children, but since they are children of axes objects they can have line, surfaces, text, and other patch objects as siblings. You have already seen these objects created in Chapter 4 using **fill** and **fill3**.

Surface objects are used to visualize data in a 3-D perspective. A surface is generated with a set of colored quadrilaterals, where each individual quadrilateral is very similar to a patch object. The edge and face colors can also be defined as solid, interpolated, or transparent. Usually the colors are related to the height of the object, but this need not be the case. These objects have no children, but may accompany other line, patch, surface, and text objects in their axes parent. They can be created with commands that create mesh and surf type graphics, in addition to **pcolor** (which will be discussed later).

Rectangle objects are 2-dimensional objects that have four sides and corners that have a specific roundness or "curvature." This type of object, created with the function **rectangle**, includes square-cornered rectangles, rounded rectangles, and ellipses. Rectangle objects, like image objects, can only be viewed in 2-D.

Image objects can be viewed only in a 2-D perspective; if you attempt to view them from any other perspective, they will not appear. Images are graphical representations of matrix data, where each matrix element value defines the color of a particular rectangle in the image. More specifically, the element is an index that points to one color within a list of colors (usually referred to as a color map) stored in the figure object. These objects are children of axes objects and therefore can be visualized with other axes children as long as the axes object is viewed from a 2-dimensional perspective. The image objects are displayed with the **image** or **imagesc** function.

Light objects are created using the **light** function, but are not objects that can be viewed in a plot directly. Instead, light objects affect the appearance of other objects in a plot. Specifically, light objects affect how surface and patch objects look, and have properties that include color, style, position, etc.

Finally, there are text objects. These character strings provide descriptive information to the plot. They can be used as axis labels or titles that are restricted in terms of their location with respect to the axes object. They can be character strings that are placed interactively with the plot editing tools available in the Figure Window, automatically by high-level commands such as **clabel** or **legend**, or they can be manually placed by either defining their location with a MATLAB command such as **text** or even with the mouse pointer such as by using **gtext**. These objects are children of axes objects and have no children themselves. Figure 7.2 shows a Figure Window with a typical collection of graphics objects.

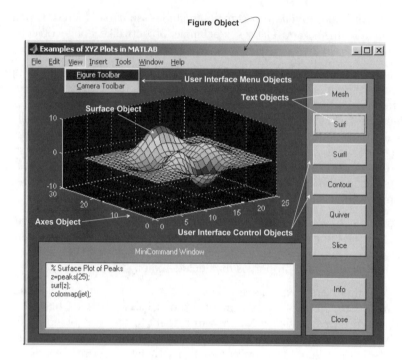

Figure 7.2   A typical collection of graphics objects.

In order to manipulate the characteristics or properties of these graphics objects with low-level graphics commands, you need to have some way of addressing each object. Previously, it was mentioned that The MathWorks coined the term "Handle Graphics" for the graphics system used by MATLAB. Handles provide the user with a way of identifying the graphics object that you want information about or whose information you want to alter. The next section further discusses the relevance of graphics object handles and how they can be obtained.

7.3   Graphics Objects Handles

To put it simply, a graphics object handle is a unique identifier assigned to every single graphics object. The term "handle" is appropriately descriptive in that it is analogous to handles of everyday objects found around the home (such as doors, frying pans, luggage, etc.). Just as these common *handles* provide you with means of gaining access, holding on to, and opening up everyday objects, graphics handles provide a means of both specifying and viewing the contents of MATLAB generated graphics objects.

Depending on the complexity of the graphics you create, there may be hundreds of objects resulting in hundreds of handles. (Minimally there will always be at least one graphics handle since if you are running MATLAB the root object must exist.) Keeping track of these handles, by assigning them to variables, provides easy access to the properties of the graphics objects.

However, even if you do not wish to keep track of these handles at the time the objects are created, as we shall see MATLAB provides a handy means to quickly acquire these handles. MATLAB's handle and property features give you a great deal of flexibility and freedom to arrive at your desired goal, but always remember that there is no one correct or best path that you can take.

Handles can be either integers or floating point numbers that MATLAB automatically generates for an object. Once assigned, the number, which is the handle, cannot be changed and will remain viable until the graphics object is deleted. Although MATLAB generates handles as needed, there are a number of objects for which the handles will always be the same from session to session: the first one of these is the screen or root object, which always has the handle of zero; secondly, if using the **figure** command, a figure's handle will always be an integer and the first figure created will have a handle of one. In MATLAB's default mode of operation, this integer is displayed in the border at the top of every Figure Window. For example, the Figure Window shown on the previous page has "Figure No. 1" displayed in the border and therefore, its handle is the integer 1. The figure objects are numbered consecutively from 1 to the number of figures you have displayed on your screen. The only exception to this rule is when you close or quit a Figure Window that is not the last one created. In this case, the next created figure will use the next lowest available number. For example, if you have figures 1 through 5 displayed, and then close figures 2 and 3, the next new figure generated will be assigned a handle number of 2 and will also be labeled as such ("Figure No. 2"). All other graphics objects will be assigned a floating-point number.

Before we discuss how these handles are used to change the properties of graphics objects, we will discuss some techniques that can be used to obtain the graphics handles. Generally speaking, you can either obtain the handles of objects at the time they are created, or you can get the handles from objects that already exist.

7.3.1 Determining Handles at Creation

All low-level and most high-level plotting functions return the handles to the objects that they create if an output argument is supplied during their execution. For example, we can create a figure, an axes, and a line object and store their handles respectively in the variables figure_handle, axes_handle, and line_handle with something like the following code:

```
figure_handle = figure;
axes_handle = axes;
line_handle = plot(exp(-([-3:3].^2)));
```

If more than one graphics object of a particular type is created, as when you create multiple lines with one **plot** statement, the handles will be returned as a column vector. The caveat with the high-level functions (such as **plot**) is that they only return the handles to objects that are created by the function if a figure and axes object were already present when, in fact, we know that the figure and axes object will automatically be generated if one currently does not exist. Therefore, even though the third line in this example would generate the desired plot all by itself, we could not have obtained the figure and axes handle at the time of their creation without the first and second lines (later you will learn about other low-level functions and techniques that can be used to obtain object handles after creation).

Many high-level graphics functions can be called in ways that will return the data being plotted. Some of these functions, such as **ezplot**, will return an error if called with an output argument. Many of these commands, e.g., ellipsoid, stairs, can be forced to suppress plot creation and will return other information, if called with an output argument, that can be plotted, such as the coordinates of lines vertices. When called in this manner they do not return graphics handles. The following table is a list of high-level MATLAB commands that generate plots but do not return graphics object handles when an output argument is supplied.

Graphics Functions That Do Not Return a Handle When Called with Outputs		
bar	ezplot	rose
compass	feather	stairs
cylinder	fplot	quiver
ellipsoid	hist	sphere
errorbar	polar	cylinder

For these commands, you could always create the plot in two steps: the first step would be to create the coordinates with the command, and the second step would be to pass the coordinates to the functions **plot** or **surf**. As an example, the **stairs** function can return a handle to the line objects that make up its plot, however if you wanted to get the actual values making up the plot you must use the function calling two output arguments. When you do this, you are not given the handles to the line objects. Consider this example where we use the **stairs** function in conjunction with **plot** to look at the histogram of some normal-randomly distributed numbers.

```
x=randn(1,100);
hist_x=hist(x);
[X,Y]=stairs(hist_x) %this is step one
plot(X,Y) %this is step two
```

This gives you access to the data used to create the plot. Still it would be very helpful to access this data, which we know must be part of the object, at a lower level. This is exactly what we will discuss in the next section as we venture into the properties of objects, but first we will present how to get handles of current objects when they were not determined at creation time.

7.3.2   Getting Handles of Current Objects

Several functions are useful for obtaining handles of current objects. (Remember that the current figure is the active figure which subsequent plotting commands would affect.) To determine which figure is the current one, you can use

```
figure_handle = gcf;
```

where **gcf** is a function that "gets current figure." The variable figure_handle could have been named anything you like. Although this is not very useful if you work with only one Figure Window since **gcf** will always return the number 1 if only one figure is displayed. However, **gcf** can be useful when working with multiple Figure Windows. At the very least, it can be used as a way of double-checking to assure that subsequent plotting commands appear in the figure that you expect.

But what about what's inside the figure? You can use the "get current axes" function **gca** to obtain the graphics handle of the current axes. The current axes are the axes in the current figure to which subsequent plotting commands would be sent. When you switch between figures, the **gca** command will return the handle to the current axes within the active figure. If there are multiple axes in the figure, the function will return the handle either to the one that was most recently created or the one that was most recently plotted, whichever event occurred most recently.

Finally, there is also a graphics object within each figure referred to as the *current object*. This is the object in the current figure that was last touched in some way; either most recently created, manipulated, or clicked on with the mouse pointer. To obtain the handle to this object, the function **gco**, for "get current object," is available.

So, why is it useful to store the handles or use handle requesting functions? For one, they will provide you with a means of determining or modifying an object's properties. Secondly, the commands **gcf** and **gca** can be used as a means of switching between multiple figures or axes so that the next plotting command you issue will appear in the figure and location as you intended. Finally, once the objects' handles are stored in some set of variables, you can make any figure or axes current. This could be accomplished by respectively executing the commands **figure**(figure_handle) or **axes**(axes_handle), where figure_handle and axes_handle are the object handles to the figure and axes you want to make current.

Consider that when developing your own custom M-files it is a good idea to assume that at the time your function is invoked there may be other figures and axes objects already displayed that you or a user do not want altered. Therefore, it may be wise not to use programming styles that depend on a particular figure number being available in which to display graphics. As an example, consider a situation in which you want to plot some data to two different figures, and then, after you looked at the data, you would like to reuse the two figures. Do not use a form in your code that does the following:

```
figure(1);
plot(...);
figure(2);
plot(...);
disp('Press any key to continue');
pause
figure(1);
plot(...);
```

where the "..." represents legal plotting arguments. Rather, assume that figures 1 and 2 already contain some graphics that you (or the user) do not want destroyed and use a form such as

```
fig_handle1 = figure;
plot(...);
fig_handle2 = figure;
plot(...);
disp('Press any key to continue');
pause
figure(fig_handle1);
plot(...);
```

We will also recommend that you use variable names for your handles that are a little more descriptive than fig_handle1 and fig_handle2. This will make your programs more readable and make it easier for other individuals to look at your code and determine how the handles are being used!

Handles and the functions that return handles can be used as arguments to commands that make use of handles. In the next section, you will learn about the functions **get** and **set** that can be used to query and specify a graphics object's property values. There are also the commands **clf**, **cla**, and **delete** that can be used to clear the current figure, clear the current axis, or delete an object by using its handle. The **delete** function is used in conjunction with **gca** and **gcf** to remove graphics objects that are displayed in the current MATLAB work session. For example, the objects in the current axes, but not the axes itself, can be deleted with the **cla** command; using **delete(gca)** will not only remove the contents of the axes, but will delete the axes as well. Likewise, **clf** will clear the contents of the current figure, whereas **delete(gcf)** will delete the current Figure Window. You can also delete just a single object with **delete(gco)**. Remember that if an object is deleted, all children of that object will also be deleted. The thought might have already occurred to you that quite a lot of graphics objects can exist in a Figure Window, and you are right. It would be very inconvenient if we had to rely solely on a user touching a graphics object, or always keeping track of our handles at creation. After we discuss the properties of objects, we will introduce another technique for obtaining the handles of graphics objects by making use of the function **findobj**.

7.4   Properties

Every graphics object has a set of properties associated with it, i.e., named values, that contain all the information needed for display. At the time a graphics object is created, the properties that you do not explicitly specify are initialized to their default values. Property values can be in the form of strings, vectors, or matrices, but are always used to define a characteristic or attribute related to an object.

In this section, you will be introduced to the Property Editor; a GUI included with MATLAB that lets you browse graphics objects and their properties. Then, and more importantly, you will learn M-File programming techniques that will let you determine what the names of the properties are, how their values can be determined and manipulated, and how you can alter their default values.

By the end of this chapter, you will know how to create objects, find objects, and change their properties in just the manner that best works in your specific application. In fact, from here on out in this book, object properties will be key in almost all the discussions, so much so that we will indicate object property names in italics.

7.4.1 The Property Editor

The Property Editor is a convenient graphical utility that allows you to quickly navigate around objects and edit most of an object's properties. (We say "most" because not all properties are visible via the Property Editor.) The greatest benefits of the Property Editor are its convenience of accessing object properties and its ability to depict the organization of objects in a figure. It is very handy for one-time "tune-ups" of plots and for quickly seeing object properties and values.

You invoke the Property Editor either by going into property editing mode in a Figure Window, achieved by clicking on the arrow in the toolbar and then double-clicking on an object in the figure, or by typing the command **propedit** in the Command Window. If you type **propedit** without a Figure Window in existence, the Property Editor will open to the root object properties and looks something like Figure 7.3.

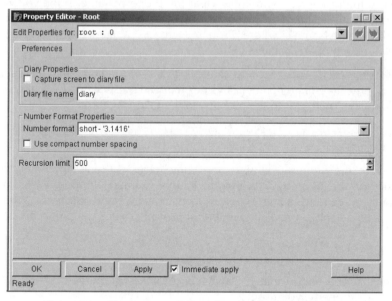

Figure 7.3   The Property Editor showing the root object.

If you have any Figure Windows open, when you invoke the Property Editor from the Command Window, it will open with the current figure in the "Edit Properties for:" box.

7.4.2 Manipulating Properties

Manipulating object properties requires that you first know the property names for the various objects. From using the Property Editor you might think that this is a daunting requirement, however it is not necessary to memorize the names nor must you continually look them up in the MATLAB documentation.   MATLAB provides two functions, namely **get** and **set**, that allow you to list all available or settable properties for a given object. Even though you do not need to memorize all the properties, after you become familiar with them you will see that the property names are fairly intuitive or at least become so fairly quickly. If you think about the English words that describe the characteristic of the graphics object that you want to change, you will very likely come up with the property name. For instance, if you want to change the color of a line from yellow to blue, you would look at a line object property called *Color*. If you want to change the style of the line from solid to dashed line, you will need to look at the *LineStyle* property of the line object.

To list all of the properties for a graphics object with handle h, use **get**(h). In addition to listing the properties, **get** will list the corresponding property values next to the property. If you know what the property name is and you wish to determine the value that is currently assigned to that property, you can use **get**(h,'PropertyName'), where PropertyName is a string of characters that spellout the property's name.

To see which properties you can specify the properties value, use **set**(h). If a property has a finite number of values that can be specified for that property, they will be listed next to the property in the following form

```
PropertyName : [ PropertyValue1 | {PropertyValue2} |
              PropertyValue3 | PropertyValue4 | ... ]
```

where the property between the curly-braces ("{" and "}") is the default value for the property.   Therefore, in this example, PropertyValue2 is the default or current setting of the objects PropertyName property. Properties that are not limited to a finite number of possible values will simply be listed as

```
PropertyName
```

so that you realize that you may specify these properties. If you want to specify a property, you use

```
set(h,'PropertyName',PropertyValue)
```

where h is the handle to the object that you are manipulating. Although we use the full property name in the examples found in this book, there are a couple of things you can keep in mind that can save time in specifying and querying object property values. The first is that the capital letters are not required, but are used only to make the properties easier to read. The second is that the full name of the property does not need to be used, but only enough characters to uniquely identify a given property. As you will see, for many of the object properties, you need to pass only the first three characters of the property name.

When you use **get** to see the properties of an object, MATLAB returns a structure array of the object's property information. For example, if you used

```
MyObjectInfo = get(h);
```

you could then determine the value of any of the properties of object h in the following manner:

```
MyObjectInfo.PropertyName
```

Note that *PropertyName* must be the name of one of the valid object properties. Also, be aware that *PropertyName* must be used with the correct capitalization since MATLAB structure field names are case sensitive. So as an example, let's say you have a just created a Figure Window (this could have been done with a high-level command) and you want to determine the setting for the orientation of a printed page. You might do this in this way:

```
MyObjectInfo=get(gcf)
MyObjectInfo.PaperOrientation
```

which will return,

```
ans =
portrait
```

Just as with **get**, **set** will return a structure array if called with a return value. You can use **set** in this manner when you want to put all the possible PropertyValues for an object with handle h into a structure array.

```
PosPropVals = set(h);
```

With this method you can look at the possible property values that could be used for a particular property, say *PropertyName*, with

```
PosPropVals.PropertyName
```

Say you wanted to know what the possible units are for a figure; assuming you have a Figure Window with handle h you could use:

```
PosPropVals=set(h);
```

then look at the *Units* property with,

```
PosPropVals.Units
```

This will return something like,

```
ans =

    'inches'
    'centimeters'
    'normalized'
    'points'
    'pixels'
```

```
'characters'
```

which are the possible value specifications for the *Units* property.

7.4.3 Universal Object Properties

There are a number of properties that are common to all graphics objects in MATLAB and there are some other properties that are common among several types of graphics objects; however, there are fifteen *documented* properties that are common to all graphics objects. We are careful to emphasize documented properties since those are the ones that are either found in the MATLAB documentation or listed with **get** when MATLAB is running with the factory default settings. Undocumented properties will be discussed later and as you will learn, there are undocumented properties that are common to all graphics objects as well. If you are running a version of MATLAB before version 6, you might notice some differences. The common properties that are documented are listed in the table below along with the specific attributes related to their values.

The tables in this and the next few sections that summarize the graphics properties are organized and sorted alphabetically by the property name in the first column. The second column tells you whether or not you can use the function **set** to specify the property value. If the property is read only (indicated with a "Yes" in this column), you cannot modify the property value. The third column lists the type of information or the possible options that you can set or get. Limited property value options are indicated where the contents within this column are surrounded by square brackets "[ ]" and the choices are separated with the separator character "|". The factory default value is indicated by the option that is surrounded by braces "{}". These values are passed to and from the **get** and **set** functions as strings. For example, to set the Selected property to *on* for an object with handle h, you could use **set**(h,'Selected', 'on'). Entries in the third column that are surrounded by square brackets, yet have no separator characters, indicate that there is a strict format for the value matrix and that MATLAB expects the individual elements to be in a particular order. Finally, if brackets do not surround entries in the third column, this indicates that their values can be numbers, integers, handles, strings, or characters. The format column indicates whether the property values are stored as a limited number of elements, an unrestricted (in length) row vector, a column vector, or a matrix.

Some of these properties do not affect all of the graphics objects; nonetheless, each of these properties still exists and therefore is listed here as a common property. For example, the *ButtonDownFcn*, Clipping, Interruptible, and Visible properties of the root object will not alter any aspect of MATLAB's graphics and interface operations.

The *Type* property is read only; in other words, you can only use the **get** function with them. If you try to set this property, you will get an error message like,

```
??? Error using ==> set
```

```
Attempt to modify read-only figure property: 'Type'.
```

The "ValueType/Options" column within the table shows whether a string or number can be specified or retrieved from the property with the **set** and **get** functions. Entries containing brackets indicate that there are only a limited number of options available for those object's properties. Even though string quotes are not shown, they would need to be used when setting those properties. For example, if you want to set the *Visible* property to "off" for an object whose handle has been stored in the variable h, you can use

```
set(h,'Visible','off')
```

The property "Format" column indicates how the property values must be passed to or from the object. The entry "row" means that the value can be either a row vector of numbers or characters. The entry "column" means that the value can be either a column vector of numbers or characters. Both row and column formatted property values may also be a single element or the empty number ([ ]) or string ([") matrix when appropriate. Only the *UserData* property is unrestricted (in the same sense that any other variable within the MATLAB work space is unrestricted) with regard to the data that it can store.

Property	Read Only	ValueType/Options	Format
BusyAction	No	[{queue}	cancel]	row
ButtonDownFcn	No	string	row
Children	No*	handle(s)	column
Clipping	No	[{on}	off]	row
CreateFcn	No	string	row
DeleteFcn	No	string	row
HandleVisibility	No	[{on}	callback	yes]	row
HitTest	No	[{on}	off]	row
Interruptible	No	[no	{yes}	off	{on}]	row
Parent	No	handle	one element
Selected	No	[{off}	on]	row
SelectionHighlight	No	[{no}	yes	{off}	on]	row
Tag	No	string	row
Type	Yes	string	row
UserData	No	number(s) or string	matrix
Visible	No	[{on}	off]	row

* Although you cannot create new handles in the *Children* property, you can change the order of the handles and so change the stacking order of the objects.

7.4.3.1 *ButtonDownFcn, BusyAction,* and *Interruptible*

The *ButtonDownFcn* (button down function), *BusyAction*, and *Interruptible* properties will be fully addressed in Chapter 10 when we discuss using objects as a mechanism for the user to interface with MATLAB. Briefly, *ButtonDownFcn* is used to specify a single or set of legal MATLAB commands that perform some action when the user clicks the mouse button in an area that is near, in, or on (depending on the graphics object type) a graphics object. The *BusyAction* property controls what should happen to the events that are spawned from actions taken by the user directed at the object when some other event is currently being executed. Let's say you have two objects that have their *ButtonDownFcn* property defined and you click on one object and then the other object before the commands stored in the first object's *ButtonDownFcn* have completed. If the *BusyAction* property is set to queue, the commands in the second object's *ButtonDownFcn* property will be executed after the first object's commands finish. If the *BusyAction* property is set to cancel, the commands in the second object's *ButtonDownFcn* property will be ignored. The *Interruptible* property specifies whether the sequence of events that are programmed to occur when a user interacts with graphics objects can be interrupted by the execution of additional event-driven sequences.

7.4.3.2   *Children* and *Parent*

The *Children* and *Parent* properties of an object will contain the graphics object handles to the children and parent of that object. The root object never has a parent. And therefore, the value of the *Parent* property of the root object will be the empty matrix. All other objects will have a single number that corresponds to their parent stored in the *Parent* property. Objects such as lines, text, patches, and surfaces have no children and therefore, their Children properties will be the empty matrix. Figure and axes objects will always have a parent object and may have children, depending on whether or not children have been created. As a historical note, in versions before MATLAB 5, these properties were read-only. With the modern versions of MATLAB you can reassign an object to another object of the same *Type*. For example, you can move an axes object from one figure to another just by setting the axes object's *Parent* property to the figure handle to which you want to move the axes object. You can also reorder the handles in the *Children* property, i.e., you can set the property to any permutation of the current handle values that are stored in the object's *Children* property, with the result of changing the stacking of the objects on the display. In this manner the *Children* property sets the order in which certain objects (figure, axes, uicontrol, and uimenu) are displayed when their screen positions overlap. The lower the index number of the handle that is stored in the *Children* property's column matrix, the closer the object will be drawn to the user's viewpoint to the screen. By manipulating this property, you can force an object that might otherwise be hidden from view to move to the front of the screen.

7.4.3.3   *Clipping*

Although an available property of all objects, the *Clipping* property only affects line, patch surface, image, and text objects. Line, patch, surface, and image objects by default have their *Clipping* property set to "on", while text objects have this property set to "off". For these objects, when the *Clipping* is set to "on", portions of the object that lie beyond the region of space defined by their parent axes will not be seen. When *Clipping* is set to "off", portions of the object will be seen even if they are outside of the axes object perimeter. You can illustrate this property with the following code.

```
x = -5:15;
LineHandles = plot(x,x+5,'--r',x,x-3,'g');
TextHandles(1) = text(6.5,5,...
        'This String will have clipping off');
TextHandles(2) = text(-1,3.5,...
```

```
                'This String will have clipping on');
      axis([0 10 0 10]);
```

You should have the result shown in Figure 7.4.

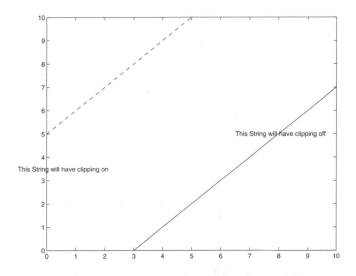

Figure 7.4 **Line objects default to *Clipping* "On", but text objects default to *Clipping* "Off".**

Now we shall set *Clipping* "off" for the red dashed line (this object's handle is stored as the first element in the LineHandles variable because it is defined by the first three arguments in the **plot** command) and "on" for the text object whose handle is stored in TextHandles(2) by typing

```
      set(LineHandles(1),'Clipping','off');
      set(TextHandles(2),'Clipping','on');
```

The result is shown in Figure 7.5. Notice how the dashed line now extends beyond the boundaries of the axes object since we set its *Clipping* property to "off". Also notice that the characters that previously lay outside the axes boundaries are no longer visible now that the *Clipping* property has been set to "on" for that text object. The *Clipping* property affects how the graphics are displayed, but does not affect the data or characters of the object; the characters that are no longer displayed still reside within the text object, so that if at some later time you decide to set *Clipping* "off" for this object, the characters will reappear.

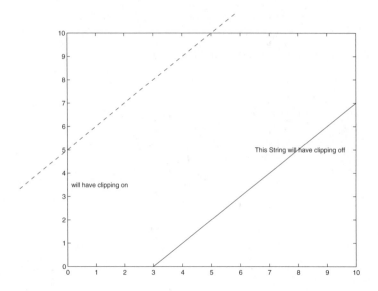

Figure 7.5 Here *Clipping* is set to "off" for the dashed line and "on" for one of the text objects.

7.4.3.4   *CreateFcn* and *DeleteFcn*

The *CreateFcn* (create function) property can be assigned a string containing legal MATLAB commands that will be executed during an event in which a duplicate of the object is being created. If that sounds a little confusing, consider an example where, if you use the **copyobj** function on a graphics object that has the *CreateFcn* property set, the string in this property will be evaluated as if it were typed on the command line. (**copyobj** makes a copy of a graphics object along with all of its children.)

Try the following:

```
h = figure;
set(h,'CreateFcn','display(''Cloning a figure.'')');
copyobj(h,0);
```

You should see a result like,

```
ans =
Cloning a figure.
```

after you type the second command. The only other way this property can really have an affect is if the root's DefaultObjectTypeCreateFcn is set so that whenever an object of ObjectType is created, the string will be evaluated. For instance, if you typed,

```
set(0,'DefaultFigureCreateFcn',...
     'display(''OK, here''s a figure.'')')
```

every time a new figure was created, the display string would be shown on the command line output.

The *DeleteFcn* (delete function) property's value may be assigned a string containing legal MATLAB commands that will be executed during an event which destroys or removes the object from existence. For example, if you executed

```
delete(object_handle)
```

the object whose handle had been assigned to the variable *object_handle* will execute its destroy function. In addition, if the object that is being deleted has children, the children's respective destroy functions will also be executed. Other events that will execute destroy functions are when you close a Figure Window or perform another graphics command that replaces any currently displayed graphics objects.

7.4.3.5 *HandleVisibility*

The *HandleVisibility* property has nothing to do with what you see on your screen; rather, it has to do with one manner in which objects are grouped by MATLAB and accessed from the command line, functions, and callback routines (which we will discuss in Chapter 10). The default value of this property is "on", and means that the object's handle is available to the command line, any function, and callback routine. If the value is set to "off'", the object's graphics handle is not visible to functions, and callback routines assume you have not set the root's *ShowHiddenHandles* (see the next section of this chapter) property to "off". For instance, if you set figure 1's *HandleVisibility* to off with,

```
figure(1);
set(1,'HandleVisibility','off');
figure(2);
```

and then get the *Children* property of the root object, the value 1 (a figure's handle is in most cases the figure number) will not be in the returned list, as you can see with,

```
get(0,'Children')

ans =
     2
```

In addition, if you execute a **close**('all') command, only figures with their *HandleVisibility* set to "on" will be closed. With the value set to "off", functions like **findobj**, **gco**, **gcf**, and **gca** will not return the figure's graphics handle. If the *HandleVisibility* property is set to "callback", then only a callback routine (such as that defined in a figure's *ButtonDownFcn*) will have visibility to the handle. For example, continuing the simple example from above, type

```
set(1,'HandleVisibility','callback');
set(2,'ButtonDownFcn','get(0,''Children'')');
% Note that Children is surrounded by 2 single quotes.
```

Now, every time you click in figure number 2, MATLAB will return

```
ans =
      2
      1
```

But if you typed

```
get(0,'Children')
```

MATLAB would only return

```
ans =
      2
```

7.4.3.6   *HitTest*

This property is used to control access to graphics objects when a mouse click has occurred on an object. Specifically, the value assigned to an object's *HitTest* property determines if the object can become the current object, i.e., its handle be returned by the **gco** command and a figure's *CurrentObject* property (see the next section). If *HitTest* is set to "off", clicking on the object will select the object below it, typically the figure containing it. Although all graphics objects have this property, only axes, figures, images, lines, patches, rectangles, surfaces, and text will respond to it. The default is "on" which means that the object clicked on will be set to the current object.

7.4.3.7   *Selected* **and** *SelectionHighlight*

When the *Selected* property of an object is in its "on" state, the object will be circumscribed with a *selection box* that has *little handles* in the corners. The selection box can either be dashed or solid, depending on the object type (for instance, text objects will have a solid selection box, while axes objects will have a dashed box). Although all objects have this property, root, light, uicontrolmenu, and uimenu objects are not affected by it. Lines, surfaces, images, and uimenus do not have a visible selection box with either setting; instead, the presence of the little handles is all that is affected. Figure 7.6 illustrates the *Selected* property "off" and "on" states.

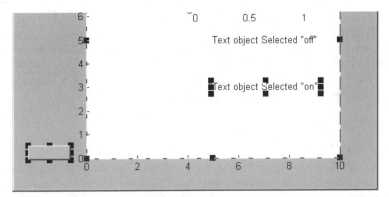

Figure 7.6   The effects of setting *Selected* "on" and "off".

Working closely with the *Selected* property is the *SelectionHighlight* property. The *SelectionHighlight* property is "on" by default for every object, thereby letting you see the selection box when the *Selected* property is set to "on". You also have the option of having an object be selected while keeping its selection box invisible by setting the *SelectionHightlight* to "off" and the *Selected* property "on". This combination of properties makes it convenient to search for a set of objects using **findobj** without having the visual indication that those objects are selected, such as with

```
findobj('selected','on')
```

7.4.3.8   *Tag* and *Type*

The *Tag* property is an extremely useful graphics property that allows you to store any string vector in any of the graphics objects. For instance, you may assign a meaningful name to one or more of the graphics objects you create. Later you could then use **findobj** with the *Tag* property as a simple way of obtaining the handle to a particular object.

The *Type* property just identifies the object's type as being one of the 13 possible types (root, figure, axes, line, rectangle, patch, surface, image, light, text, uimenu, uicontextmenu, or uicontrol). This is a read-only property that comes in handy when searching for particular objects.

7.4.3.9   *UserData*

The *UserData* (user specified data) property allows you to store number or string matrices in a graphics object. It provides a good place to put information that you want to associate with the object, but does not necessarily have to be related to the object. One advantage of storing information in an object's *UserData* property is that even if you clear the workspace with the function **clear**, the data will remain with this object and can be retrieved at any time. The graphics object does not alter or use the information stored in this property. However, we will learn that since you can design an event driven graphics system with MATLAB, you can program MATLAB to make use of this data storage location during the occurrence of some event. For example, you could have the contents displayed to the user or mathematically manipulated if the user clicked on a graphics object. The use of the *UserData* property will become clear when we discuss user interfaces and provide illustrative examples.

7.4.3.10   *Visible*

The *Visible* property allows you to determine whether or not a graphics object is displayed (visible) or hidden (not visible). This property has no effect on the root object. If the *Visible* property of a figure is specified as "off", the Figure Window and its contents will be invisible. For all other objects, the *Visible* property affects only the visibility of the object whose property is specified; making the axes object invisible will not make its children invisible. By default, this property is set to "on" when an object is created.

7.5   Object Specific Properties

The properties covered in the previous section are universal or common for every graphics object in the MATLAB graphics environment. In this section we will examine the properties that are specific to the root, figure, axes, line, rectangle, patch, surface, light, and text objects. The properties of those objects that are related to color (color, light, and transparency) will be quickly mentioned in this section but left for detailed discussion in the next chapter. Image object properties are looked at again in this section, while user interface control, menu, and context menu object properties will be discussed in Chapter 10. This section has been designed to give you informative tables, illustrations, and simple examples that make use of the object properties. The Appendix contains duplicates of these tables for quick reference.

At this time it is important to point out that the order in which the commands discussed in this section are performed and whether executing them from an M-File or the command line has different implications on the final results. This is because MATLAB does not update the display or render graphics with every command when running an M-File. Rather, an event queue is established to store consecutive graphics statements so that they may be more efficiently executed. There are four events that cause MATLAB to flush out the queue so that these stored commands can generate graphics objects or modify an object's property alterations:

1.   a return of control to the MATLAB Command Window prompt,

2. a **pause** or **waitforbuttonpress** statement,

3. the execution of the **getframe** function, and

4. the execution of a **drawnow** command.

The **pause** command waits for a key to be pressed, while the command **waitforbuttonpress**, which will be discussed in Chapter 10, waits for the mouse button to be clicked. Since these two commands suspend execution of MATLAB code, the graphics environment is updated so that the objects in the Figure Windows accurately represent their current attributes. The **getframe** function requires the objects to represent their present state because this function can be used to take a snapshot of a Figure Window. The **getframe** and **drawnow** functions will be discussed in detail in Chapter 9.

The examples in this section assume that you are reading along and perhaps typing in commands so that you see their effects on your monitor. By typing the commands, the graphics events are getting flushed immediately so that you will see the same results presented in the figures shown in this book after you have executed the given sequence of commands.

7.5.1 Root Properties

In addition to the universal properties that the root has which were discussed in the previous section, the root object contains quite a few properties, some of which are not even related to graphics. These properties can be categorized into properties about the display, properties related to the state of MATLAB, and properties related to the behavior of MATLAB. The nature of these categories will become clear in the following discussions. The following table summarizes the documented properties of the root object.

Property	Read-Only	ValueType/Options	Format
Display Related			
FixedWidthFontName	No	string	row
ScreenDepth	Yes	integer	1 element
ScreenSize	Yes	[left bottom width height]	4-element row
Related to the State of MATLAB			
CallbackObject	Yes	handle	1 element
CurrentFigure	No	handle	1 element
ErrorMessage	No	string	row
PointerLocation	No	[x-coordinate,y-coordinate]	2-element row
PointerWindow	Yes	handle	1 element
ShowHiddenHandles	Yes	[on | {off}]	row
Behavior Related			
Diary	No	[on | {off}]	row
DiaryFile	No	string	row
Echo	No	[on | {off}]	1 element
Format	No	[short | long | {shortE} | longE | hex | bank | + | rat]	row
FormatSpacing	No	[{loose} | compact]	row
Language	No	string	row
RecursionLimit	No	integer	1 element
Units	No	[inches | centimeters | normalized | points | {pixels}]	row

7.5.1.1   Display Related Root Properties

The first display related property we will present is *FixedWidthFontName*. This property takes a string that specifies what fixed-width font MATLAB will use for axes, text, and uicontrols whose *FontName* property is set to FixedWidth. The advantage given by *FixedWidthFontName* is that you do not need to independently code font names in MATLAB applications and thereby enables these applications to run without modification in locales where non-ASCII character sets are used; MATLAB attempts to set the value of *FixedWidthFontName* to the correct value for a given locale. In general you will not be changing this property since you should create axes, text, and uicontrols with their *FontName* properties set to FixedWidth when you want to use a fixed-width font for these objects. You can also change this property to set a different font for the fixed-width font. In most cases, the default for the value of *FixedWidthFontName* is 'Courier'. Here's an example.

```
get(0,'FixedWidthFontName')
```

```
ans =

Courier
```

On startup, MATLAB determines the value that is assigned to the *ScreenDepth* property. The value assigned to *ScreenDepth* specifies the number of bits that correspond to the number of colors that the display system of your computer is configured to display. The number actually corresponds to the exponent of a power of 2, i.e., the number of bits used for color. For example, if you have your monitor set up for 256 colors, then *ScreenDepth* will be 8 (2^8 = 256); for 16K colors, the value will be 24 (2^{24}=16,777,216).

The *ScreenSize* property contains the size of the screen as a four-element vector that specifies the lower left corner coordinate (left, bottom) and the width and height as

```
[left bottom width height]
```

The left and bottom elements of this vector are both zero for all root unit specifications except pixels. When the root's *Units* property is set to "pixels", the left and bottom elements will both be the number one. The width and height elements will depend on the monitor size and units used. So, for an example, if you are using a 1024 x 768 display,

```
get(0,'ScreenSize')
```

will return
```
ans =

        1         1      1024       768
```

7.5.1.2 Root Properties Related to the State of MATLAB

The next six properties are related to the *state* of MATLAB, i.e., they contain information that can be used to determine what is going on in a MATLAB session. The first we will discuss is the *CallbackObject* property. Although this will make more sense when we get to Chapter 10, at this time let us be satisfied with knowing that *callbacks* are simply the code that is executed when a user interface is invoked. This property of the root, when accessed by the command line, will contain an empty matrix. It is only when an object's callback routine (e.g., **ButtonDownFcn**, **Callback**, **DeleteFcn**, **CreateFcn**, etc.) is being executed that this property will contain a value, namely, the handle of the object whose callback routine is currently being executed. This property provides the best way for a callback routine to determine which object it is executing from (particularly if multiple objects execute the same callback routine) so that the routine can, for instance, access its own *UserData* property to get information that has been stored there. As a simple example, try typing

```
figure; figure;
set(findobj('Type','figure'),...
    'ButtonDownFcn','get(0,''CallbackObj'')')
```

Notice that CallbackObj is surrounded by two single quotes.

Then click in one and then the other Figure Window; you should see the figure number appear in the Command Window corresponding to the figure you clicked in. However, if at the command line you type

```
get(0,'callbackobj')
```

MATLAB will return

```
ans =

    []
```

The *CurrentFigure* property will contain the handle of the current Figure Window, i.e., the Figure Window that was most recently created, clicked in, or made current with

```
figure(h)
```
or
```
set(0,'CurrentFigure',h)
```

where h is the handle of an existing Figure Window. Note that **figure** will restack the Figure Windows if multiple ones exist, while **set** does not. If there are no figure objects,

```
get(0,'CurrentFigure')
```

returns the empty matrix. However, **gcf** will always return a figure handle, and creates one if no figure objects exist.

The *ErrorMessage* property contains a string consisting of the last error message issued by MATLAB or the last value to which you set this property. The **lasterr** function retrieves the value of this property by executing

```
get(0,'ErrorMessage')
```

The content of this property can be useful in routines that could result in a MATLAB error message if the user of your function or graphical user interface were to do something incorrectly. For example, to execute functions you know could result in an error, you can use the **eval**('*try*','*catch*') where *try* is the function you would like to execute but know may result in an error if used incorrectly, and *catch* is a function that will get the string stored in *ErrorMessage*, (e.g., error_string = **get**(0, 'ErrorMessage');) parse the string, and perform some action that is dependent on the error message that was found to have occurred.

The *PointerLocation* (current pointer location) property can be used to report the position of the mouse pointer. Executing

```
get(0,'PointerLocation')
```

will return a 2-element vector that contains the x- (horizontal) and y-coordinates (vertical) of the pointer with respect to the lower left corner of the computer screen (not the Figure Window). This will be a useful property in some graphical user interface applications, such as when creating functions that allow you to use the mouse to define the position of a graphics object by clicking and dragging the object. You can also use it to place the mouse pointer in a particular location on the screen at the occurrence of a particular user action or event. For example, you can use *ScreenSize* and *PointerLocation* to place the pointer in the middle of the screen with

```
ScreenSize = get(0,'ScreenSize');
set(0,'PointerLocation',ScreenSize(3:4)/2);
```

The *PointerWindow* property reports the handle of the Figure Window that contains the mouse pointer. This is another property that is particularly useful in callbacks. If there are no Figure Windows being displayed at the moment or the mouse pointer is not within a Figure Window, this property will be set to zero. By itself, this property may not seem too useful, but you will learn about potential uses for it in Chapter 10 when we discuss graphical user interfaces.

The *ShowHiddenHandles* property is related to the universal property of *HandleVisibility* and by default is set to "off" which allows the *HandleVisibility* of each object to dictate whether its handle will be visible at the command line, during a callback type routine, or not at all. If you set *ShowHiddenHandles* to "on", the object property is overridden and all object handles will be visible.

7.5.1.3   Behavior Related Properties of the Root

The remaining eight root properties (*Diary, DiaryFile, Echo, Format, FormatSpacing, Language, RecursionLimit,* and *Units*) have no effect on MATLAB in a graphics context, but they nevertheless are properties of the root object. The *Diary* and *DiaryFile* properties are directly related to the **diary** command, which as you know can be used to keep a record of your MATLAB command line entries and outputs. (See Chapter 1.) Using this property for instance, you could use the following in an M-File;

```
set(0,'diary','on')
```

which would be equivalent to typing

```
diary on
```

at the command prompt. Likewise **diary off** is an alias for **set**(0,'diary','off'). The **diary** command also lets you specify the name of the diary file. If, for some reason, instead of using the diary command, you wanted to make use of the root property, it would be just as effective to use

```
set(0,'DiaryFile','filename')
```

The MATLAB commands **echo on** and **echo off** are just setting the root property *Echo*, respectively, to "on" or "off". Setting *Echo* to "on" forces each line of a script file to be displayed as it is executed.

The *Format* property can be used to affect how MATLAB displays numbers in the Command Window. Values for *Format* are the strings "short", "long", "shortE", "longE", "shortG", "longG", "hex", "bank", "+", and "rational". Similarly, *FormatSpacing* affects the line spacing of output to the Command Window. Its possible values are "loose" and "compact". For example, using the **format** command at the command prompt to set the Command Window to longE and compact format with

```
format long e
format compact
```

can be achieved using **set** with,

```
set(0,'Format','longE')
set(0,'FormatSpacing','compact')
```

Please recognize that you would typically control the format of output using formats in the **printf** function; see the MATLAB helps and documentation for more about how to produce formatted output.

The property *Language* is a system environment setting that reports the language your version of MATLAB is designed for. The code

```
get(0,'Language')
```

will return,

```
ans =
english
```

The property *RecursionLimit* tells MATLAB how deep in recursion, that is how deep nested M-File calls can go, before MATLAB will terminate it. The default value is 500.

The property *Units* specifies the units MATLAB uses to interpret size and location data of your computer screen. Possible values are "pixels", "normalized", "inches", "centimeters", "points", and "characters". All units are measured from the lower left corner of the screen. "Normalized" units map the lower left corner of the screen to (0,0) and the upper right corner to (1.0,1.0). "Inches", "centimeters", and "points" are absolute units. One point equals 1/72 of an inch. "Characters" are units defined by characters from the default system font, specifically, the width of one unit is the width of the letter x, the height of one character is the distance between the baselines of two lines of text. The default value for *Units* is "pixels".

7.5.2   Figure Properties

You have already seen lots of figure objects. Figure objects are those objects created with the **figure** function, or by other functions that invoke **figure** such as **plot**, **surf**, etc.. Each figure object will be a window and we often refer to these as Figure Windows. The properties that are found with every figure object, except those that are universal properties, are listed in the following table. The table uses the same structure as the previous property tables.

Property	Read Only	ValueType/Options	Format
		Positioning the Figure	
Position	No	[left bottom height width]	4-element row
Units	No	[inches | centimeters | normalized | points | {pixels}]	row
		Style & Appearance	
Color	No	[Red Green Blue] or color string	RGB vector
MenuBar	No	[{figure} | none]	1 element
Name	No	string	row
NumberTitle	No	[{on} | off]	row
Resize	No	[{on} | off]	row
WindowStyle	No	[{normal} | modal]	row
		Colormap Controls	
Colormap	No	M RGB number triplets	M-by-3 matrix
Dithermap	No	N RGB number triplets	M-by-3 matrix
Dithermapmode	No	[auto | {manual}]	row
FixedColors	No	N RGB number triplets	N-by-3 matrix
MinColormap	No	number	1 element
ShareColors	No	[no | {yes}]	row
		Transparency	
Alphamap	No	default is 64 values progression from 0 to 1	M-by-1 vector
		Renderer	
BackingStore	No	[{on} | off]	row
DoubleBuffer	No	[on | {off}]	row
Renderer	No	[{patinters} | zbuffer | OpenGL]	row
RendererMode	No	[{auto} | manual]	row
		Current State	
CurrentAxes	No	handle	1 element
CurrentCharacter	No	character	1 element
CurrentObject	No	handle	1 element
CurrentPoint	No	[x-coordinate, y-coordinate]	2-element row
SelectionType	Yes	[normal | extended | alt | open]	row

continued on next page

Property	Read Only	ValueType/Options	Format
		Properties that Affect the Pointer	
Pointer	No	[crosshair | fullcrosshair | {arrow} | ibeam | watch | topl | topr | botl | botr | left | top | right | bottom | circle | cross | fleur | custom]	string
PointerShapeHotSpot	No	(row index, column index)	2-element row
PointerShapeCData	No	1s where black, 2s where white, NaNs where transparent	16-by-16
		Callback Execution	
CloseRequestFcn	No	string, function handle, or cell-array {'closereq'}	string, 1-element, cell-array
KeyPressFcn	No	string	string
ResizeFcn	No	string	string
UIContextMenu	No	Number	1 element
WindowButtonDownFcn	No	string	string
WindowButtonMotionFcn	No	string	string
WindowButtonUpFcn	No	string	string
		Controlling Access to Objects	
IntegerHandle	No	[{on} | off]	string
NextPlot	No	[{add} | replace | replacechildren]	string
		Properties that Affect Printing	
InvertHardcopy	No	[{on} | off]	string
PaperOrientation	No	[{portrait} | landscape]	string
PaperPosition	No	[left bottom width height]	4-element row
PaperPositionMode	No	[{auto} | manual]	string
PaperSize	No	[width height]	2-element row
PaperType	No	[{usletter} | uslegal | A0 | A1 | A2 | A3 | A4 | A5 | B0 | B1 | B2 | B3 | B4 | B5 | arch-A | arch-B | arch-C | arch-D | arch-E | A | B | C | D | E | tabloid | <custom>]	string
PaperUnits	No	[{inches}|centimeters| normalized | points]	string
		General	
FileName	No	A name of a FIG-File to be used with GUIDE; see Chapter 10.	string

7.5.2.1   Figure Properties Affecting Position

The first set of figure object properties, *Position* and *Units*, deals with location of the Figure Window on the screen. The *Position* property of a figure object contains a vector that specifies the left bottom, width, and height ([left bottom width height]) in the current units (as specified by the contents of *Units*). The creators of MATLAB give a special name to this four-element vector – *rect*. Typically we will see the rect vector whenever we discuss the position of a graphics object. For example, if you want to place the current figure in the exact center of the screen and resize it so that it is 200 pixels wide by 50 pixels high, you can do the following:

```
set(0,'units','pixels');
set(gcf,'units','pixels');
screenrect = get(0,'screensize');
screenwidth = screenrect(3);
screenheight = screenrect(4);
figwidth = 200;
figheight = 50;
figposition = [(screenwidth/2-figwidth/2)...
               (screenheight/2-figheight/2)...
               figwidth figheight];
set(gcf,'position',figposition);
```

The first two lines make sure that the root and figure are both using pixel units, just in case you (or the user if you are designing a MATLAB routine that others may use) have changed one or both of these object's units from their factory default values. This code assumes that there is a current figure available. It still works if there is no existing figure object; however, a figure will be generated when you type the second line, and this figure will have the default Position property value. Then only when you type the last line will the figure be sized as desired. Most likely, you will want to specify the desired dimensions when the figure object is created so that you can get the desired end result immediately. To achieve this, you can modify the above code by removing the second line and replacing the last line with

```
figure('position',figposition)
```

or with

```
figure('position',figposition,'units','pixels')
```

to safeguard against alterations in the default unit values.

7.5.2.2   Style and Appearance Properties of the Figure Object

The next set of figure object properties we will look at closely follows the previous discussion and has to do with the style and appearance of the Figure Window.

When you call up a figure, it is by default a pleasing shade of gray (specified by the RGB vector [0.8 0.8 0.8].) However, you can specify any color by making use of the figure's *Color* property. For example, if you want to set figure 1's color to yellow, you can type

```
set(1,'Color','yellow')
```

or

```
set(1,'Color',[1 1 0])
```

Speed!

In order to give Figure Windows more meaning, you can assign a name to them. The name can be specified by setting the *Name* property to any string of characters. These characters will then appear in the top or title bar of the Figure Window next to the "Figure No. X" (where X is the handle of the Figure Window). The "Figure No. X" in the title bar can be suppressed by setting the *NumberTitle* property of the Figure Window to "off". By setting the *Resize* property to "off", you can prohibit the user from changing the dimensions of the figure with the mouse. (The default value for this property is "on" so that the user can resize the window.)   Although properly part of "Callback Execution", it is helpful to mention the *ResizeFcn* (resize function) property here. *ResizeFcn* allows you to specify any legal MATLAB commands as a string which will be evaluated when the user attempts to modify the height and width of the figure with either the mouse or the **set** function.

The property *MenuBar* allows you to either hide or show the menu bar placed at the top of a Figure Window. For those of you who are developing MATLAB code for use on a Macintosh operating system, this property will let you suppress the display of the default menus that appear when the figure is selected. In the MATLAB version for the Macintosh, the default menus for Figure Windows are "File," "Edit," "Options," and "Window." If the *MenuBar* property is set to "none" for a given figure, these menus will not be visible when that figure is active. The default value for this property is "figure", however if you are using GUIDE for your GUI development the default is "none". If you are just creating plots and not defining any custom user interface menus, it is probably best to leave this figure property in its default mode. Note that this property affects only built-in menus; menus defined with the **uimenu** command, which will be discussed in Chapter 10, are not affected by this property.

The figure object property *WindowStyle* can be set to either "normal" (the default) or "modal". With this property, you can direct a Figure Window to trap all keyboard and mouse events that occur, essentially forcing the user to deal with the Figure Window in some way before any other action can take place. With *WindowStyle* set to "modal" the user will not have access to any other MATLAB window (including the Command Window). In addition, a modal Figure Window remains "stacked" on top of other MATLAB windows until it is deleted, at which time focus returns to the window that last had the focus.

7.5.2.3   Figure Properties that Control the Colormap

From the previous table you can see that several properties have to do with controlling the color map. We will only examine three of those properties here, namely *FixedColors*, *Dithermap*, and *Dithermapmode*; the remaining three properties, *Colormap*, *MinColormap*, and *ShareColors*, affect the color of surface, image, and patch objects that are displayed in the figure and are deferred to a more detailed discussion in Chapter 8. Essentially, you can use any RGB triplet or any of the legal color strings to define the color of your choosing. *FixedColors* keeps track of the colors that are being used by axes, line, and text objects within the figure. As an example, create a quick plot and see what colors are being used.

```
ezplot('sin(x)')
get(gcf,'FixedColors')

ans =

        0        0        0
   1.0000   1.0000   1.0000
   0.8000   0.8000   0.8000
        0        0   1.0000
        0   0.5000        0
   1.0000        0        0
        0   0.7500   0.7500
   0.7500        0   0.7500
   0.7500   0.7500        0
   0.2500   0.2500   0.2500
```

Notice that each row of the returned array specifies an RGB triplet.

Two other color-related figure properties discussed here, *Dithermap* and *DithermapMode* are used by MATLAB if you are using a low-color display (typically 8-bit color). By default, the *DithermapMode* is set to "auto" and MATLAB creates a dithermap using the Floyd-Steinberg algorithm that contains colors from the entire color spectrum until something is drawn in the figure. Every time MATLAB renders a figure (like when you add something new to a figure), MATLAB regenerates the dithermap when the *DithermapMode* is set to "auto". To speed up the amount of time it takes MATLAB to render a figure, once you are done adding objects that contain new colors to a figure, you should set the *DithermapMode* to "manual". Remember that once you add new colors after having set the mode to "manual", combinations of colors in a 6-pixel group will be selected to approximate any colors that you add that do not exist in the *Dithermap*, and therefore, you will lose some accuracy in the color content shown on your display.

7.5.2.4   Figure Properties that Affect Transparency

The only property in this category is one called *Alphamap*. We will see this property in Chapter 8 when we discuss object transparency. In this chapter suffice it to say that by default the *Alphamap* property contains a row vector of 64 elements and is used in conjunction with the rendering of surface, image, and patch objects, but not other graphics objects.

7.5.2.5   Properties that Affect How Figures are Rendered

The *BackingStore* property in its default mode is set to "on". This property specifies whether or not the Figure Window must be redrawn every time you switch between the figure in question and another window. In its "on" setting, the figure will be redrawn or refreshed every time you switch between figures, in addition to when the figure is resized or another graphics object is added to the figure. In its "off" state, the figure is redrawn only when resizing or adding additional graphics objects. When there are simple line plots in a figure, there may not be a noticeable delay when the figure is being refreshed; however, if

3-dimensional surface plots or figures that contain a large number of objects are refreshed, it can be an annoyance to have *BackingStore* set to "off". We recommend that you leave this property in its "on" state and make use of the command **refresh** to force a complete redrawing, i.e., refreshing, of a Figure Window.

The properties *Renderer* and *RendererMode* are related to rendering speed and accuracy of displayed plots. By default, the *RenderMode* property is set to "auto" which is usually desirable since MATLAB will then determine which is most likely the best way to render your graphics; however, there can be advantages such as when printing (see Chapter 6) to overriding the default to achieve the results you need.   Whenever you set the *RenderMode* to "manual", or set the *Renderer* property (which will also set *RenderMode* to "manual"), MATLAB will no longer use what it thinks is the best rendering algorithm for the figure.   The three rendering methods that MATLAB supports are Z-buffering ("zbuffer"), Painters ("painters"), and OpenGL ("OpenGL"). The first two methods are algorithmically based, while the third, OpenGL is a hardware-based rendering method that is available on many computer systems. The Painters method is MATLAB's original rendering method and is typically faster when the figure contains only simple or small graphics objects. However, the Painters algorithm will not work if you are displaying image, surface, or patch objects in a figure using RGB specifications. The Z-buffering algorithm determines which graphics object is closest to the viewer (you) at each pixel and draws the front-most portion of the virtually closest object, i.e., if you can't see it, it won't be drawn.   This can be the fastest rendering algorithm when the figure is built up of many complex graphics objects. Z-buffering can draw graphics object faster and more accurately because objects are colored on a per pixel basis and MATLAB renders only those pixels. Although fast, this method can consume a lot of system memory, especially if the scene is complex. OpenGL, if available on your system, is generally faster than Painters or Z-buffer, especially if your computer has a video card that offers its own OpenGL processing.   If you have a simple figure, such as a line plot, the Painters algorithm will generally display it more accurately and quicker.

The property *DoubleBuffer* applies to animations, which is the subject of Chapter 9, and can take the values "on" or "off" which is the default. Double buffering can be used to speed up the process of animating an image as it first draws to an off-screen *pixel buffer* and then *blits* (think of it as throwing the whole image out at once) the buffer contents to the screen once the drawing is complete. You would typically want to take advantage of double buffering to produce flicker-free rendering for simple animations, such as those involving lines. It is not as effective for objects containing large numbers of polygons. We will revisit these ideas again in Chapter 9 where we will discuss how to determine the best rendering methods for the type of animation you want to produce.

7.5.2.6   Properties Related to the Current State of a Figure

The figure object keeps track of things like which axes within the figure will be plotted to next, what was the last keyboard character pressed within the Figure Window, and even the location of the pointer when the last mouse button press and release occurred. The figure object has several properties starting with the word "Current" that are used to store these handles, characters, or locations. The *CurrentAxes* property stores the handle to the current axes, where the current axes are the axes that will be the parent to graphics objects created by subsequent plotting commands. If you have a figure that contains no axes objects, this property will contain the empty matrix; however, the moment you query this property with the **get** command, an axes object will be created and its handle will be returned. Earlier, we learned that you can use the command **gca** to get the handle of the current axes; now, you may realize that this command is merely

```
get(gcf,'CurrentAxes')
```

If there are multiple axes objects within the figure, one will always be the current axes. You can also specify any of these axes objects to be current by setting the *CurrentAxes* property to the handle of the axes with

```
set(figureHandle,'CurrentAxes', axesHandle)
```

so that your next plotting command creates the object in the axes.

MATLAB has many features that facilitate the creation of user interfaces. Depending upon the type of interface you design, you may come across a need for the *CurrentCharacter* property. When a figure is active and you press a key (or key combination such as shift + a character), the corresponding string will automatically be stored in the *CurrentCharacter* property. This property is read-only and is often used in callback routines in conjunction with the *KeyPressFcn* property. We will show this in an example in the subsection dealing with callback execution.

The *CurrentObject* property will contain the empty matrix until you press the mouse pointer somewhere within the Figure Window. If you click the mouse on top of or in the region which is very close (usually called the hot zone) to a graphics object, that object's handle will be placed in the *CurrentObject* property until you select another object with the mouse. If there are multiple objects located under the mouse pointer location at the time the user presses the mouse button, the object which is closest to the top of the graphics object stack will be selected. The object stack is initially determined by the order of object creation. The most recent object created will be at the top of the stack. However, the stacking order changes once objects are clicked on with the mouse. The object most recently selected will be moved to the top of the graphics stack. The stacking order is kept track of with the *Children* property of the figure. For instance, the object whose handle is the first element in the column of handles stored in the *Children* property will be at the top of the stack. The *CurrentObject* property can be set by passing the handle of an object that exists within the figure with

```
set(figurehandle,'CurrentObject',objecthandle)
```

If the current object is deleted, the *CurrentObject* property will be the empty matrix. Finally, you may query this property with

```
get(figurehandle, 'CurrentObject')
```

or by using **gco** which will get the current object within the current figure, whereas when you use the **get** function and explicitly specify the figure handle, the current object for that figure will be returned.

Another useful property of figure objects is that they have the ability to keep track of the last location that the mouse button was either clicked down or released within the figure. The x- and y-coordinates of the most recent of these two events are stored in the *CurrentPoint* property and the coordinates' values are in units specified by the *Units* property of the figure. If the mouse pointer is moved while the button is held down (a click and drag), the *CurrentPoint* will be updated as the pointer is moved. These x- and y-coordinates are always measured with respect to the lower left corner of the figure and therefore are independent of the figure's location within the screen. This property is useful when you want a user to have the ability to provide information to MATLAB with the mouse. For instance, if you want the user to specify the corner points of an object, you might make use of this property and one of the properties presented in the next subsection, e.g., *WindowButtonDownFcn*, that is designed to execute as a result of a mouse event.

The *SelectionType* property's value depends on either the way that the mouse button is pressed (single or double click), the button that is pressed (for a multi-button mouse), or which key was held down when the mouse button was pressed. The following tables present the actions that are required to set the *SelectionType* value to "normal", "open", "alt", or "extend". When a key + button combination is given, it means to hold down the specified key and then press the mouse button. Since the value for *SelectionType* is system dependent, values for Windows, Macintosh, and Unix X-Windows operating systems are listed in the first table. The second table lists the values possible from either two-button or three-button mice.

Selection Type	Windows	Macintosh	X-Windows
"normal"	Single click	Single click	Single click
"open"	Double click	Double click	Double click
"alt"	Alt + click	Option + click	Ctrl + click
"extend"	Shift + click	Shift + click	Shift + click

Selection Type	2-Button Mouse	3-Button Mouse
"normal"	Left button	Left button
"open"	Double click*	Double click*
"alt"	Right button	Right button
"extend"	Right + Left button	Center button

* Note: A double click with a multi-button mouse must be performed with the same button.

To experimentally determine how your system's mouse operations affect the *SelectionType* property, try the following:

```
fighandle = figure;
windowbuttondownstr =
['disp([get(gcf,''selectiontype'')])'];
set(fighandle,'windowbuttondownfcn',windowbuttondownstr);
```

Now click down in the Figure Window using several of the techniques described in the previous tables.

7.5.2.7   Figure Properties that Affect the Pointer

This set of properties belonging to figure objects allows you alter the appearance of the mouse pointer. The *Pointer* property specifies the symbol type that is used to identify where the pointer is located within the figure. By default this is the "arrow" symbol that is most likely similar to the arrow that you are accustomed to seeing when you select from menus within your operating system. However, when the pointer is within the Figure Window, you can specify that it use any one of the 17 symbol-names that MATLAB offers. The 17 symbols with their corresponding names are shown in the following table, however your pointers might look different based on the pointer "scheme" which is active on your system.

Pointer Names and Symbols			
crosshair	+	left	⟨↔⟩
fullcrosshair	crosshair lines extend full horizontal & vertical	top	 ⇕
arrow		right	
ibeam		bottom	
watch	(busy)	circle	○
topl		cross	 ╬
topr		fleur	
botl		custom	16 x 16 pixels contained in PointerShapeCdata
botr			

If the standard pointer styles do not provide exactly what you need, you can generate a custom pointer using the "custom" option of the *Pointer* property and define your custom pointer with the *PointerShapeCData* property and the *PointerShapeHotSpot* properties. By default the custom pointer is the 16-by-16 pixel face shown in the above table. Altering the 16-by-16 matrix in the *PointerShapCData* will allow you to make whatever pointer you need. The elements of the matrix can be either 1's (corresponding to black pixels), 2's (corresponding to white pixels), or NaNs (corresponding to transparent pixels). The (1,1) element of the matrix specifies the upper left corner pixel, while element (16,16) specifies the lower-right corner. The *PointerShapeHotSpot* defines which pixel in the custom pointer is used to determine the location of the pointer (or the value stored in the *CurrentPoint* figure property) and is by default (1,1). In Chapter 10 we will demonstrate some GUI techniques by developing a convenient GUI-based pointer editor called **ptredit**.

7.5.2.8   Figure Properties that Affect Callback Execution

The functions that get invoked whenever an action is taken on a MATLAB graphics object, e.g., moving the mouse over an object, clicking on an object, etc., are called *callbacks*. We will discuss callbacks in great detail in Chapter 10. For this section you need to realize that there are a number of figure properties that affect the execution of callbacks. The first of these, *CloseRequestFcn* (close request function), is a property to which you can assign a set of MATLAB commands that will be executed any time the Figure Window is closed. The value of this property can be a string of the commands, a handle to a function, or a cell-array containing the commands that you want executed when the figure is closed such as when you issue the **close** command with either

```
close(figure_handle)
```

or

```
close all
```

and whenever you close a Figure Window from the computer's window manager menu, or when you quit MATLAB. Consequently, this string must contain valid MATLAB commands, just as you would with using the **eval** function. For example, if you wanted to display a message before a particular window was closed, you could use

```
handle = figure;
set(handle,'CloseRequestFcn',...
    ['display([''You have closed figure #''',...
    ',num2str(get(0,''CallbackObject''))]);closereq']...
    );
```

which will display a message in the Command Window when the figure is closed.

The **closereq** function is the default value of the *CloseRequestFcn* property and executes the following code,

```
shh=get(0,'ShowHiddenHandles');
set(0,'ShowHiddenHandles','on');
currFig=get(0,'CurrentFigure');
set(0,'ShowHiddenHandles',shh);
delete(currFig);
```

which will unconditionally delete the current figure and destroy the Figure Window.

Power!

So as you can see, the *CloseRequestFcn* property is typically used when you want to query the user before finalizing the closing of a figure. The following code that uses the function **questdlg**, will prompt the user with a pop-up window before closing the figure.

```
selection = questdlg(['Do you really want to close ',...
              'Figure #',int2str(gcf),'?'],...
              'Close My Figure',...
              'Yes','No','Yes');

switch selection,
    case 'Yes',
       delete(gcf)
    case 'No'
       return
end
```

To use this code, save it to an M-File named *closemyfig.m* (or any other name you like) then

```
set(gcf,'CloseRequestFcn','closemyfig')
```

The **qestdlg** function takes a string or cell-array of strings, and creates a modal dialog box (one that must be answered before other action can take place) that automatically wraps the cell array or string (vector or matrix) to fit an appropriately sized window. It will return the name of the button (in this case "Yes" or "No") that is pressed. The second string argument ('Close Figure Function') provides the title to the modal window. The following strings, 'Yes', 'No', and 'Yes', specify each button and the last being the default. If you tried the above code as described, you should see a result similar to Figure 7.7.

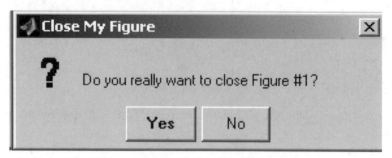

Figure 7.7   Using the *CloseRequestFcn* property.

As you might imagine, you could easily have a nondestructive function assigned to the *CloseRequestFcn* property that will not allow the figure to be closed, such as with

```
set(gcf,'CloseRequestFcn','disp(''I'll Never Die!'')')
```

Most likely, you will create such a situation inadvertently during some code development. Fortunately, with your command of handle graphics, you can always type **delete**(*figure_handle*) at the command line.

Later in this chapter we will discuss *default properties,* but until then, you might find it of value to know that you can apply a custom function like *closemyfig.m* to all figures without having to use **set** with each one. To do this, you can set a default value on the root level with

```
set(0,'DefaultFigureCloseRequestFcn','closemyfig')
```

MATLAB will now use this setting for the *CloseRequestFcn* of all subsequently created figures.

As you will see in Chapter 10, MATLAB has many features that facilitate the creation of user interfaces. As you have seen already, the *CurrentCharacter* property will contain a string representing the key that was pressed while in the active figure. (Recall that this property can only be queried.) Complimentary to the *CurrentCharacter* property is the *KeyPressFcn* property. The *KeyPressFcn* property allows you to specify a string of MATLAB commands that will be executed every time a key is pressed in that Figure Window. There are some rules to keep in mind when creating strings for this property, but they will be discussed in Chapter 10. Basically, if the string that you want to use can be executed with the **eval** function without any errors (as with the *CloseRequestFcn*), it will most likely work when executed at the occurrence of a key press. As a simple illustration try the following:

```
figurehandle = figure;
keypress = ['disp([''The current character is: '' '...
            'get(gcf,''CurrentCharacter'')])'];
set(figurehandle,'KeyPressFcn',keypress);
```

After executing these commands, click your mouse in the Figure Window and type a few characters. If you type the character "b", the message "The current character is now: b" will appear in the Command Window.

There are several points to remember when creating executable or *evaluatable* strings; first, if you want a string to contain a quote within it, then you need to use two single quotes in a row. The second is that the string will be passed to the **eval** function and evaluated in the base MATLAB workspace even if these properties are specified within a function that you create (as if they were script files). Therefore, the string can only make use of variables and information that are available in the workspace and not the local variables within a function (unless they are specified as global variables). A complete elaboration on this point will be discussed in Chapter 10.

In addition to the *Resize* property that we presented earlier, there is a property named *ResizeFcn* (resize function) that, just like the previous property, allows you to specify any legal MATLAB commands as a string which will be evaluated when the user attempts to modify the height and width of the figure with either the mouse or the **set** command.

The *UIContextMenu* property of a figure keeps a value which is a handle to a user interface context menu that is to be associated with the figure. We will examine this property in detail in Chapter 10.

The *WindowButtonDownFcn*, *WindowButtonMotionFcn*, and *WindowButtonUpFcn* are three figure properties that can be used to evaluate and execute a string containing MATLAB commands at the occurrence of a mouse driven event. The *WindowButtonDownFcn* is used to store a string that will be evaluated whenever a mouse button is pressed down within that Figure Window. The *WindowButtonMotionFcn* is used to store a string that will be evaluated whenever the mouse pointer moves within the Figure Window. Finally, the *WindowButtonUpFcn* string will be evaluated when the mouse button is released.

To further illustrate what we have just presented about figure properties that affect callback execution, let's look at an example that uses several of the mouse event driven properties. In this example, we shall have MATLAB identify the type of selection and the current location of the pointer when the user clicks down the mouse button in the Figure Window. In addition, we shall use the crosshair cursor instead of the arrow. If the user then holds down the mouse button and moves the pointer around, the location of the pointer relative to the initial location will be displayed. Finally, when the user releases the mouse button, the current point will be displayed and the cursor should once again become an arrow. For now, we will display these quantities in the Command Window, however, once you learn about the text object properties, you will see how easy it is to display these values within the figure object itself.

```
fighandle = figure;

bdfcnstring =
['selecttype=get(gcf,''selectiontype'');'...
    'firstpnt = get(gcf,''currentpoint'');'...
   'figunits = get(gcf,''units'');'...
   'set(gcf,''pointer'',''crosshair'');'...
```

```
   'disp([''The selection type is:'' selecttype]);'...
   'disp([''First X: '' num2str(firstpnt(1)) '' ''
figunits]);'...
   'disp([''First Y: '' num2str(firstpnt(2)) '' ''
figunits]);'...
   'set(gcf,''windowbuttonmotionfcn'',bmfcnstring,'...
   '''windowbuttonupfcn'',bufcnstring);'];

bmfcnstring = [...
   'currentpnt = get(gcf,''currentpoint'');'...
   'offset = currentpnt-firstpnt;'...
   'disp([''X-Offset: '' num2str(offset(1)) '' ''
figunits]);'...
   'disp([''Y-Offset: '' num2str(offset(2)) '' ''
figunits]);'];

bufcnstring = ['set(gcf,''pointer'',''arrow'');'...
   'lastpnt = get(gcf,''currentpoint'');'...
   'disp([''Last X: '' num2str(lastpnt(1)) '' ''
figunits]);'...
   'disp([''Last Y: '' num2str(lastpnt(2)) '' ''
figunits]);'...
   'set(gcf,''windowbuttonmotionfcn'','''');'];

set(fighandle,'buttondownfcn',bdfcnstring);
```

The variable *budfcnstring* is set up to determine the *SelectionType*, *CurrentPoint*, and *Units* values of the figure and change the cursor into a crosshair when the user first clicks anywhere within the figure. Since this string is evaluated in the base workspace, the *firstpnt* and *figunits* variables will be available when the *bmfcnstring* and *bufcnstring* strings are evaluated. In addition, the *WindowButtonMotionFcn* and *WindowButtonUpFcn* are specified when the user clicks down on the mouse button. In this example, it was not absolutely necessary to set the *WindowButtonUpFcn* at the time when the user clicks down (it could have just as easily been defined in the last line of the code). However, the *WindowButtonMotionFcn* needs to be specified at that time, because after the user releases the button, we will clear the motion property so that further pointer motion does not display X- and Y-Offset values until the mouse button is once again pressed. The variable *bdfcnstring* resets the cursor to the default and displays the final location point. (Notice, as was stated at the beginning of this chapter, that the property names are not case sensitive, however we keep the cases in the discussion to avoid confusion.)

Power!

7.5.2.9   Figure Properties that Control Access to Objects

The *IntegerHandle* property is, by default, set to "on". This means that the figure's handle will match its figure number shown in the title bar of the figure. For example, figure 1's handle will be the number 1, which means you can use **set**(1,...) to change figure 1's properties. However, if you set the *IntegerHandle* property of a figure to "off", you will no longer be able to access the figure using the figure number, since MATLAB will reassign a new floating point number as the figure handle. You can always determine this handle (assuming the *HandleVisibility* property has not been set off) by clicking in the figure and then performing a **gcf** or using the **findobj** function. This property can be useful when you want to reduce the chances of the user accidentally changing a figure's properties, since integers are much more easily typed than some random floating point handle number.

By default, if you create a plot with any of the high-level graphing routines such as **plot**, **surf**, **contour**, etc., the plot will be drawn in the current figure (if there is no current figure, a figure will be created). This can happen because the current figure's *NextPlot* property is in the "add" state. In this state, all of the figure's properties will remain the same. You can also set this property to "replace", which will force any subsequent plotting function to reset all of the figure's properties (except the position property) and remove all figure contents before creating the new plot. If you do not want to reset all of the figure's properties, but do want to remove all figure contents before creating a new plot, set the contents of *NextPlot* to "replacechildren". However, if you want to put a level of protection on the contents of the figure, you can do so by setting the figure's *HandleVisibility* to "off"; this will force any following high- or low-level plotting functions to create a new figure (or make use of the next figure that does not have its *HandleVisbility* property set to off) before it can draw its graphics objects.

7.5.2.10   Figure Properties that Affect Printing

Seven properties affect how a figure is printed. These are *InvertHardCopy*, *PaperOrientation*, *PaperPosition*, *PaperPositionMode*, *PaperSize*, *PaperType*, and *PaperUnits*.

PaperOrientation determines whether the figure is oriented in a portrait or landscape fashion on the printed page (yet you can always use the command **orient** *landscape* or **orient** *portrait* to specify how the figure is to be printed).

PaperPosition takes a four-element vector that defines a rectangle of the form [left, bottom, width, height] specifying the location of the figure on a printed page. The element *left* specifies the distance from the left side of the paper to the left side of the rectangle and *bottom* specifies the distance from the bottom of the page to the bottom of the rectangle. Together these distances define the lower left corner of the rectangle. The elements *width* and *height* define the dimensions of the rectangle. The units of these values is defined by the *PaperUnits* property and in most cases can be left in its default setting of "inches". The *PaperPosition* property is used to enforce "What you see is what you get" (WYSIWYG); when set to "manual" the figure will be printed using the value specified by the *PaperPosition* property. In the default "auto" mode the figure will be printed the same size as it appears on the computer screen and centered on the page, i.e., WYSIWYG.

PaperSize and *PaperType* are used to specify the size of the paper on which to print the figure. *PaperType* lets you choose from many standard paper sizes. If you set the value of *PaperType* to one of the standards, the size of that standard, in the units specified in *PaperUnits*, will then be contained in *PaperSize*. Here's an example to illustrate.

```
set(1,'PaperType','A0')
get(1,'PaperSize')
```

```
ans =
    33.1354    46.8466

set(1,'PaperUnits','centimeters')
get(1,'PaperSize')

ans =

    84.1000   118.9000
```

7.5.3 Axes Properties

The Axes object has more properties than any other object, in part because many of the properties are duplicated for each axis. All these properties give you a great deal of freedom in specifying exactly how you want your plot or graphics to appear. First, we will look at the properties that affect the appearance of the axes object itself and then we will look at those properties that affect the children of axes objects.

The following table lists alphabetically all properties specific to the axes object. Notice the axis-specific properties beginning with *XColor*; each of these can take either X, Y, or Z depending on the axis you wish to affect. For instance, *XColor* is the property whose value determines the color of the x-axis, while *YColor* is the property affecting the color of the y-axis. Where properties are duplicated for each axis, we have denoted this by underlining the axis identifier in this table. In the discussions we will simply use the appropriate property name.

Property	Read Only	ValueType/Options	Format
Properties Affecting Transparency and Lighting			
ALim	No		
ALimMode	No	[{auto} | manual]	
AmbientLightColor	No		
Properties Controlling Boxes and Tick Marks			
Box	No	[on | {off}]	row
TickLength	No	[2-Dticklength 3-Dticklength]	2-element row
TickDir	No	[{in} | out]	
TickDirMode	No	[{auto} | manual]	
XMinorTick	No	[on | {off}]	row
XTick	No	numbers	
XTickLabel	No	string	matrix
XTickLabelMode	No	[{auto} | manual]	row
XTickMode	No	[{auto} | manual]	row
continued on next page			

Property	Read Only	ValueType/Options	Format
Properties Affecting Character Formats			
FontAngle	No	[{normal} | italic | oblique]	
FontName	No	name of desired font	string
FontSize	No	number	1-element
FontUnits	No	[inches | centimeters | normalized | {points} | pixels]	string
FontWeight	No	[light | {normal} | demi | bold]	string
Properties Determining Axis Location and Position			
Position	No	[left bottom width height]	4-element row
Units	No	[inches | centimeters | {normalized} | points | pixels | characters]	
XAxisLocation	No	[top | {bottom}]	string
YaxisLocation	No	[{left} | right]	row
CurrentPoint	No	mouse click near and far x, y, z axis locations	2-by-3 matrix
Title	No	handle of text object	1 element
Properties Affecting Grids, Lines, and Color			
Color	No	[Red Green Blue] or color string	
ColorOrder	No	M RGB number triplets	M-by-3 matrix
CLim	No	[cmin cmax]	2-element row
CLimMode	No	[{auto} | manual]	string
DrawMode	No	[{normal} | fast]	
XGrid	No	[on | {off}]	
GridLineStyle	No	[- | -- | {:} | -. | none]	string
Layer	No	[top | {bottom}]	string
LineStyleOrder	No	string array of linestyle symbol(s)	matrix
LineWidth	No	number	1element
MinorGridLineStyle	No	[- | -- | {:} | -. | none]	
XColor	No	[Red Green Blue] or color string	row
Xform	No	4 x 4 Perspective Transformation	4 x 4 matrix
XLabel	No	Handle of text object	1 element
XMinorGrid	No	[on | {off}]	row
NextPlot	No	[add | {replace} | replacechildren]	string

continued on next page

Property	Read Only	ValueType/Options	Format
Properties Affecting Axis Limits			
DataAspectRatio	No	[x y z] relative ratio of axis lengths	2-element row
DataAspectRatioMode	No	[{auto} | manual]	string
PlotBoxAspectRatio	No	[x y z] relative ratios of box lengths	3-element row
PlotBoxAspectRatioMode	No	[{auto} | manual]	
XDir	No	[{normal} | reverse]	row
XLim	No	[xmin xmax]	2-element row
XLimMode	No	[{auto} | manual]	row
XScale	No	[{linear} | log]	row
Axes Properties Related to Viewing Perspective			
CameraPosition	No	[x y z] numbers	3-element row
CameraPositionMode	No	[{auto} | manual]	string
CameraTarget	No	[x y z] numbers	3-element row
CameraTargetMode	No	[{auto} | manual]	string
CameraUpVector	No	[x y z] numbers	3-element row
CameraUpVectorMode	No	[{auto} | manual]	
CameraViewAngle	No	number	1 element
CameraViewAngleMode	No	[{auto} | manual]	string
Layer	No	[top | {bottom}]	string
Projection	No	[{orthographic} | perspective]	
View	No	[DegreesAzimuth DegreesElevation]	2-element row

The first three properties in this table, *ALIm*, *ALimMode*, and *AmbientLightColor* deal with image, surface, and patch objects, and how lighting and transparency is affected by them. *ALim* and *ALimMode* specifically affect transparency, while *AmbientLightColor* deals with the color of light. Since Chapter 8 deals specifically with light and transparency, we shall leave these properties for that discussion.

7.5.3.1   Axes Properties Controlling Boxes and Tick Marks

The *Box* property specifies whether the axes region should be enclosed within a box in its 2-D view or by a cube in its 3-D view. Figure 7.8 illustrates the differences between the different perspectives when the Box property is set to "off" or "on".

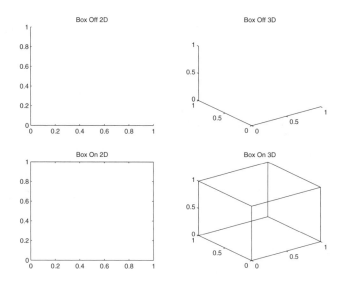

Figure 7.8   The effect of the Box property in 2-D and 3-D views.

In addition to the box attribute of the previous figure, look at the tick markers in the 2- and 3-D perspectives: the tick markers are, by default, 1% (0.01) of the width and height of the axes object in 2-D perspectives and 2.5% (0.025) in 3-D views. In addition, notice that the tick markers face inward in the 2-D plots and outward in the 3-D plots; both of these attributes can be controlled with the *TickLength*, *TickDir*, and *TickDirMode* properties to suit your personal preferences. The *TickLength* property value is a 2-element vector where the first element specifies the length as a percentage of the axes object width and height in 2-D perspectives, and the second element specifies the length as a percentage of the axes object in 3-D perspectives. By default, *TickDir* is set to "in" for 2-D graphs and "out" for the 3-D graphs as long as the *TickDirMode* is set to "auto". Once you change the value of *TickDirMode* to "manual", which can also occur by setting the *TickDir* property, 2-D and 3-D graphs will have their ticks pointing in the direction defined by *TickDir*. Yet, as with all settable properties, you can override the default properties by using the **set** command or by specifying the property values upon object creation. You can also specify whether or not to show minor tick marks with the *XminorTick* property. Setting the value of *XMinorTick* to "on" will show tick marks between the major tick marks.

One of the most common questions that people ask is

"How can I specify the values that will be displayed on the axis?"

To do this for the x-axis, you will make use of the *Xtick* and *XTickLabel* properties (for the y- and z-axes, just substitute Y or Z for the X in the property names). The *XTick* property is used to identify the locations on the axis where tick marks will be placed. It also ends up automatically specifying which numbers, called tick labels (since they can be forced to include characters), are displayed along the axis. The *XTickLabel* property lets you specify a string of characters for each tick mark on the x-axis. Let's say we want to plot the sine function from 0 to 4π; typing

```
x = 0:(pi/16):(4*pi);
plot(x,sin(x));
axis([0 4*pi -1 1]);
```

will yield the plot shown in Figure 7.9.

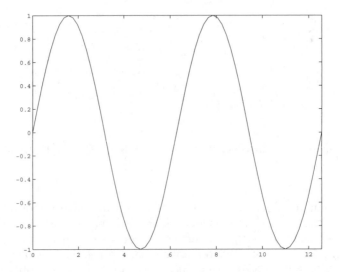

Figure 7.9   Default X-axis tick mark labels and locations.

As you can see, the tick marks and labels are automatically generated. However, we can specify them manually. For instance, we can force a tick mark at π/2 increments with

```
set(gca,'XTick',[0:(pi/2):4*pi])
```

to get Figure 7.10.

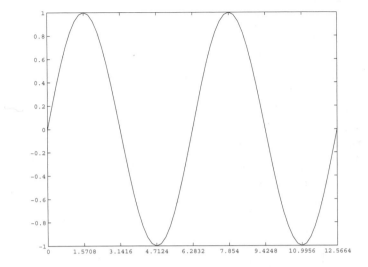

Figure 7.10   Specifying x-axis tic marks with XTick.

Unfortunately, the tick mark labels are floating point representations instead of symbolic representations. However, we can specify that MATLAB replace the number 1.5708 with 'pi/2', 3.1416 with 'pi', etc. by typing

```
set(gca,'XTickLabel',...
    ['0|pi/2|pi|3pi/2|2pi|5pi/2|3pi|7pi/2|4pi'])
```

To create the value string we recommend making use of the separator character "|" (usually typed with the shift+\ key on most keyboards) instead of using the **str2mat** function or manually typing in a string matrix. Generally, using the "|" is easier to type and read once you realize that the separator character can be used in this fashion. We show the other methods here in order to be thorough.

```
set(gca,'XTickLabel',...
        str2mat('0','pi/2','pi','3pi/2','2pi',...
        '5pi/2','3pi','7pi/2','4pi'));
```

or

```
set(gca,'XTickLabel',...
        ['0    ';'pi/2 ';'  pi ';'3pi/2';' 2pi ';...
        '5pi/2';' 3pi ';'7pi/2';' 4pi '])
```

Both of these approaches require you to set up the string matrix and force you to pay particular attention to the number of spaces within each string. Modern MATLAB has some very powerful character string manipulation capabilities that make such tasks easy, foremost of which is the cell-array. You can also use cell-arrays to store your *XtickLabels* value as shown in the following code.

```
s={'0','pi/2','pi','3pi/2','2pi',...
                '5pi/2','3pi','7pi/2','4pi'}

set(gca,'XTickLabel',s)
```

No matter which technique you end up using, they will all provide you with the result in Figure 7.11.

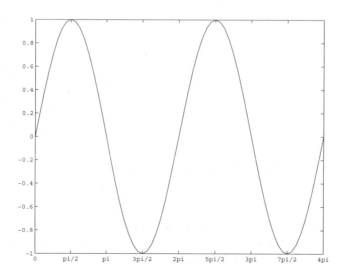

Figure 7.11   Specifying x-axis tic mark labels with XTickLabel.

Unfortunately, tick labels do not interpret TeX character sequences like *Title* and *XLabel* properties do; e.g., **xlabel**({'Units of \pi.' }) would yield 'Units of π' as the x-axis label. Therefore we cannot simply specify symbols like the Greek letter "π" in a cell-array. However, if your computer has a symbolic font installed on it, you can take advantage of the *FontName* property for an axis object as shown here:

```
set(gca,'FontName','Symbol')
t=['0|p/2|p|3p/2|2p|5p/2|3p|7p/2|4p']
set(gca,'XTickLabels',t)
```

This will produce the attractively labeled x-axis of Figure 7.12.

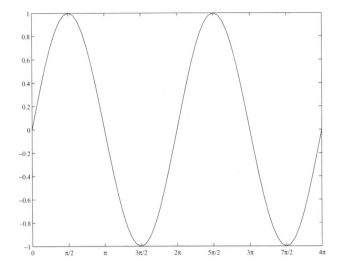

Figure 7.12   Changing the FontName property of the x-axis.

If you specify fewer numbers of tick mark labels than there are tick marks, the labels will be reused in a cyclical fashion. Once you specify either the location of the tick marks or their labels, the *XTickMode* and *XTickLabelMode* will respectively be set to their "manual" mode of operation. The manual mode keeps MATLAB from automatically determining the tick locations or labels that should be used to account for the data being plotted within the axes object. For example, plot a simple line with

```
figure
plot(1:10)
hold on
```

and then set the *Xtick* property to the manual setting and plot another line which extends beyond the x-axis limits with

```
set(gca,'XTickMode','manual')
plot(6*ones(1,15))
```

You can see from the result shown in Figure 7.13 that the labels stop after just 10. This happened because we forced MATLAB not to automatically calculate new tick mark locations.

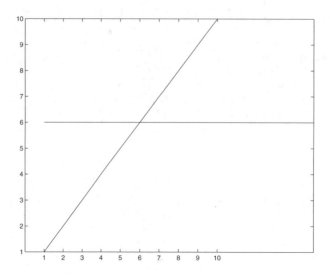

Figure 7.13   The result of setting XtickMode to "manual" before adding a second line.

Now create the plot in Figure 7.14 by typing

```
figure
plot(1:10)
hold on
set(gca,'XTickLabelMode','manual')
plot(6*ones(1,15))
```

The lines in this plot are identical; however, the labels "1" through "10" are now spread over the x-axis, whereas the data really runs from 1 to 15. As you can see, this is dangerous, because the plot labels misrepresent the data that was plotted (the 1, 2, 3, and 4 on the x-axis really correspond to the numbers 1, 5, 10, and 15).

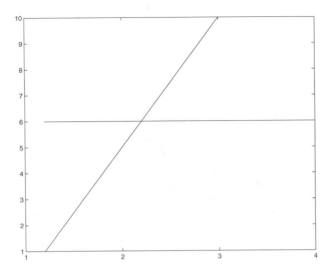

Figure 7.14   Setting XtickLabelMode incorrectly leads to incorrect results.

7.5.3.2   Properties Affecting Axes Character Formats

In addition to specifying which tick marks and labels are displayed, the character format of the x-, y-, and z-axes tick labels can be specified with the *FontAngle*, *FontName* (which you have already seen), *FontSize*, *FontWeight*, and *FontUnits* properties. If you want the tick labels of the current axes to be both bold and italic, you can use

```
set(gca,'FontAngle','Italic','FontWeight','Bold')
```

You need to be aware that these font properties are applied to all of the axes' tick mark labels. You do not have control over the font used on an axis-by-axis basis. For example, you cannot assign a bold Helvetica font for the x-axis tick mark labels, while using a normal Times font for the y-axis tick mark labels. Also note that the text objects generated with the commands **xlabel**, **ylabel**, **zlabel**, and **title** will use the font properties of the current axes when they are created; however, you do have individual control over their attributes and can alter them to your pleasing (see text object properties). The *FontUnits* property is particularly useful if you want the tick mark labels to scale proportionately with the size of the axes and figure. If you want the tick mark labels to scale, set the axes *FontUnits* to normalized.

7.5.3.3   Axes Properties Determining Axis Location and Position

The *Units* property of an axes object only affects the value and interpretation of the axes *Position* property. The value of *Units* can be specified in "inches", "centimeters", "normalized", "points", "pixels", or "characters". The position rect vector of a single axes object that is created either by a high-level graphics function (other than **subplot**) or by the low-level

axes command will default to [0.130 0.110 0.775 0.815] (normalized units). This position tends to keep it visually centered in the figure; however, you can reposition the current axes with **set**(gca,'Positon',*newposrect*), where *newposrect* is a four-element rect vector ([left bottom width height]) that defines the lower left corner coordinate, (left, bottom), with the first 2 elements and the width and height with the last 2 elements. The (left, bottom) coordinate measurement is made with respect to the lower left corner of the figure object within which the axes object exists.

In Chapter 3 you learned how to use the **plotyy** command to plot different y-axis limits against the x-axis. Now with handle graphics you can do even more. The axis location properties *XaxisLocation* and *YaxisLocation* give you the ability to specify whether the x-axis labels are on "top" or "bottom" of the plot and the y-axis labels are on the "left" or "right" side of the plot. By default the x-axis label will be on the "bottom" and the y-axis labels on the "left". Now you can quickly create overlaying plots with different x-axis and y-axis limits in the same graph by modifying the *XAxisLocation* and *YAxisLocation* properties. The following example will illustrate this and generate the plot in Figure 7.15.

```
figure;
plot(0:9,[0:9].^2);
a1=gca;
a2=axes;
plot(-10:10,[-10:10].^3);
set (a2,'xaxislocation','top','yaxislocation','right',...
       'color','none');
```

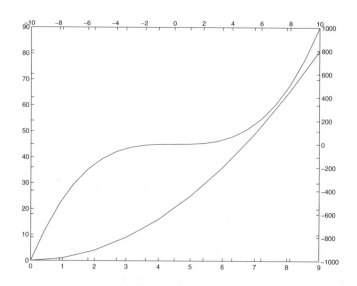

Figure 7.15   Superimposed axes objects.

Notice that this figure has extra tick marks. To turn off the extra tick marks, set both axes' *Box* properties to "off".

```
set([a1 a2],'box','off')
```

Now, if you wanted grid marks, you can turn on the grids with

```
axes(a1); grid on;
axes(a2); grid on;
```

Unfortunately, this does not look good because there are a different number of tick marks on the left as compared to the right and the bottom as compared to the top. Here's one solution that will produce Figure 7.16.

M-File

```
numxticka1=length(get(a1,'xtick'));
xlima2=get(a2,'xlim');
xincr=(abs(diff(xlima2))/(numxticka1-1));
newxtks = [xlima2(1):xincr: xlima2(2)];
set(a2,'xtick',newxtks);

numyticka1=length(get(a1,'ytick'));
ylima2=get(a2,'ylim');
yincr=(abs(diff(ylima2))/(numyticka1-1));
newytks = [ylima2(1):yincr: ylima2(2)];
set(a2,'ytick',newytks);

% Rounding may not always be appropriate, but is done
% on the next two lines to make the graph look cleaner.
set(a2,'xticklabel',round(get(a2,'xtick')))
set(a2,'yticklabel',round(get(a2,'ytick')))
```

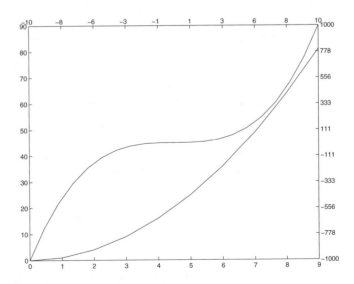

Figure 7.16   Matching tick marks on superimposed axes.

If you are using a color printer, Figure 7.16 could be improved by changing the color of one of the lines along with its associated x-axis and y-axis. This can be done using the *XColor* property presented in the next discussion.

7.5.3.4   Axes Properties Affecting Grids, Lines, and Color

In Chapter 3 we showed you that a grid could be created with the **grid** command.   This command essentially sets the *XGrid, YGrid,* and *ZGrid* properties to their "on" state.   With these axes properties you have the prerogative of specifying that the grid be displayed for the axis you want. Furthermore, you can specify the line type of the grid to something other than the default dotted lines (':') with the property *GridLineStyle* and the line width to something other than the default of 0.5 with the *LineWidth* properties. The line width affects all axes property lines (grid lines and the full or partial box drawn around the axes object). For example, we can have solid horizontal grid lines for a plot with

```
figure;
plot(randn(1,10))
set(gca,'YGrid','on','GridLineStyle','-','Linewidth',3)
```

to obtain Figure 7.17.

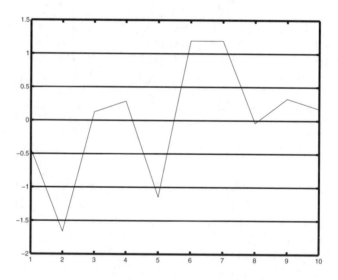

Figure 7.17   Altering the axes' LineWidth and GridLineStyle properties.

In addition to specifying line styles and widths, we can also specify the color of the axes object. By default, the *Color* property is set to the string "none"; however, you can use any of the legal color strings or any RGB color intensity triplet. For example, if you want the current axes to be red, you can use

```
set(gca,'color','red')
```

or

```
set(gca,'color',[1 0 0])
```

The x-, y- and z-axis line colors can be individually specified, respectively, with the *XColor*, *YColor*, and *ZColor* properties. These properties do not only specify the color of the actual axis lines, they also define the color used for tick marks, tick mark labels, and grid lines. If you want to make parts of the axis "invisible," you will need to set the component's color to the color of the axes. For instance, let's continue with the last example, but specify that the x-axis components should not appear in the figure (see Figure 7.18). Typing,

```
set(gca,'YColor',[.3 .3 .3],'Xcolor',get(gca,'color'))
```

almost accomplishes what we want. However, you will notice that the x-axis labels, which are now white, are visible against the default gray ([0.8 0.8 0.8]) Figure Window. We can overcome this by forcing the Figure Window's color to white as well with,

```
set(gcf,'Color','white')
```

which results in Figure 7.18.

You might notice that on some platforms the top corners of the y-axis lines have a white spot on your screen instead of the dark gray that was specified. This has to do with the order in which lines are rendered and stacked upon one another by MATLAB. These two dots are the end-points of the z-axis lines. Even though we are viewing this plot in a 2-dimensional perspective, the z-axis lines are still drawn orthogonal to the screen. To make sure that these appear in the same color, we can define the color of the z-axis lines with

```
set(gca,'ZColor',[.3 .3 .3]);
```

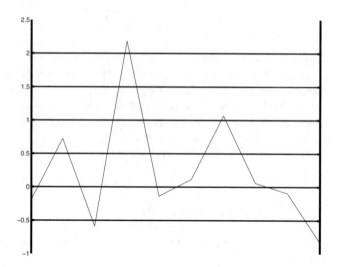

Figure 7.18   Making an axis invisible.

Along the same line of thought, let's look at this plot in three dimensions and turn the *ZGrid* property on with

```
view(-37.5,35)
set(gca,'ZGrid','on')
```

If you have been following along by typing in the examples, you should get something like the plot shown on the left side of Figure 7.19.

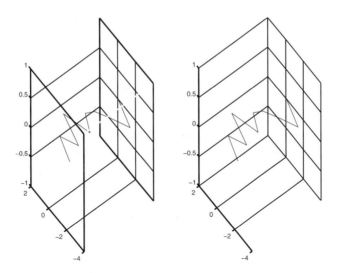

Figure 7.19   Black box lines created a blanked-out strip.

Notice that the lines of this plot have gaps in them and the grid lines in the z-axis at -1 and 1 are obscured. The reason for this is that the lines created by the axes *Box* property in the x-axis direction are black and drawn on top of the plot because they are closer to the observer in the perspective shown. For this example, it would be best to keep the *Box* property set to "off" as shown above on the right-hand side. It is important to recognize that there is a stacking order in terms of a viewer's perspective of 3-D graphics and that even if the object is colored to make it "invisible," the object still exists and may be drawn in front of other objects. In 2-D views of the plot, you can control whether the axis lines are drawn above or below the graphics objects in the plot with the *Layer* property. By default, this property will be set to "bottom" which will force the axis lines to be drawn below the axes' children. If you want the axis lines to be drawn over the children, set the value of *Layer* to "top".

The axes property *DrawMode*, in its default setting of "normal", will make sure that objects are drawn in such a manner so that those that are farther from the viewer are rendered before those that are closer. This property can be set to "fast", which disables the 3-dimensional sorting of objects and forces MATLAB to draw the objects in the order that they were originally created. The figures will be drawn quicker; however, the plot may be misleading in 3-dimensional perspectives as to the true order of the object's location with respect to one another.

You have already been using color in your plots to distinguish multiple data sets, and you already know how to specify the color for each line as you plot it, or to allow MATLAB to automatically assign colors as it plots. However, you are probably asking yourself the following question.

"How do I change the order in which colors are used when plotting multiple lines?"

By default MATLAB uses a predetermined set of colors to cycle through when plotting more than one line at a time (e.g., using **plot**(X,Y) where X and Y are matrices or **plot**(x1,y1,x2,y2,x3,y3,...)). This default order is yellow, magenta, cyan, red, green, and blue. The number of colors, the color values, and the order in which the colors are used can be predetermined and set as desired with the axes *ColorOrder* property. The *ColorOrder* property is an M-by-3 matrix containing M RGB triplets. For example, the default is the following 6-by-3 matrix:

RGB Triplets			Corresponds to the color
1	1	0	yellow
1	0	1	magenta
0	1	1	cyan
1	0	0	red
0	1	0	green
0	0	1	blue

If you want a particular plot to contain several lines that cycle between the colors red, green, and blue (shown in Figure 7.20 with slight variations in shading since this book is printed in black and white), you could do the following:

```
figure;
colorordermatrix = [1 0 0; 0 1 0; 0 0 1];
axes('ColorOrder',colorordermatrix,'NextPlot','add');
xdata = [1:10];
ydata = xdata'*[1:5];
plot(xdata,ydata);
```

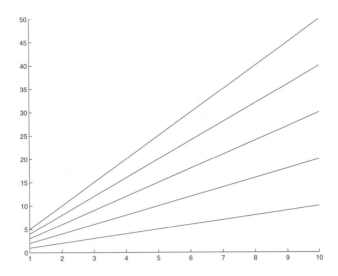

Figure 7.20   Controlling the order of automatic color assignment.

The *NextPlot* property must be specified as "add" instead of "replace", so that, when the **plot** command is executed, the axes object that has the desired *ColorOrder* value is not deleted.

There are two techniques that do not require the *NextPlot* property to be set to "add" that can be used to get the line colors to cycle through the desired set of colors. One technique is to use the low-level command **line** to generate the line objects, and the other is to set the default *ColorOrder* value to the one you want. The **line** command leaves the axes properties alone and, therefore, will use the colors in the order specified by the *ColorOrder* property. We will learn more about setting the default properties later in this chapter, but for now, if you want to do this, use the command

```
set(0,'DefaultAxesColorOrder',colorordermatrix)
```

where colorordermatrix is the variable that contains your M-by-3 color matrix.

Another frequently asked question is

"Since I use a black-and-white printing device, I would rather have MATLAB cycle through various line style types instead of colors. How can I do this?"

This is accomplished in a manner similar to the one used for colors, except in this case we need to make use of the *LineStyleOrder* and *ColorOrder* properties. The *ColorOrder* property should be set to one color and the *LineStyleOrder* should contain a matrix in which each row defines a legal line style. By default, the *LineStyleOrder* property is set to the solid line character string "-". Using the same *xdata* and *ydata* variables from the previous example, we can create the same plot, except this time, we will require that all of the

generated lines are colored white and that they cycle through several line
types (solid, dashed, dash-dotted, dotted, and the "x" marker). Type

```
figure;
% Specify black color
colorordermatrix = [0 0 0];
% Specify Line Styles
linestylematrix = ['- ';'--';': ';'-.';'x '];
axes('ColorOrder',colorordermatrix,...
     'LineStyleOrder',linestylematrix,...
     'NextPlot','add');
plot(xdata,ydata)
```

to obtain Figure 7.21.

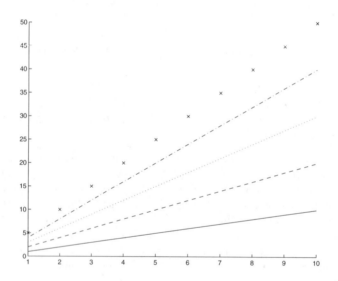

Figure 7.21   Cycling line styles automatically with *LineStyleOrder*.

Note that the specification for the *LineStyleOrder* can be composed of any
valid line style or marker type.   You could also have specified the
linestylematrix using the form

```
linestylematrix = ['-|--|:|-.|x '];
```

where the "|" is used to separate each style or marker. Here again, we do
not necessarily need to specify the *NextPlot* property. We could just as easily
replace the **plot**(xdata,ydata) with **line**(xdata,ydata). We could also specify the
default values for the *ColorOrder* and *LineStyleOrder* with

```
set(0,'DefaultAxesColorOrder',colorordermatrix)
set(0,'DefaultAxesLineStyleOrder',linestyleordermatrix)
```

where the *colorordermatrix* is a variable that contains a single RGB color intensity triplet (e.g., [0 0 0] in the example above) and *linestyleordermatrix* is a string matrix containing the desired line styles (e.g., ['- ';'–';': ';'-.';'x '] in the example above).

If both the *ColorOrder* and *LineStyleOrder* axes properties have multiple rows, lines will be created in a manner such that the first line style will be used while the colors cycle through their possibilities, then the second line style will be used while the colors again cycle through their possibilities, and so on. For instance, if you used

```
set(gca,'ColorOrder',[1 0 0; 0 1 0],...
    'LineStyleOrder ,['--';'-.';': ']);
```

and you plotted seven lines at once, the color of each of the lines would appear as stated in the following table:

Line Number	Line Attributes
1	Red dashed
2	Green dashed
3	Red dash-dotted
4	Green dash-dotted
5	Red dotted
6	Green dotted
7	Red dashed

The *CLim* property affects the color attributes of surface and patch graphics objects. A complete discussion of this property will be left for Chapter 8. For now, it will suffice to understand that the *CLim* property defines how MATLAB maps the colors that are stored in the figure's *Colormap* property to the data values of the surface and patch objects found within that axes object. By default, this property is a 2-element vector ([cmin cmax]) that contains the smallest and largest z-axis data values of surface and patch children of the axes object. This allows MATLAB to map the entire spectrum of *Colormap* colors to the data values. However, you have the ability to set the limits to your liking. This gives you the freedom of specifying either that only a given portion of the color map will be used across your plotted data or that the portions of the surface or patch objects falling above or below the two limits will, respectively, be colored with the maximum or minimum color in the *Colormap*.

Here again, there is a high-level command equivalence to setting this property.

The high-level command...	*is equivalent to...*
`caxis([cmin cmax])`	`set(gca,'CLim',[cmin cmax])`

Once you set the *CLim* property, the *CLimMode* property will be changed from "auto" to "manual". If later you want MATLAB to automatically define the *CLim* limits, you can set the *CLimMode* property back to "auto".

We can look (on your display) at three examples that make use of both the *CLim* and the *View* properties (additional examples related to color maps and the *CLim* property will be provided in Chapter 8) by plotting with the **peaks** function as follows:

```
[x,y,z] = peaks;
surf(x,y,z);
shading interp;
set(gca,'view',[90 0]);% You could also use view([90 0]);
```

to see how the full range of colors from the current color map are being used.

Now, if you type

```
get(gca,'CLim')
```

you will see that

```
ans =
      -6.5466    8.0752
```

is returned and is identical to the result returned from the command

```
[min(min(z)) max(max(z))]
```

Next, redefine the *CLim* limits with

```
set(gca,'CLim',[-3 3]);
```

You could also have used

```
caxis([-3 3]);
```

so that the color of data points in the surface above (in the z-axis direction) the upper limit are colored with the last color defined by the value of the figure's *Colormap* property, whereas those in the surface below (in the z-axis direction) the lower limit are colored with the first color in the *Colormap*.

Finally, once again redefine the *CLim* limits with

```
set(gca,'CLim',[-12 10]); %You could use caxis([-12 10])
```

so that only a portion of the color map is used to color the surface.

7.5.3.5   Properties Affecting Axis Limits

In addition to the *DrawMode* property, several other axes object properties have an affect on the children of that axes in one way or another. For one, the upper and lower data limits of the individual axis lines can be defined with the *XLim*, *YLim*, and *ZLim* properties. Normally, these are automatically specified because the *XLimMode*, *YLimMode*, and *ZLimMode* properties are set to "auto". However, if at some time you specify any of these limits, either directly with **set** or with the **axis** command, the respective *LimMode* property will be set to "manual".

The command...	is equivalent to...
`axis([5 10 3 7])`	`set(gca,'XLim',[5 10],'YLim',[3 7])`
`axis([1 4 -10 10 5 6])`	`set(gca,'XLim',[1 4],'YLim',...` `[-10 10],'Zlim',[5 6])`
`axis('axis')`	`set(gca,'XLimMode','manual','YLimMode',...` `'manual','ZLimMode','manual')`

Once a particular axis mode has been placed in its manual setting, the limits of that axis will not change (even if you add other graphics objects with values that fall outside the data limits of the axes) until you place the mode into the auto setting. This is particularly useful when there is a region of interest to which you plan to add plots. You can always define the axis after all the plots have been added; however, you will see that when you are adding plots from the command line or when you are working with animated plots, these properties come in handy. Typing

```
x = [-5:.5:7];
plot(x,x.^2)
```

will generate the plot shown below on the left of Figure 7.22. If you then use

```
set(gca,'xlim',[-2 3])
```

the x-axis will run between -2 and 3, and the y-axis will automatically be adjusted with new limits as shown with the figure on the right of Figure 7.22.

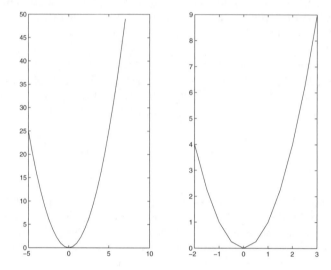

Figure 7.22   Using XLim to adjust x-axis limits.

If we now want to hold the y-axis limits constant so that we may add another plot to the figure without altering the limits, we can type

```
set(gca,'YLimMode','manual')
```

Previously, we learned that the command **hold on** could be used to set the axes object in a mode that will keep all existing objects in the plot when subsequent graphics statements are executed.   This command sets the property *NextPlot* to "add", instead of the default of "replace", so that new graphics objects are added as children to the axes object. The "replace" mode will delete (or replace) the existing children of the axes object and clear all of the axes properties to their default values before creating the new children. There is also a mode called "replacechildren" that removes all axes children, but does not reset the other properties before adding the new children.

So let's add another line to the existing figure by setting the *NextPlot* property to "add". Since both the *YLimMode* and *XLimMode* properties are set to manual, only the portion of the next plot that falls within the existing limits will be seen, as in Figure 7.23.

```
set(gca,'NextPlot','add')
x2=[-10:10];
y2 = 2*x2+6;
plot(x2,y2,'b');
```

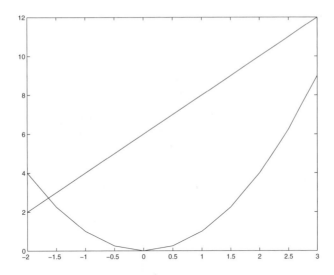

Figure 7.23 Adding plots while keeping current axis limits.

If the X and YLimMode had been set to automatic (auto) either before or after the last plot, the limits would be automatically determined so that both of these lines would be seen in their entirety.

In addition to being able to specify the upper and lower limits of the x-, y-, and z-axis, you can also specify the direction of increasing values for each of the axis lines. This is defined with the *XDir*, *YDir*, and *ZDir* properties. By default, the directions are all set to "normal", meaning that the axes object forms a standard right-handed coordinate system. However, under certain circumstances, you may wish to have one or even all of the directions reversed. This is accomplished by setting the direction property to "reverse" for the desired axis lines. The command **axis**('ij') is a high-level command that alters the direction properties so that you can put the 2-D coordinate system origin in the upper left corner. Its handle graphics equivalent is

```
set(gca,'YDir','reverse')
```

If you have generated the last plot you can demonstrate the affects of the *XDir* and *YDir* properties with,

```
set(gca,'XDir','reverse','YDir','reverse')
```

which should look like the plot in Figure 7.24.

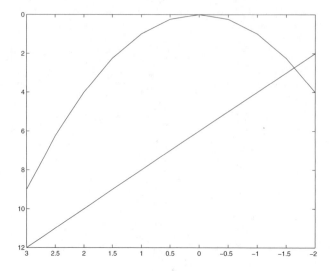

Figure 7.24   Reversing axis direction with XDir and YDir.

The axes properties also give you control over whether or not an axis is scaled linearly or logarithmically. This can be individually specified for any of the x-, y-, and z-axes with the *XScale*, *YScale*, and *ZScale* properties of the axes. For example, we can plot the series 1:100 in a logarithmic scale with the following code,

```
plot(1:100);
set(gca,'YScale','log');
grid on;
```

which produces the plot shown in Figure 7.25.

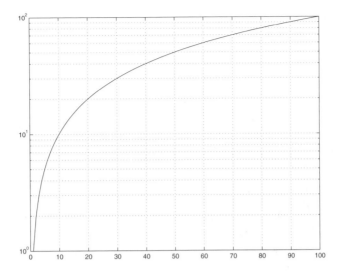

Figure 7.25   Setting YScale to "log".

The *DataAspectRatio* property is a three-element vector that defines the relative ratios of a unit of length along the x-, y-, and z-axis. The *PlotBoxAspectRatio* property is also a three-element vector that defines the relative ratios of the box that contains the axes object. By default, both of the mode properties will be set to "auto", thereby letting MATLAB try to display the objects within the axes with the highest possible resolution in the space defined by the axes object's *Position* property. *Stretch-to-fill* is the term associated with this default behavior of MATLAB. MATLAB attempts to create the largest axes it can in the region specified by the *Position* property with a data aspect ratio that best fits the x- and y-axis limits. Some of the axis command inputs that specify values for these ratios are shown in the following table.

The high-level command...	*is equivalent to...*
axis('equal')	set(gca,'DataAspectRatio',[1 1 1])
axis('square')	set(gca,'PlotBoxAspectRatio',[1 1 1])
axis('normal')	set(gca,'DataAspectRatioMode','auto')
	set(gca,'PlotBoxAspectRatioMode','auto')

The best way to get an idea of how these two ratios affect the apparent size of the axes object and the data within them is to look at several examples that use various settings. Run MATLAB's aspect ratio demo by typing **ardemo** or try the following examples. Figures 7.26 through 7.31 use the plot of a square and a circle that were created with

```
x = [-1 -1 1 1 -1]; y = [-1 1 1 -1 -1];
x2 = cos((0:5:360)*pi/180);
```

```
y2 = 2*sin((0:5:360)*pi/180);
plot(x,y,x2,y2)
axis([-2 2 -2 2])
```

followed by the appropriate form of **set**, e.g.,

```
set(gca,'DataAspectRatioMode','auto',...
    'PlotBoxAspectRatioMode','auto')
```

for Figure 7.26. The title on top of each plot shows the value of the aspect ratio properties of the axes object.

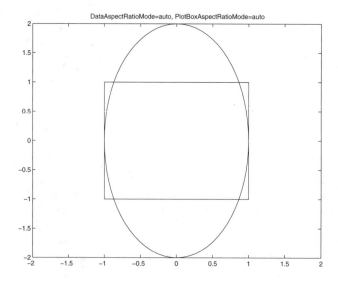

Figure 7.26

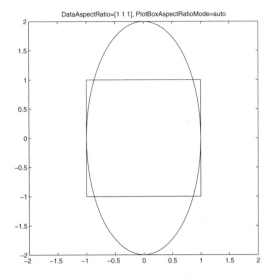

Figure 7.27

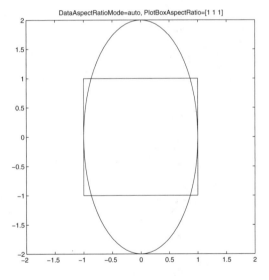

Figure 7.28

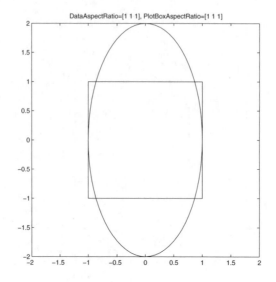

Figure 7.29

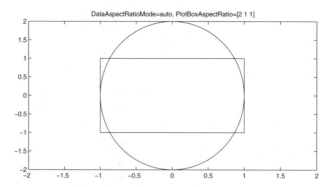

Figure 7.30

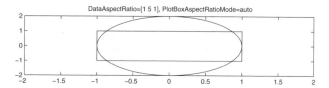

DataAspectRatio=[1 5 1], PlotBoxAspectRatioMode=auto

Figure 7.31

As you can see, each of these settings changes the way the visualized data is perceived. You should always be careful to look at the axis tick mark labels to obtain a true appreciation and understanding of the data's relevance.

7.5.3.6   Axes Properties Related to Viewing Perspective

The *View* and *Xform* are two closely related axes properties that control the manner in which 3-dimensional graphics objects are drawn on the 2-dimensional plane. The *View* property stores a 2-element vector ([azimuth elevation]) that defines an observational viewpoint in terms of the number of degrees in azimuth and elevation just as was described with the **view** function in Chapter 4. Any point in 3-D space can be defined with the azimuth and elevation angles and some measure of the distance (or range) from the observer to the origin. The origin is not necessarily the point (0,0,0); rather it is the point (*xmin,ymin,zmin*) defined by the lower limits of the *XLim*, *YLim*, and *ZLim* properties. Specifying an azimuth, elevation, and range would allow the observer to swing around within the 3-D space to view an object from any desired location. However, the *View* property does not require a range, since MATLAB will automatically determine a range that allows the object being viewed to be as large as possible while under the constraint of remaining within the axes object's position boundaries.

The next properties we will discuss prescribe how the objects in the axes (which we might just as well refer to as the *scene*) are viewed as if you were looking through a camera. These properties are the *CameraPosition*, *CameraTarget*, *CameraUpVector*, and *CameraViewAngle*. Along with each of these is a corresponding mode property (i.e., just add "Mode" to the end of the property names mentioned) that by default is set to "auto". The *CameraPosition* property specifies the position in data (x,y,z) coordinates from which you are looking through the camera, while the *CameraTarget* specifies the location in data coordinates at which you are pointing your camera. The default settings force the *CameraTarget* to be the center of the axes containing your graphics objects, the *CameraUpVector* to the y-axis direction in 2-D views and the z-axis direction in 3-D views, and the scene to fill as much as possible of the axes position rectangle.

These properties give you many controls over the way objects are viewed, particularly when it is important to view a scene from different angles without resizing the scene. In other words, this lets your perspective revolve around or move through a scene without changing the apparent relative distance at which you are looking at the axes, which makes it easier to do data comparisons between different views.

The most useful point to remember is that if you want to keep MATLAB from resizing the axes object, use

```
set(gca,'CameraViewAngleMode','manual')
```

After you have done this, you can revolve around a scene by changing the view with the **view**([*az el*]) command where *az* and *el* refer to the azimuth and elevation from which you want to view the objects in the axes (see Chapter 4). You can also move through the scene by changing the values in *CameraPosition* and *CameraTarget*. Making a movie (see Chapter 9) by combining snapshots of a scene that you revolve and move through is easy and can produce a great presentation!

The *CameraUpVector* property allows you, in a sense, to define the relative tilt of the camera with respect to the line defined by the camera and camera target locations.

A closely related property is the axes *Projection* property. This lets you define either an "orthographic" or a "perspective" display of your graph. These projections were introduced in Chapter 4, but we shall discuss them again. Orthographic should be used when trying to maintain the relative x-, y-, and z-axis data units. For example, when you are plotting 3-D views of mathematical functions, you should use the orthographic projection mode:

```
set(gca,'projection','orthographic')
```

If you are plotting objects that you want to have shrink in size the farther they lie from the camera's position, you should use perspective mode:

```
set(gca,'projection','perspective'))
```

7.5.4 Line Properties

The following table summarizes the properties that every line object has in addition to those that are common to all graphics objects.

Property	Read Only	ValueType/Options	Format
Color	No	[Red Green Blue] or color string	RGB row
EraseMode	No	[{normal} | background | xor | none]	row
LineStyle	No	[{-} | -- | : | -. | none]	row
LineWidth	No	number	1 element
Marker	No	[+ | o | * | . | x | square | diamond | v | ^ | > | < | pentagram | hexagram | {none}]	row
MarkerSize	No	number	1 element
MarkerEdgeColor	No	[none | {auto}] -or- a ColorSpec	row
MarkerFaceColor	No	[{none} | auto] -or- a ColorSpec	row
XData	No	numbers	vector
YData	No	numbers	vector
ZData	No	numbers	vector

Line objects are children of a single axes object and therefore must have some property that defines their relative position within their parent. The *XData*, *YData*, and *Zdata* properties are just for this purpose. These three properties store the data values that you are plotting with a line when you issue either a high-level command such as **plot**(x,y) or the low-level graphics command, **line**(x,y). Every line can be thought of as a bunch of connected dots, where the ith dot is at a coordinate specified by (*XData*(i),*YData*(i), *ZData*(i)). To render a line, MATLAB requires that the *XData* and *YData* property values be the same length vectors. *ZData*, however, must either be an empty matrix, [ ], or a vector that is the same length as *XData* and *YData*. In the event that you use a plotting command, such as **plot**(x,y) or **line**(x,y) for 2-D plotting, i.e., specifying only x and y, *ZData* will contain the empty matrix and it is assumed that the *ZData* coordinate is the number zero for each *XData*, *YData* element pair. Only when you create a 3-D line, such as with **plot3**(x,y,z) or **line**(x,y,z), or when you specifically set this property value to some vector, perhaps with

```
set(line_handle,'ZData',vector_of_zvalues);
```

will the *ZData* property contain numeric values.

Most of the other line properties are used to specify the visual features of the line object. So in order to illustrate the effects of these properties, we shall create a simple line object and keep track of its handle.

```
figure;
x = [1:6];
y = sin(x);
line_handle = plot(x,y);
```

This last line could have been replaced with any one of the following:

```
line_handle = plot(y);
```

or

```
line_handle = line('XData',x,'YData',y);
```

or

```
line_handle = line(x,y);
```

The *Color* property contains a single RGB intensity vector that defines the color of the line. To set this property, you may pass either a legal color string or a 3-element vector. For instance, to make our line green, we can use

```
set(line_handle,'color','green')
```

or

```
set(line_handle,'color','g')
```

or

```
set(line_handle,'color',[0 1 0])
```

Previously, when we specified the colors of lines by passing a color string along to the **plot** command, the routine was essentially setting this property for you.

This was also the case with the *LineStyle* property. For instance, if you use the command **plot**(x,y,'g:') to create a green dotted line, the plot routine will separate the string into its two subcomponents, 'g' and ':'. The 'g' is used to set the *Color* property and the ':' is used to set the *LineStyle* property. The *LineStyle* property can be any one of the five types identified in the previous table.

The *Marker* property lets you choose from one of 14 markers (13 styles and "none"). Markers are placed at every data point specified by the *X*, *Y*, and *ZData* coordinate vectors. Since you can choose a line style and marker type at the same time, you are probably plotting too many lines in a graph if you find that you have run out of combinations!

The *LineWidth* property, by default, is set to 0.5 points (1 point = 1/72 inch). To illustrate some of the various line thicknesses, we can run the following script:

```
figure;
axes('XLim',[0 6],'YLim',[0 7],'Box','on');
x = [1:4]; y = ones(size(x));
thicknessrange = [0.25 0.5 1 2 4 10];
for thicknessindex = 1:length(thicknessrange)
    line('XData',x,'YData',y*thicknessindex,...
        'LineWidth',thicknessrange(thicknessindex));
    text(5,thicknessindex,...
        num2str(thicknessrange(thicknessindex)));
end
title('LineWidths indicated next to line')
```

This script will generate the plot shown in Figure 7.32.

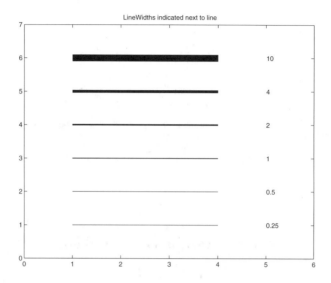

Figure 7.32 **Controlling line widths.**

Since the width of a line is specified in terms of points, values of 1 and less will all look identical on the screen; however, when you print them out, you will see the difference.

Please note that the *LineWidth* will also change the widths of the markers, but not the size of the markers. To change the marker size, you need to use the *MarkerSize* property.

By default, the *MarkerSize* is six points. As a quick exercise, see if you can generate a similar script (before looking at the code) to the one we just used above to generate the figure shown in Figure 7.33.

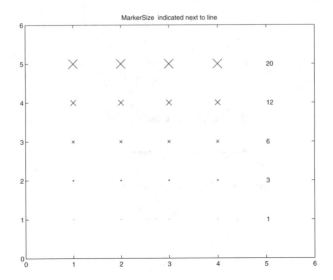

Figure 7.33   Using *Marker* and *MarkerSize*.

```
figure;
axes('XLim',[0 6],'YLim',[0 6],'Box','on');
x = [1:4]; y = ones(size(x));
markersizerange = [1 3 6 12 20];
for markersizeindex = 1:length(markersizerange )

    line('XData',x,'YData',y*markersizeindex ,...
        'LineStyle','none','Marker','x',...
        'MarkerSize',markersizerange (markersizeindex ));
    text(5,markersizeindex ,...
        num2str(markersizerange (markersizeindex )));
end
title('MarkerSize  indicated next to line')
```

A line's marker also has edge (*MarkerEdgeColor*) and face (*MarkerFaceColor*) color properties. A marker's face is the region within the boundary defined by the marker's edge. The "+", ".", "x", and "*" markers do not have faces, and therefore, their color is only affected by the *MarkerEdgeColor* property. To illustrate, the following code will create hexagrams that have a yellow face, red edge, and are connected by a blue dashed line as shown in Figure 7.34:

```
figure;
l=plot([-.5 .5 .5 -.5 -.5],[-.5 -.5 .5 .5 -.5]);
set(l, 'linestyle','--',...
    'color','blue',...
    'linewidth',2,...
    'marker','hexagram',...
    'markersize',15,...
    'markeredgecolor','red',...
    'markerfacecolor','yellow');
axis([-1 1 -1 1]);
```

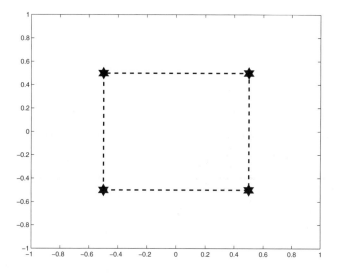

Figure 7.34   Using marker and line properties.

Of course, you will have to try this on your computer to see the colors. As you can see, using various combinations of the *Color, LineStyle, LineWidth, Marker, MarkerEdgeColor,* and *MarkerFaceColor* properties will give you quite a bit of freedom in defining many line object appearances.

The *EraseMode* property of line objects is used to give some level of control over the manner in which a particular line object is erased and/or redrawn. This property is primarily manipulated when animating graphics objects, of which we will make extensive use in Chapter 9 when we discuss animation. For now, we will simply point out that this property is set to "normal" by default so that objects are rendered in the figure to provide an accurate presentation of the objects that currently exist in their relative order in relation to the perspective of the observer. The price paid for the accurate figure representation is speed. The other three modes of erasing are much faster, but have certain implications with regard to what is shown in the figure. The "none" setting will keep MATLAB from updating the region of the figure where the object was found before it was either deleted or moved. The "xor" mode allows the particular object to be moved or deleted without affecting the objects that are rendered below it. However, since the object is xored with the color of the objects below it, its color will be influenced by other objects and can be guaranteed only when the object is located on top of the figure object. Finally, the "background" setting will make sure that the object is drawn with the right color. However, when an object with an *EraseMode* set to "background" is deleted or erased, any other object located below it will temporarily be damaged with an *imprint* of the erased object drawn in the figure's background color. All of these inaccuracies are removed at the time that either a refresh is issued or another graphics object which has its *EraseMode* property set to "normal" is created, moved, or deleted.

7.5.5 Rectangle Properties

In the previous example, we created what looked like a rectangle by using the **plot** function and specifying data that defined the sides of the rectangle. Although this looks like a rectangle, it is not an actual rectangle as far as MATLAB is concerned. A rectangle in MATLAB is a unique graphics object and therefore has properties that specify it. The following table lists those properties unique to rectangle objects.

Property	Read Only	ValueType/Options	Format
Curvature	No	[x, y]	1 or 2 element
EraseMode	No	[{normal} | background | xor | none]	row
FaceColor	No	ColorSpec | {none}	row
EdgeColor	No	{ColorSpec} | none	row
LineStyle	No	[{-} | -- | : | -. | none]	row
LineWidth	No	number	1 element
Position	No	[x,y,width,height]	vector

As you would expect, rectangle objects have a *Position* property, specified by the same rect vector format we have previously seen. Also, since a rectangle is made of a line, there are the properties *LineStyle* and *LineWidth*. Similar to what we have seen with the *Marker* property of line objects, we see that rectangle objects have *FaceColor* and *EdgeColor* properties as well. One property that rectangles have that you probably did not anticipate is *Curvature*. This property takes either a one- or two-element vector as its value where the vector specifies the curve into the corners. If there is only one value specified, then both the vertical and horizontal segments of the rectangle take the curve; if two elements are provided, then the first affects the horizontal segment, and the second the vertical segment. The range of these values are 0 to 1 where 0 is no curvature (corners would meet at right angles) and 1 is maximum curvature. The properties of a rectangle object are perhaps best understood by example. The following code will produce the result shown in Figure 7.35.

```
figure;
curvesize=[0 0.2 0.5 0.8 1];
axis([1 20 1 20]);
for inc=1:5
 rect_h(inc)=rectangle;
 set(rect_h(inc),'Position',[2,3*inc,2,2],...
     'Curvature',curvesize(inc));
 text(5,3*inc, num2str(curvesize(inc)));
end
inc=inc+1;
rect_h(inc)=rectangle
set(rect_h(inc),'Position',[9 6 6 6],...
     'Curvature',[0.3 0.7],'LineStyle',':',...
     'LineWidth',2,'EdgeColor','blue',...
```

```
     'FaceColor',[1 0 0]);
text(10,4, {'Curvature = [0.3 0.7]',...
     'EdgeColor = blue','FaceColor = red'});
axis equal;
```

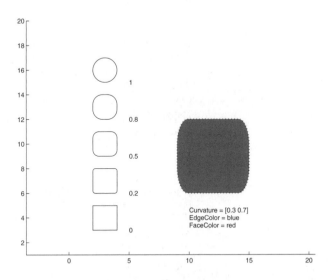

Figure 7.35   Rectangle objects.

As you can see, when the value of *Curvature* is 1 (or [1 1]), the rectangle becomes a circle.

Although convenient and potentially very useful, rectangle objects are somewhat limited as far as graphics control. For instance, you might have noticed that rectangles do not have *XData* or *YData*, so you can not rotate rectangle objects with the **rotate** command, nor can you specify their transparency since they don't have *AlphaData*. However, you will learn in the next section about a much more robust object that will allow you to manipulate its appearance in practically every way imaginable.

7.5.6   Patch Properties

A patch object is made up of one or more polygons. It is defined by the coordinates of its vertices. Each patch can have its own color, transparency, etc., and can be either 2-D or 3-D. The following table lists all the patch properties that are not common to all graphics objects.

Property	Read Only	ValueType/Options	Format
Properties Defining Patch Objects			
Faces	No	permutation of 1:M	N-by-V matrix
Vertices	No	numbers x-, y-, z-coordinates	M-by-3 matrix
XData	No	coordinates of the points at the vertices	vector or matrix
YData	No	coordinates of the points at the vertices	vector or matrix
ZData	No	coordinates of the points at the vertices	vector or matrix
Properties Specifying Lines, Color, and Markers			
CData	No	numbers	vector
CDataMapping	No	[direct | {scaled}]	row
EdgeColor	No	[none | {flat} | interp] or [Red Green Blue] or color string	row
FaceColor	No	[none | {flat} | interp] or [Red Green Blue] or color string	row
FaceVertexCData	No	RGB per patch, face, or vertex	matrix
LineStyle	No	[{'-'} | '--' | '-.' | ':' | 'none']	row
LineWidth	No	number	1 element
Marker	No	['square' | 'diamond' | 'v' | '^' | '>' | '<' | '.' | 'pentagram' | 'hexagram' | 'o' | 'x' | '+' | '*' | {none}]	row
MarkerEdgeColor	No	[none | {auto} | [R G B] | color_string]	row
MarkerFaceColor	No	[{none} | auto | [R G B] | color_string]	row
MarkerSize	No	number	1 element
continued next page			

Property	Read Only	ValueType/Options	Format
Properties Affecting Lighting and Transparency			
AmbientStrength	No	numbers	vector
BackFaceLighting	No	[unlit | lit | {reverselit}]	row
FaceLighting	No	[none | {flat} | gouraud | phong]	row
DiffuseStrength	No	number	1 element
EdgeLighting	No	[{none} | {flat} | gouraud | phong]	row
SpecularColorReflectance	No	number ranging from 0 to 1	1 element
SpecularExponent	No	number > or = to 1	1 element
SpecularStrength	No	number ranging from 0 to 1	1 element
VertexNormals	No	numbers	M-by-3 matrix
NormalMode	No	[{auto} | manual]	row
EraseMode	No	[{normal} | none | xor | background]	row
AlphaDataMapping	No	[none |direct | {scaled}]	row
EdgeAlpha	No	[{scalar = 1} | flat | interp]	1 element or string
FaceAlpha	No	[{scalar = 1} | flat | interp]	1 element or string
FaceVertexAlphaData	No	transparency data	1 element or M-by-1 matrix

7.5.6.1   Properties Defining Patch Objects

Just as line objects used *XData*, *YData*, and *ZData* properties to store data that defines the coordinates that are connected sequentially to form a line, patch objects use these three properties to store data that defines the locations of its vertices. Here again, if the *ZData* property contains the empty matrix, it is assumed that the patch object lies in the xy-plane (the z-axis coordinates are assumed to be zero). In addition, if the first and last vertex coordinates do not form a closed path, MATLAB automatically joins these two vertices.

Try not to think of patches as a single polygon; patches can have as many faces as you want. Each column of the *X*, *Y*, and *ZData* properties refers to a face of the patch object. Additional properties, *Vertices* and *Faces*, are part of the patch object to make it easier to define patches with more than one face. With the *Vertices* property, you can define all the possible vertices you want to use (and additional ones if it makes your life easier) as an M-by-3 matrix, where each of the M rows represents a vertex's x,y,z coordinates. Then you define groups of the vertices that are to be connected with an N-by-V matrix in the *Faces* property, where N is the number of faces and V is the maximum number of vertices you want in any single face. The faces are drawn by connecting the vertices in the order specified by going from column 1 to column V.

You might be wondering,

What if I want different type polygons for some of the faces in the patch?

If you wanted a mix of quadrilateral faces and triangular faces, the maximum number of rows you need is four. For the rows defining triangles, you need only three vertices, so just put a NaN in the fourth column of those rows. As an example, we can create a patch object with different polygons.

```
figure;
vertex = [-0.5 -0.5  0;     % Vertex 1
           0.5 -0.5  0 ;    % Vertex 2
           0.5  0.5  0;     % Vertex 3
          -0.5  0.5  0;     % Vertex 4
            0    0  -1];    % Vertex 5
faces = [1 2 3 4;           % Face F1
         1 2 5 NaN;         % Face F2
         2 3 5 NaN;         % Face F3
         3 4 5 NaN;         % Face F4
         4 1 5 NaN];        % Face F5
p=patch('vertices',vertex,...
        'faces',faces,...
        'facecolor',[.5 .5 .5]);
axis([-1 1 -1 1 -1 0]);
view(3);
```

to produce the upside-down pyramid shown in Figure 7.36.

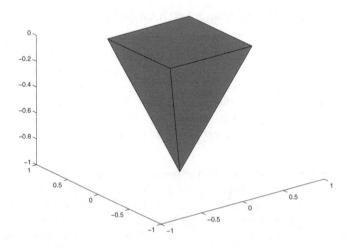

Figure 7.36   A single patch object with 5 faces.

7.5.6.2   Properties Specifying Lines, Color, and Markers

Each face of a patch object can be thought of as being composed of three subcomponents. The first subcomponent is the edge or line that connects each of the vertices in the sequence in which they are specified. The second subcomponent is the patch's face or region that lies within the vertices. And the third, which is by default not displayed, is the set of markers that can be located at the vertices.

You can define the edge color, style, and width of a patch object with the *EdgeColor, LineStyle,* and *LineWidth* properties.

Markers are drawn only if you set the *Marker* property with a valid setting. Just as with line objects, a marker's edge color, face color, and size can be altered with the *MarkerEdgeColor, MarkerFaceColor,* and *MarkerSize* properties. The only difference with a patch's marker edge and face color properties is that they have an "auto" setting. If in "auto" mode, their colors will depend on the value specified in the *EdgeColor* property. This allows the marker colors to be interpolated. Finally, remember that the marker's edge width is determined by the *LineWidth* property.

To better understand how these properties can be used, let's look at a simple example. First, we will define the coordinates of a single polygon. Then we will create translated patch objects that illustrate the use of the different properties. The results are shown in Figure 7.37.

```
x = [-1 -1 1 1 -1];
y = [-1 1 1 -1 -1];
figure;
axes('XLim',[-4 4],'YLim',[-4 4],'box','on')
p1 = patch('XData',x,'YData',y,'FaceColor','blue');
p2 = patch('XData',x+2,'YData',y+2,...
'FaceColor',[1 0 0],'Edgecolor',[0 1 0],...
'linewidth',3, 'marker','o');
p3 = patch('XData',x-2,'YData',y+2,'FaceColor','none',...
    'Edgecolor',[.3 .3 .3],'linewidth',6);
p4 = patch('XData',x+2,'YData',y-2,...
'FaceColor',[0 1 1],'Edgecolor','none',...
'linewidth',3,'marker','hexagram',...
'markeredgecolor','yellow','markerfacecolor','red',...
'markersize',20);
p5 = patch('XData',x-2,'YData',y-2,'FaceColor',[0 1
1],...
    'Edgecolor',[0 0 0],'linewidth',40);
```

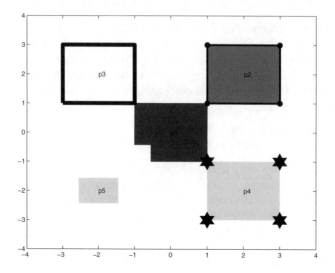

Figure 7.37   Demonstrating patch properties.

The last two patch objects, p4 and p5, are created to illustrate that there is a difference between specifying the color of the axes object (black on most platforms) or none for the *EdgeColor* and *FaceColor* properties. If a subcomponent's color value is set to "none", that component of the patch object will not be rendered. On the other hand, a white component will get rendered and may partially hide another object (see p1 in the example) or even its complement component (as seen by the fact that p5's face component is partially obscured by its edge component). This is particularly evident when the figure or axes object is not white.

You should also be aware of the fact that the order in which these patch objects were created has significance in the final result. This is because we did not specify any values for the *ZData* properties of these patch objects and, therefore, they are all in the same xy-plane. The objects most recently created will be on top of the other objects.

The *EdgeColor* and *FaceColor* of a patch object can also be set to "flat" or "interp" (interpolated). In order to use one of these options, you need to specify the color data, *CData* or *FaceVertexCData*, of the patch object. If either of these two *CData* properties is not specified and you try to define the *EdgeColor* or *FaceColor* property as "flat" or "interp", or the *MarkerEdgeColor*, or *MarkerFaceColor* as "flat",   MATLAB will return one of the following warning messages:

```
Warning: Patch FaceVertexCData length (0) must equal
Vertices length (5) for flat EdgeColor.

Warning: Color Data is not set for Interpolated shading.

Warning: Patch FaceVertexCData of size 0 cannot be used
with Flat shading.
```

```
Warning: Color Data is not set for marker flat coloring.
```

and the patch object will not be rendered.

Although you were introduced to how color maps and true color are specified in MATLAB in Chapter 5, we will visit those topics again in Chapter 8. Furthermore, we will discuss the *Cdata*, *FaceVertexCData*, and *CDataMapping* properties in greater detail in Chapter 8. For now, we will give just a brief discussion as it applies to the patch object. As a general overview, the color data (*Cdata* and *FaceVertexCData*) properties are used to specify the colors at the corners of each face. Which of these two properties you use will most likely depend on what form you used to enter in the location of the patch's corners. If you used the *X*, *Y*, *ZData* properties, then *CData* is probably the most appropriate. If you used the *Vertices* and *Faces* properties, then you will probably want to use *FaceVertexCData*. If you are using multifaceted patches you will notice that, in general, using the *FaceVertexCData* is easier because you have to define the color at a vertex that is used for multiple faces only once, where as with *CData* you need to specify the color at the corner for each face.

For now, when the *CDataMapping* is set to the default value of "scaled", you can think of the color data values as a range of values that get scaled to the size of the figure's color map to identify the color at each of the patch vertices or each face. By scaled, we mean that the smallest color data value will be mapped to (or, in other words, drawn using) the first color in the color map, and the largest color data value will be mapped to the last color in the color map. This default setting lets you flip between different sized color maps without having to recalculate the *CData* or *FaceVertexCData* values. When the *CDataMapping* is set to "direct", you can think of the color data values as indices to the figure's color map that are used to identify the color at each of the patch vertices or each face.

Remember that the size of the matrices you assign to either of color data value properties and the setting of the *FaceColor* and *EdgeColor* properties will specify how MATLAB should interpret your coloring intentions.

If you want...	set the patch's ...
a single true color for all faces	*FaceColor* to an RGB value or *FaceColor* to "flat" and *FaceVertexCData* to an RGB value
a single color for all faces that are indexed to the color map	*FaceColor* to "flat", *CDataMapping* to "direct", and *FaceVertexCData* to a single colormap index containing the color you want.
a single color for all faces that are scaled to the color map	*FaceColor* to "flat", *CDataMapping* to "scaled", and *FaceVertexCData* to a single value that is chosen in relation to the axes object's *CLim* property limits value.
one true color for each face	*FaceColor* to "flat" and *FaceVertexCData* to an M-by-RGB (i.e., M-by-3) matrix, where M is the number of faces.

continued next page

If you want...	set the patch's ...
one color for each face that is indexed to the color map	*FaceColor* to "flat", *CDataMapping* to "direct", and *FaceVertexCData* to a column of M colormap indices containing the color you want, where M is the number of faces.
one color for each face that is scaled to the color map	*FaceColor* to "flat", *CDataMapping* to "scaled", and *FaceVertexCData* to a column of M (the number of faces) values that are chosen in relation to the axes object's *CLim* property limits.
interpolated true color for each face	*FaceColor* to "interp" and *FaceVertexCData* to an V-by-RGB (i.e., V-by-3) matrix, where V is the number of vertices.
interpolated color for each face that is indexed to the color map	*FaceColor* to "interp", *CDataMapping* to "direct", and *FaceVertexCData* to a column of V colormap indices containing the color you want, where V is the number of vertices.
interpolated color for each face that is scaled to the color map	*FaceColor* to "interp", *CDataMapping* to "scaled", and *FaceVertexCData* to a column of V (the number of vertices) values that are chosen in relation to the axes object's *CLim* property limits.

7.5.6.3   Properties Affecting Lighting and Transparency

This last set of patch properties have to do with how patch objects are affected by lighting and transparency. Although we will cover these properties in greater detail in Chapter 8, we will give a brief overview here. *VertexNormals* are automatically calculated by MATALB when the *NormalMode* property is set to "auto". We will see these properties with surface objects as well. These normals are used by MATLAB to perform calculations that determine the visual effects of lighting models on the patch object.   Modifying the *VertexNormals* sets the *NormalMode* property to "manual" and prohibits MATLAB from recalculating the normals. Providing your own set of normals in lieu of MATLAB automatically determining them can lead to very interesting lighting effects.

Additional properties that give you control over how lighting affects the patch objects are *SpecularStrength*, *DiffuseStrength*, *AmbientStrength*; properties used to specify the respective intensity components of light objects that are reflected off of the patch object. The *SpecularExponent* property determines the size of the highlight spot due to a light source shining on the patch object. The default value is 10. Increasing this value makes the spot smaller and decreasing it makes the spot larger. The *SpecularColorReflectance* property controls the color of the reflected light spot emanating from the patch object. It can vary between 0 and 1, where values approaching 0 indicate that MATLAB should reflect more of the patch's color, and values approaching 1 indicate that MATLAB should reflect more of the light object's color. The *FaceLighting* and *EdgeLighting* properties specify the method that MATLAB should use to calculate the effect of light on the patch object. You will find that setting the *EdgeLighting* and *FaceLighting* to "flat" for patch objects is generally optimal. "Gouraud" and "phong" lighting is usually best used when viewing curved surfaces. Finally, there is a property called *BackFaceLighting*. *BackFaceLighting* by default is set to "reverselit", which means that patch objects whose *VertexNormals* point away from the camera are illuminated. *BackFaceLighting* can be used to highlight only the patch objects whose *VertexNormals* are facing the camera by setting it to "unlit". The "lit" setting can be used if you are seeing strange lighting effects along the edges of the patch objects.

The *EraseMode* property of patch objects has the same meaning as was briefly discussed for line objects. This property will also be elaborated when animation is presented in Chapter 9.

The remaining properties, *AlphaDataMapping*, *EdgeAlpha*, *FaceAlpha*, and *FaceVertexAlphaData* determine how patch objects present transparency data, a.k.a., alpha data. We will discuss this in greater detail in Chapter 8, but for now we will present a brief overview as an appetizer. The properties *EdgeAlpha* and *FaceAlpha* specify the transparency of the edges and faces of patches respectively. Both take values of either a single scalar value between 0 and 1, where 0 is fully transparent and 1 (the default) is completely opaque, or "flat" or "interp" in which cases the property *FaceVertexAlphaData* must contain valid alpha data. Note that you must have supplied alpha data when you select "flat" or "interp" or MATLAB will report a warning like,

```
Warning: Patch FaceVertexAlphaData of size 0 cannot be
used with Flat Alpha..
```

or

```
Warning: Alpha Data is not set for Interpolated shading.
```

The *AlphaDataMapping* property specifies the method MATLAB will use to map the transparency, i.e., "none" such that alpha data is "clamped" between 0 and 1, "scaled" (the default) where alpha data is mapped linearly to span the portion of the alphamap indicated by the axes *ALim* property, or "direct" in which case alpha data in the *FaceVertexAlphaData* property is taken directly. As we said, we will present using these properties fully in Chapter 8.

7.5.7 Surface Properties

Power!

Surface objects have properties which are an assembled mix of properties from both the line and patch graphics objects. You have already become somewhat familiar with surface objects by virtue of the **surf** function in Chapter 4. However, you will wield great power over surface objects once you go beneath the surface (so to speak) and become familiar with the properties which are listed in the following table. We will only briefly describe these properties in this section, but see them again in the following chapter on color, light, and transparency. Also, many of these properties are already familiar to you, having been presented with line and patch objects.

Property	Read Only	ValueType/Options	Format
Properties that Define a Surface			
XData	No	coordinates of the points at the vertices	vector or matrix
YData	No	coordinates of the points at the vertices	vector or matrix
ZData	No	coordinates of the points at the vertices	vector or matrix
Properties that Specify Lines, Colors, and Markers			
CData	No	numbers	vector
CDataMapping	No	[direct | {scaled}]	row
LineStyle	No	[{'-'} | '--' | '-.' | ':' | 'none']	row
LineWidth	No	number	1 element
EdgeColor	No	[none | {flat} | interp] or [Red Green Blue] or color string	row
FaceColor	No	[none | {flat} | interp | texturemap] or [Red Green Blue] or color string	row
Marker	No	['square' | 'diamond' | 'v' | '^' | '>' | '<' | '.' | 'pentagram' | 'hexagram' | 'o' | 'x' | '+' | '*' | {none}]	row
MarkerEdgeColor	No	[none | {auto} | [R G B] | color_string]	row
MarkerFaceColor	No	[{none} | auto | [R G B] | color_string]	row
MarkerSize	No	number	1 element
Properties Affecting Lighting and Transparency			
AmbientStrength	No	numbers	vector
BackFaceLighting	No	[unlit | lit | {reverselit}]	row
DiffuseStrength	No	number	1element
EdgeLighting	No	[{none} | {flat} | gouraud | phong]	row
FaceLighting	No	[none | {flat} | gouraud | phong]	row
NormalMode	No	[{auto} | manual]	row
SpecularColorReflectance	No	number ranging from 0 to 1	1 element
SpecularExponent	No	number > or = to 1	1 element
SpecularStrength	No	number ranging from 0 to 1	1 element
VertexNormals	No	numbers	M-by-3 matrix
AlphaData	No	default = 1 (opaque)	M-by-N matrix of double or uint8
AlphaDataMapping	No	[none |direct | {scaled}]	row
EdgeAlpha	No	[{scalar = 1} | flat | interp]	1element or string
FaceAlpha	No	[{scalar = 1} | flat | interp]	1element or string

Surface objects store matrices in the *XData*, *YData*, and *ZData* properties. These three properties define the vertices of the quadrilaterals which make up the surface object. The *XData* and *YData* properties do not need to be matrices. They can be vectors, provided that there are as many elements in

XData as there are columns in *ZData*, and as many elements in *YData* as there are rows in *Zdata*. If you store an M-by-N matrix in *ZData*, then you must store either a 1-by-N vector or an M-by-N matrix in the *XData* property. Likewise, you must store either a 1-by-M vector or an M-by-N matrix in the *YData* property.

In addition to the three vertex position properties, the color data property, *CData*, must also be defined. Most high-level commands set *ZData* and *CData* to the same matrix so that color will be proportional to the height of the surface. However, we will see that if you set the *FaceColor* property to "texturemap", *CData* can be any size you desire. Under this circumstance, the CData will be treated like an image that you want to have "wrapped" across or made to fit within the surface. Normally, the FaceColor property is set to flat and the color of a particular quadrilateral within the surface will be identified by the upper left-hand element of the set of four elements defining the color. Consider for a moment that the following MATLAB code

```
sx = [1 2 3 ];
sy = [4 5 6];
sZ = [1  2  3;
      4  5  6;
      7  8  9];
figure
axes('view',[-37.5, 30])
surface('XData',sx,'YData',sy,'ZData',sZ,'CData',sZ);
% We could also use surface(sx,sy,sZ,sZ).
```

will generate Figure 7.38.

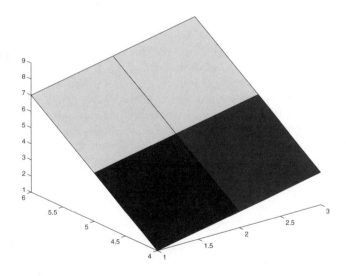

Figure 7.38   Creation of a simple surface.

As you can see, the 3-by-3 matrix, sZ, creates four quadrilaterals.

The lines that surround each quadrilateral are called the edges, and are by default black. These can be defined in several different ways using the *EdgeColor* property, just as the patch edge lines can be specified. For example, if you want the surface with handle, h_surface, to have red edges, you can use

```
set(h_surface,'EdgeColor','red')
```

or

```
set(h_surface,'EdgeColor',[1 0 0]);
```

If you do not want edges, you can set the *EdgeColor* to "none". This property may also be set to "flat" or "interp". The value "flat" will use a single color for the edge of a particular quadrilateral that is based on the upper left most element of the group of four elements in *CData* that are associated with that face. On the other hand, "interp" will linearly interpolate the edge line segments between the face's four *CData* elements.

You can also set the *LineStyle* of the edge lines to something other than the default solid lines ('-'). In addition to the style of the line, you can also set its thickness with the property *LineWidth* (default = 0.5). If you have decided to use one of the standard set of markers, you may set the *MarkerSize*, *MarkerEdgeColor*, and *MarkerFaceColor* properties to alter the marker's attributes. These have the same affect that they had on patch objects. The "auto" value of the *MarkerEdgeColor* and *MarkerFaceColor* lets the *EdgeColor* property dictate the color of the markers. Just as with patch objects, remember that the surface's marker's edge width is determined by the *LineWidth* property.

Finally, the last property you can manipulate to affect the appearance of the surface quadrilateral edges is the *MeshStyle* property. When the *LineStyle* is set so that the edges are solid lines, you can specify which edges are drawn. By default the edges are on all four sides of each quadrilateral, or in other words, down each row and column of the *ZData* matrix. However, if you want the edges to run only along the columns of the *ZData*, you can set the *MeshStyle* to "column", and if you want them to run along the rows of the *ZData*, set this property to "row". Figure 7.39 shows the effect of *MeshStyle* set to "row" as well as some line width alteration applied to the previous example with

```
set(h_surf,'MeshStyle','row','EdgeColor',[1 0 1],...
    'LineWidth',4)
```

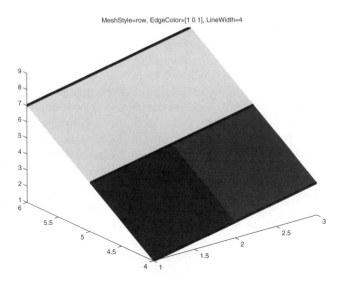

MeshStyle=row, EdgeColor=[1 0 1], LineWidth=4

Figure 7.39   Manipulating the MeshStyle 'row.'

Figure 7.40 shows the effect of using

```
set(h_surf,'MeshStyle','column','EdgeColor',[1 0 1],...
    'LineWidth',4)
```

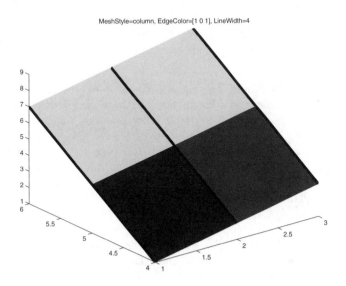

MeshStyle=column, EdgeColor=[1 0 1], LineWidth=4

Figure 7.40   Manipulating the MeshStyle 'column.'

The manner in which the color of the faces of the surface's quadrilaterals can be specified is similar to the way that an individual patch object's face color is specified. You may specify that all quadrilaterals be a certain color by setting the *FaceColor* property to either a legal color string or RGB triplet vector. By default, this property is set to "flat" so that each quadrilateral face is a solid color that usually corresponds to the height of the upper left corner of the quadrilateral. In Chapter 8 you will learn how to set the quadrilateral face colors so they are proportional to the height of the quadrilateral's center. You may also set the *FaceColor* to "interp" so that the color is interpolated through the four elements of *CData* that are associated with the vertices of each quadrilateral. This will give a smooth blend of color between adjacent quadrilaterals and depending on the values, *CData* can be used to give an accurate representation of surface height across the entire surface.

Just as with patch objects, when the *CDataMapping* is set to the default value of "scaled", you can think of the color data values as a range of values that get scaled to the size of the figure's color map to identify the color at each of the corners of a face in the surface. This default setting lets you flip between different sized color maps (see Chapter 8) without having to recalculate the *CData* values. When the value of *CDataMapping* is set to "direct", you can think of the color data values as indices to the figure's color map that are used to identify the color at each corner of each face of the surface object.

Remember that the size of the matrices you assign to either of color data value properties and the setting of the *FaceColor* and *EdgeColor* properties will specify how MATLAB should interpret your coloring intentions.

Just as with patches, surface *VertexNormals* are automatically calculated by MATLAB when the *NormalMode* property is set to "auto". The normals are used to perform calculations that determine the visual effects of lighting models on the surface of each object. Modifying the *VertexNormals* sets the *NormalMode* property to "manual" and keeps MATLAB from recalculating the normals.

Although we will discuss lighting and transparency in detail in Chapter 8, you have already seen an introduction to these properties with the patch object discussion. So we will dispense with repeating that brief introduction, and merely point out that the properties of *SpecularStrength*, *DiffuseStrength*, and *AmbientStrength* are, just as with patch objects, used to specify the respective intensity components of light objects that are reflected off of the surface object. The defaults and uses are the same as with patch objects. The *SpecularExponent* property determines the size of the highlight spot due to a light source shining on the surface object. The *SpecularColorReflectance* property controls the color of the reflected light spot emanating from the surface object. The *FaceLighting* and *EdgeLighting* properties specify the method that MATLAB should use to calculate the effect of light on the surface object. *BackFaceLighting*, by default, is set to reverselit, which means that patch objects whose *VertexNormals* point away from the camera are illuminated. Just as we stated with patch objects, *BackFaceLighting* can be used to highlight only the surface objects whose *VertexNormals* are facing the camera by setting it to "unlit". The "lit" setting can be used if you are seeing strange lighting effects along the edges of the surface object.

Finally, just as patch objects can have transparency, so can surfaces. The same properties you saw with patch objects exist with surface objects as well. All we will say here, is that just like with patch objects, alpha data must exist for the surface object before using *EdgeAlpha* or *FaceAlpha*. If alpha data has not been defined, you will get warnings like,

```
Warning: size(AlphaData) must equal size(ZData) or
size(ZData)-1 for flat alpha.
```

or

```
Warning: size(AlphaData) must equal size(ZData) for
interpolated alpha.
```

We will cover alpha data thoroughly in the next chapter.

7.5.8 Image Properties

Although Chapter 5 dealt with images, we intentionally left out any discussion of handle graphics. Image objects are created anytime you invoke the **image** or **imagesc** commands. As you have seen, MATLAB lets you do quite a lot without relying on handle graphics and object properties. But by now you have gained a great deal of familiarity with object properties, and are probably quite comfortable using the **set** and **get** commands. Image objects in MATLAB are children of axes objects, just like lines, patches, and surfaces. Knowing the properties that affect images will give you a greater command over what you can do with them. The following table lists the properties that pertain to image objects.

Property	Read Only	ValueType/Options	Format
General Properties of the Image Object			
CData	No	numbers	matrix or M-by-N-by-3 array
CDataMapping	No	[{direct} | scaled]	row
XData	No	[min, max] default = [1, size(CData,2)]	2-element vector
YData	No	[min max] default = [1, size(CData, 1)]	2-element vector
Properties Affecting Transparency			
AlphaData	No	default = 1 (opaque)	M-by-N matrix of double or uint8
AlphaDataMapping	No	[{none} |direct | scaled]	row

The *CData* property of an image contains the actual data that makes up the image. The dimension of the data in *CData*, either MxN, or MxNx3, determines if MATLAB displays the image using colormap colors, or as an RGB image. If the *CData* contains a 2-D array, then the image is either an indexed image or an intensity image (see Chapter 5). In either case the image is displayed using colormap colors. If the data in *CData* is 3-D, then the image is a truecolor image.

The properties *XData* and *YData* specify the coordinate system for the image. For MxN images, the default for *XData* is [1 N] and the default for *YData* is [1 M]. This means that for an image object, the left column of the image has an x-coordinate of 1, the right column an x-coordinate of N, the top row a y-coordinate of 1 and the bottom row an y-coordinate of M. We can demonstrate this with the following code,

```
X = [1  2  3  4;
     5  6  7  8;
     9 10 11 12];
h_image = image(X);
colormap(colorcube(12));
xlabel('x-coordinates');
ylabel('y-coordinates');
```

which will produce the result shown in Figure 7.41.

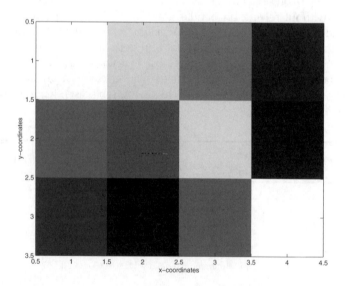

Figure 7.41   Default image object coordinates.

If you invoke the **image** function again, this time specifying a non-default value for *XData*, such as with the following code, you will get the result shown in Figure 7.42.

```
h_image2 = image(X, 'XData',[-1 2]);
colormap(colorcube(12));
xlabel('x-coordinates');
ylabel('y-coordinates');
```

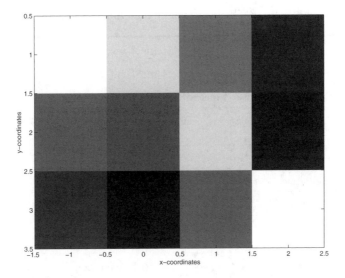

Figure 7.42 Changing XData to a non-default value.

As you can see in Figure 7.42, the x-coordinates have been changed to the range we specified ([-1 2]). Since we have changed the coordinate range, the return value of commands like,

```
get(gca,'CurrentPoint')
```

will be points in the new coordinate system.

We will discuss the properties *AlphaData* and *AlphaDataMapping* in the next chapter.

7.5.9 Text Properties

The last object we will discuss, and the last object that is a child to an axes object, is the text object. You have been adding text, either with text commands like **text**, **xlabel**, **label**, etc., or by annotating using the "Insert Text" button from the toolbar in the Figure Window. But just like the other objects we have seen, MATLAB offers the ability to alter the text properties so that you can highlight the important aspects of the graphical information. In addition to properties common to all axes children objects, the properties associated with text objects can be categorized into those that define the text object, those that position the text object, and those that specify the font. The following table summarizes the properties that are specific to text objects.

Property	Read Only	ValueType/Options	Format
Color	No	[Red Green Blue] or color string	RGB row
Editing	No	[{off} | on]	row
EraseMode	No	[{normal} | none | xor | background]	row
Extent	Yes	[left bottom width height]	4-element row
FontAngle	No	[{normal} | italic | oblique]	1 element
FontName	No	string	row
FontSize	No	numbers	1 element
FontUnits	No	[inches | centimeters | normalized | points | pixels | {data}]	row
FontWeight	No	[light | {normal} | demi | bold]	row
HorizontalAlignment	No	[{left} | center | right]	row
Interpreter	No	[{tex} | none]	row
Position	No	[x y z] coordinates	row
Rotation	No	[AngleInDegrees]	1 element
String	No	string	row
Units	No	[inches | centimeters | normalized | points | pixels | {data}]	row
VerticalAlignment	No	[top | cap |{middle}| baseline | bottom]	row

Each text object displays the text that is in the *String* property. Multiple lines of strings are displayed by using string cell-arrays. Every cell will correspond to a line in the multi-line string. For example, the following code

```
text(.1,.1,'This is single line text object.');
text(.5,.5,[{'This is line one'...
            'This is line two'...
            'This is line three'...
            'of a multiline string object'}]);
```

produces the results shown in Figure 7.43.

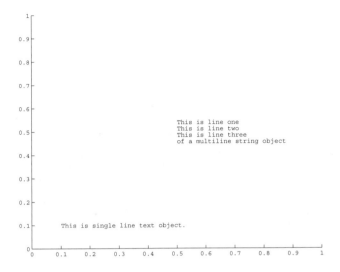

Figure 7.43   Cell-arrays allow multiple line text objects.

Each text object is anchored to a location within the figure that is specified with respect to its parent (an axes object). The *Units* of a text object are by default in data units. This makes it easy to add text at the command line by picking off a location in the graph using the axis tick mark labels as a guide or, perhaps more important, allows placing text with respect to a specific point on a plotted line, surface, or other graphics object. However, you can also specify that the *Units* be in inches, centimeters, normalized, points, or pixels. These units are all relative to the lower left corner of the axes object parent.

The *Position* property is stored as a 2- or 3-element row vector that defines a coordinate, (x,y,z), in the 3-dimensional space. If a 2-element row is specified, the z-axis coordinate is assumed zero. The position values must be specified in the units defined by the *Units* property of the text object.

In addition to defining the location of the text object, you can also define the orientation with the *Rotation* property. MATLAB allows you to specify any angular value in degrees relative to zero (which is the default value). For example,

```
axis([0 10 0 10])
text(5,5,'Text at 0 degrees');
text(5,5.5,'Text at 45 degrees', 'Rotation',45);
text(4.5,5,'Text at 90 degrees','Rotation',90);
```

creates the two text objects shown in Figure 7.44.

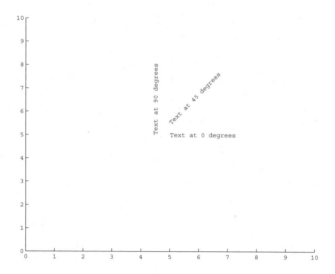

Figure 7.44   The effect of the Rotation property.

After you have experimented with the different placement properties, you may also want to look at the various properties that affect the visual aspects of the text object. Figure 7.45 shows some of the individual text item's properties as they are specified in the string of the object so that you can see their effects. Here is the code that creates the figure:

```
axis([0 10 0 10])

text(0,9,'fontweight=bold, fontname=times new roman,...
fontsize=12, fontangle=normal',...
'FontWeight','bold', 'Fontname','times new roman',...
'FontSize',12,'FontAngle','normal');

text(0,7,'fontweight=light, fontname=times new roman,...
fontsize=10, fontangle=normal',...
'FontWeight','light','Fontname','times new roman ',...
'FontSize',10,'FontAngle','normal');

text(0,5,'fontweight=normal, fontname=arial,...
fontsize=12, fontangle=normal',...
'Fontname','arial',FontSize',12,'FontAngle','normal');

text(2,3,{'fontweight=bold, fontname=brush script,...
fontsize=12','fontangle=normal, color=red'},...
'FontWeight','bold', 'Fontname','brush script',...
'FontSize',12,'Color','red');
```

You can use any of the fonts available on your system for the value of *FontName*. Only a very small fraction of the number of possibilities that you could potentially define with the text font properties has been shown.

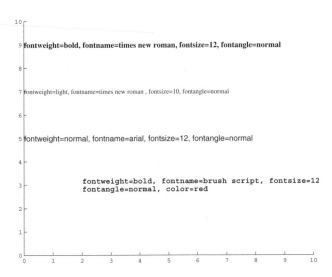

Figure 7.45   Example of text font properties.

As you can see in the above example, in addition to the font angle, name, size, and weight, you can also set the *Color* property to any legal color string or a 3-element RGB color vector. For example, if you want some blue text at the axes position (0.5,0.5,0), use any one of the following syntaxes:

```
text(0.5,0.5,'Blue Text String','color','blue');
text(0.5,0.5,'Blue Text String','color',[0 0 1]);
text(0.5,0.5,'Blue Text String','color','b');
```

The properties *HorizontalAlignment* and *VerticalAlignment* make it easy to figure out their respective effects on the text's location relative to the point defined by the *Position* property. Figure 7.46 shows several text strings that have been placed relative to the "crosshairs" drawn in the figure. The *HorizontalAlignment* property will shift the text's position relative to the point along the x-axis, while the *VerticalAlignment* property shifts the text's position relative to the point along the y-axis.

```
text(1.5,4.5,'HorizontalAlignment=left','horiz','left')
text(1.5,3.5,'HorizontalAlignment=center','horiz',...
'center')
text(1.5,2.5,'HorizontalAlignment=right','horiz',...
'right')
hold on
plot([1.5*ones(1,3)],[2.5:4.5],'+','markersize',30)

text(2.5,5,'VerticalAlignment=top','vert','top')
text(2.5,4,'VerticalAlignment=cap','vert','cap')
```

```
text(2.5,3,'VerticalAlignment=middle','vert','mid')
text(2.5,2,'VerticalAlignment=baseline','vert','base')
text(2.5,1,'VerticalAlignment=bottom','vert','bottom')
plot([2.5*ones(1,5)],[1:5],'+','markersize',30)
set(gca,'vis','off')
axis([0 5 0 6]);axis(axis)
```

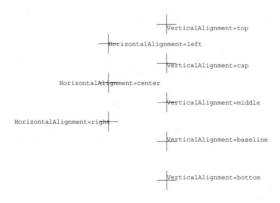

Figure 7.46   Horizontal- and VerticalAlignment properties.

The two properties *EraseMode* and *Extent* are principally applicable to animation, which will be covered in Chapter 9, however we will briefly discuss their function here. Often, it may be advantageous or desirable to have the text properties values change instead of deleting an existing object and then creating a new text object with the desired values. For example, if you want to have some text move across the top of the figure in a smooth manner, you can set the *EraseMode* to "xor" or "background" depending on your particular requirements and update the *Position* property as desired. The *Extent* property allows you to precisely determine the region covered by the characters in the string. The position along with the alignment properties define where the text will lie relative to some point and is independent of the number of characters in the string. The *Extent* property provides (since it is a read-only property) additional information that depends on a combination of other properties, specifically, *FontSize*, *FontName*, *FontWeight*, the number of characters in *String*, *Position*, and the alignment properties.

The property *Editing* when set to "on" lets you edit the *String* property contents interactively. If you have some text objects and you want to be able to quickly update them with a new value every now and then, try

```
set(findobj('type','text'),...
    'buttondownfcn','set(gco,''editing'',''on'');');
```

Try this code with the example for Figure 7.43 and you will then be able to click on a text object and edit its string. Notice that it works just as well for the text object that uses the cell-array as well as the single line. If you want to only make one of the text objects editable, simply get the object's handle (easy at creation time) and set that one's *Editing* property to "on" with something like the following code.

```
h_t=text(.5,.5,'This text object is editable!')
set(h_t,'Editable','on');
```

The last text object property we will discuss is *Interpreter*. When this property is set to its default value of "tex", all special TeX characters are interpreted as such. If you set the *Interpreter* property to "none," the special characters will be displayed literally. The special TeX characters, summarized in Chapter 3, let you mix subscript, superscripts, different fonts, and symbols.

7.6   Setting Default Properties

Now that you have a pretty good understanding of what graphic object properties are and how they can be manipulated to produce a desired visual effect, you may be asking yourself

"Do I really need to set these properties every time, when I know, for instance, that I always like to see text objects with the Times font and colored blue?"

You'll be happy to know that even though MATLAB comes with many factory-determined default values for all the properties, you can always set the defaults to your liking, so you do not need to set the same properties over and over. However, you should consider whether or not you are going to be sharing MATLAB code with colleagues and recognize that *their* default settings may not be the same as yours (especially if you alter them from the factory default).

Default properties are always set at the root level and are fairly intuitive in terms of the name that is required. For example, to set the default font for text objects, use,

```
set(0,'DefaultTextFont','Times')
```

and to set the default color and size of the text to blue and nine points, use:

```
set(0,'DefaultTextColor','blue','DefaultTextFontSize',9)
```

For other objects and properties, the concept is identical. To create the default string, just concatenate the three words "Default", the object name (such as text, figure, line, etc.), and the property name. Then set the root's property you just determined with the default value of your choosing. Experiment with some examples. Try setting the default color for line objects to "cyan" instead of the factory default of "black". What about the text that labels the tick marks? (Hint: remember that the axes object has some properties which affect the text of tick mark labels).

One of the questions that we often encounter in teaching MATLAB is

"How can I make MATLAB cycle through line styles instead of colors so that the lines are distinguishable on my black and white hardcopies?"

The answer is easily solved by setting the defaults. Remembering that the axes object contains information about the order of colors and lines that are chosen when multiple lines are plotted at once, you can put something like

```
set(0,'DefaultAxesColorOrder',[0 0 0]);
set(0,'DefaultAxesLineStyleOrder',['-|--|-.|:']);
```

You could simply type these commands at the command prompt, or you can have this be *your* default if you put these lines in your *startup.m* file. The file startup.m is looked for by MATLAB each time you start a MATLAB session and anything you put in there will be executed.

7.7   Undocumented Properties

All of the properties that have been discussed so far are referred to as *documented* properties, i.e., they are covered to some degree of detail in the MATLAB documentation such as the MATLAB Reference, User's Guide, or from the command line help. However, MATLAB has what we call *undocumented* properties, i.e., those that do not appear in any of the documentation. In fact, these properties do not even appear with **set** or **get** unless you tell MATLAB specifically that you want to make them visible. One root property that was not mentioned previously (because it is undocumented) is *HideUndocumented*, which can be set to either "on" (the default setting) or "off". In its default setting of "on," none of the undocumented properties of any of the MATLAB objects can be accessed, but when this root property's value is set to "off" by using

```
set(0,'HideUndocumented','off')
```

you will be able to access the hidden properties of any of the MATLAB objects.

You might be asking, "Why would MATLAB have undocumented properties?" There are several reasons for undocumented properties; the first is that these are experimental properties used by The MathWorks that may or may not be available for use in future versions of MATLAB. Another reason is that some properties are holdovers from previous versions of MATLAB and so are kept as aids to assist in the upward compatibility of the software package.

Warning!

You should be aware of the fact that The MathWorks does not support nor encourage the use of these undocumented properties, and it is even with hesitancy that we present how to access them in this text. However, since this book is intended to be an extensive guide to MATLAB graphics and graphical user interfaces, we feel that at least a brief mention of the existence of these properties is warranted. However, be aware that using undocumented properties, and even worse, relying on them, will most likely lead to difficulties and incompatibilities with future versions of MATLAB.

7.8 Using FINDOBJ

One of the most powerful built-in functions, as far as graphics are concerned, is the **findobj** function. The **findobj** function (which is short for "find object") relieves you of having to keep track of an object's graphics handles, such as by always having to assign an object handle to a variable name. Maintaining sets of variables for handles can be time consuming, tedious, and in general, adds a lot of overhead to MATLAB programs that create and manipulate numerous graphics objects. The **findobj** function gives you the freedom, with considerable flexibility, to quickly "search" for object handles. If MATLAB did not have a function like **findobj**, you could have created a similar routine by taking advantage of the parent-child relationship of objects. For instance, you could create a function that starts at some level in the hierarchy and searches down for an object with a particular property/value combination(s); this is essentially what **findobj** is doing. However, since this capability is already provided as a built-in function, it is quite fast at finding the object or objects that meet the specified search criteria. Although you have already seen this function in action a little, we will now formally present it.

The syntax of **findobj** in its simplest form is

```
h = findobj
```

which returns the root handle in addition to the handle of all its descendants (basically h will be a column vector containing the handle to every graphics object available in the current MATLAB session). The next form of **findobj** is to supply some search criteria such as

```
h = findobj('Property1Name','Property1Value',...);
```

which will return only the object handles of those objects that have a property named *Property1Name* which is set to Property1Value. If you do not want to start the search at the root object, you may start at any object or objects by using the syntax

```
h = findobj(ObjectHandles,...
            'Property1Name','Property1Value',...);
```

where ObjectHandles is a single element or vector containing the handles to objects from which you want the search to commence. By default, if ObjectHandles is not supplied, it is assumed to be the root handle, 0. Finally, if you have a set of object handles and want to find the subset that meets specific criteria, you can use the form

```
h = findobj(ObjectHandles,'flat',...
            'Property1Name','Property1Value',...);
```

This only checks to see if any of the handles supplied in ObjectHandles have the property/value combination specified.

Let's look at a quick example of how you might want to use **findobj** at the command line to alter the appearance or information provided by a figure. First, let's create a plot, like the one shown in Figure 7.47, and add some text in addition to the x- and y-axis labels:

```
x = 0:.1:10;
plot(x,sin(x).*exp(-.5*x));
xlabel('x'); ylabel('y')
text(4,.3,'y = sin(x).*exp(-.5x)');
text(5,-0.1,'Here''s the maximum');
```

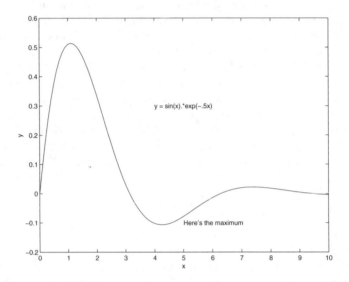

Figure 7.47   Oops!

Oops! It would appear that we didn't quite get things right. We put the wrong string in one of the text objects and we wanted the line to be dashed not solid. We could just start over, but armed with what we now know about handle graphics we can simply get the handles and alter their respective properties, as with the following code:

```
line_handle = findobj('type','line');
set(line_handle,'linestyle','--');
text_handle = findobj('string','Here''s the maximum');
set(text_handle,'string','Here''s the minimum');
```

Now our plot should look like the one shown in Figure 7.48.

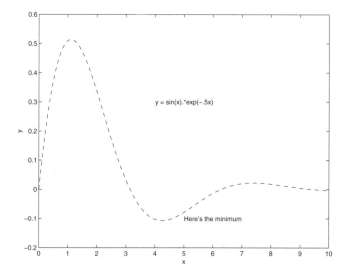

Figure 7.48 Finding handles with findobj makes changes easy!

The **findobj** function and *Tag* property are a perfect combination for finding and manipulating MATLAB graphics. Since all graphics objects have the *Tag* property, you can conceivably give each object a meaningful and unique name. For instance, if you plot several lines in the same figure, you can name them individually with

```
plot(x1,y1,'--b','tag','dataset1');
hold on;
plot(x2,y2,'tag','dataset2');
    .
    .
    .
```

You can then find the appropriate line's handle by using something like

```
line1_handle = findobj('tag','dataset1');
```

and proceed with modifying the line's properties as you see fit.

Later on, when some of the advanced graphics topics and GUIs are discussed, the usefulness of this function will become more evident.

7.9   Illustrative Problems

1. Create a plot of 10 random numbers distributed N(0,1) vs. the numbers 1 to 10. Using the **set** command along with the **gca** command turn the Ygrid to on, the GridLineStyle to – and the Linewidth to 3).

2. Continue with this example and set the y-axis color to red and the x-axis color to green.

3. Open a blue Figure Window. Plot a sine curve in solid yellow and a cosine curve in dashed green. Get the handle for the yellow curve and change its color to black.

4. Plot a sine curve as a solid line. Use the findobj command to go back and change it to a dashed line.

5. A tag is another handy concept that we only touched on in our discussions. Recall that we can use tags to keep track of objects by name. You can assign a value to the *Tag* property either with the Property Editor or with handle-graphics methods.

    For example we might execute the following command sequence.

    ```
    plot(x1,y1,'--b','tag','dataset1');
    hold on;
    plot(x2,y2,'tag','dataset2');
    ```

    Now we can find the line handle for the second curve using.

    ```
    H_line2 = findobj('tag','dataset2');
    ```

Plot a sine curve in blue and a cosine curve in green. Go back and use the first curve's tag to make it thicker.

8 USING COLOR, LIGHT, AND TRANSPARENCY

IN THIS CHAPTER...

8.1 SIMPLE COLOR SPECIFICATIONS 301
8.2 COLOR MAPS...301
8.3 MODELING OBJECT LIGHTING...338
8.4 OBJECT TRANSPARENCY 352
8.5 ILLUSTRATIVE PROBLEMS...359

8.1   Simple Color Specifications

Now that you have seen examples using both high-level plotting commands and low-level object property manipulation, it is appropriate to discuss the capabilities and features that MATLAB's functions related to color provide. As was mentioned in Chapter 2, color adds yet another dimension and therefore, an additional degree of freedom with regard to the amount of information that can be present in a figure.   Gone are the days of black and white and monochrome computer monitors. In fact, most people even have color printers in their homes and offices. The judicious use of color can greatly enhance the visibility, and information content of MATLAB graphics.

In addition to color, lighting can be used to make a three-dimensional surface look more realistic. By creating the effect of light reflecting off of the different curves of a surface, you will get more information about the data creating the plot. Light can be moved around a scene to produce different visual effects.

Finally, as if color and light (and how it is reflected) weren't enough, MATLAB also supports transparency. Transparency gives you the ability to see one set of data *through* another set. This is a very powerful capability, especially when used with volume plots.

So, let's get started and learn how to control the color, lighting, and transparency of our MATLAB graphics objects.

8.2   Color Maps

In Chapters 3 and 4 we looked at defining the color of most graphics objects with the easiest technique available, namely, using color specifications. We saw that you could use the high-level commands to specify the color of the lines by passing string arguments that contained color names as we usually refer to them when talking with other people. For example, a red line can be created by simply typing **plot**(x,y,'red'). After we learned about object handles, we saw that we can alter the properties of any graphics object. For example, to specify a green color for the current figure's background, we can set its color with **set**(**gcf**,'Color','green'). We also have seen that you don't need to spell out the whole name of a property, so you could use short names for a color, or even a single letter, as long as it is unambiguous. We also saw that we can specify a color using relative contributions of red, green, and blue in what is called RGB format, so we could specify yellow for an axes background with **set(gca**, 'Color',[1 1 0]). MATLAB has a term for the three ways you can specify colors: it's called ColorSpec and is either RGB triple, short name, or long name. The table below lists the colors that MATLAB recognizes when you use a string to represent either their long or short names.

The long name...	The short name...
blue	b
black	k
cyan	c
green	g
magenta	m
red	r
white	w
yellow	y

When we created some of our 3-dimensional surface plots we observed that the surface's color varied with the height of the surface, and we also made the color a function of the rate of curvature of the surface. In Chapter 5 we saw that MATLAB used certain built-in color maps for certain images, and three arrays representing red, green, and blue components (RGB) for others, but beyond that we didn't look very deep under the hood as to how color is controlled. The next section is aimed at teaching you about the commands that relate to these two color specification techniques and how you can use them to control an object's color.

Generally speaking, a color map is simply a three-column matrix whose length is equal to the number of colors it defines. Each row in this matrix defines a particular color by specifying the contribution of red, green, and blue components. Each component is an intensity value between zero and one, in a manner such that a zero is no intensity of the color, while a one turns on that component to full intensity. MATLAB comes with many predefined colors, many of which you have already used. The individual binary color representations are already associated with a name or a character string as shown in the following table.

R G B	Color	Character
[0 0 0]	black	k
[0 0 1]	blue	b
[0 1 0]	green	g
[0 1 1]	cyan	c
[1 0 0]	red	r
[1 0 1]	magenta	m
[1 1 0]	yellow	y
[1 1 1]	white	w

When we plotted lines and specified their colors with one of the strings above, the string was translated to the 3-element RGB vector. We could have just as easily used other colors by passing or setting the color property of the objects with some 3-element vector containing a fraction of the individual red, green, and blue color intensities. For instance, to create a dark gray color we could use a vector like [0.3 0.3 0.3] or we could create a bright gray color with [0.8 0.8 0.8], a copper color with [1 0.62 0.4], and even aquamarine with [0.49 1 0.83]. So you can see, you have the freedom of defining any color you want. Appendix A contains a list of some useful colors and their corresponding RGB values.

Color maps are just tables of colors that are organized in some desired fashion. MATLAB has many map-generating functions. The default map that is stored in the figure's *colormap* property is a 64-by-3 array of the **jet** color map. The entire list of map-generating functions is shown in the table below.

Function	Color Map Description
autumn	Smooth shades of red through yellow
bone	Gray-scale with a tinge of blue
colorcube	Regularly spaced colors with additional grays, red, green, and blue.
cool	Shades of cyan and magenta
copper	Linear copper-tone
flag	Alternating red, white, blue, and black, completely changing with each index increment
gray	Linear gray-scale
hot	Black-red-yellow-white
hsv	Hue-saturation-value, colors begin with red, pass through yellow, green, cyan, blue, magenta, and return to red
jet	Variant of hsv that is associated with an astrophysical fluid jet simulation from the National Center for Supercomputer Applications – this is MATLAB's default color map
lines	Uses the colors specified in the *ColorOrder* property of the axes object to generate a colormap

continued on next page

Function	Color Map Description
pink	Pastel shades of pink – makes grayscale images look "sepia tone"
prism	Alternating red, orange, yellow, green, blue, violet
spring	Shades of magenta and yellow
summer	Shades of green and yellow
white	All white monochrome colormap
winter	Shades of blue and green color map

You can run the MATLAB demo function **imageext** to look at a demonstration of the color maps.

In Chapter 7 we saw that *ColorMap* was a property of the figure object. To generate a matrix of RGB values, pass any one of these functions an integer that specifies the number of colors that are to be generated. For instance, to create a 32-by-3 hot color map matrix, just type something like

```
ColorMapMatrix = hot(32);
```

and if you want to place it into your current figure's *ColorMap* property, type

```
set(gcf,'colormap',hot(32));
```

or

```
colormap(hot(32));
```

The **colormap**(map) function simply performs a **set**(**gcf**, 'ColorMap', map). If you do not specify a size for the color map with an integer (e.g., the 32 in the above two examples), the matrix size will default to a 64-by-3 element matrix. This might be something to consider when creating your own color map generating functions. All of these color map generating functions can be created with simple mathematical expressions (i.e., they can be created with several lines of MATLAB code). Take a look at some of these functions in the editor and see how they work. For instance, just type **edit hsv** at the command prompt. Most of these color map generating functions return a set of RGB values that are created by sampling across three functions (i.e., one for the red, blue, and green components of the RGB vectors) between the lower and upper limits (the exception is **flag**, which cycles through red, white, blue, and black). You will see that these functions, when finely sampled, can be used to provide a nice transitional color gradation. To finely sample a color map function, just pass the function a large integer value.

8.2.1   Effects of Color Maps in General

Only surface, patch, and image objects are directly affected by the values in the *ColorMap* property of a figure. The colors of line, rectangle, text, axes, uimenu, uicontrol, and figure objects are completely independent of what lies in the figure's *ColorMap*.

This does not mean that the RGB vectors found in the figure's *ColorMap* or returned from a color map generating function are useless when you generate line, rectangle, text, axes, uimenu, uicontrol, and figures objects. Rather, you may find it convenient to obtain the colors that you want to use for these objects from one of these two sources of RGB values, particularly if you are not accustomed to defining colors with RGB vectors.

If you want to plot lines with colors other than the ones you can define with a color specification (i.e., the colors that you can specify with a string like 'red', 'green', etc.), first create a RGB matrix. Then, from this matrix, choose the colors one by one or all at once, depending on your needs. For example, if you want to generate 10 uniquely colored lines, first create a color map matrix with at least 10 colors, then use a for...end loop to plot a line with the color from this matrix using code like

```
map = hot(10);
for data_set_index = 1:10;
    plot(X(:,data_set_index),map(data_set_index,:));
end
```

You can also put the RGB matrix in the axes object's *ColorOrder* property and plot all the lines at once with something like

```
map = hot(10);
X = rand(20,10)+ones(20,1)*[1:10];
figure;
% The next line creates (since  one does not exist)an
% axes and sets its properties.
set(gca,'colororder',map,'nextplot','add','box','on');
plot(X);
title('Colored lines using colororder and...
        the hot colormap')
```

8.2.2 Color Axis Control

As you just saw, the color map generating function was used to define only the RGB values used for a set of lines. Color maps, in a more sophisticated sense, are used primarily for plotting surfaces, patches, and images. For the duration of this chapter, unless otherwise noted, the use of the word "object" will refer to one of these three.

Essentially, color maps are interpreters that are used to translate values to colors. The translated values are found in the *CData* property of each of these objects. There are two methods by which you can translate the *CData* values to colors; *direct mapping* and *scaled mapping*. These are possible values of the *CDataMapping* property.

8.2.2.1   Color Control with Direct Mapping

When an object uses direct mapping, its color data values (rounded down to the nearest integer) are used as indices to a row in the color map. For example, if you put the default color map into a matrix,

```
X=colormap;
```

you can see that size of the color map is 64-by-3. So if we had an object that had a color data value of 15 (i.e., one of its *CData* value terms was 15), the part of the object associated with that term would map to (be colored with) the color identified by the 15th row in the color map (i.e., X(15,:)). A *CData* value of 64 or greater would map to the 64th row and a *CData* value of 1 or less would map to row 1.

Image objects are similar to surface objects, except there is no *ZData* property. By default the values of an image object that are stored in the *CData* are assumed to be actual indices to the color map matrix, since an image object's default value for *CDataMapping* is direct. These indices are usually specified as integers; however, if they do have decimal portions, the values will be rounded down to the nearest integer.

8.2.2.2   Color Control with Scaled Mapping

Often the *CData* values correspond to the height of the surface or patch object. In fact, for these two objects, the *CData* property is not always specifically defined or set by the user. If the *CData* is not provided, MATLAB will automatically set this property equal to the *ZData* property values, and the *CDataMapping* property will be set to "scaled". This means that color data values will be linearly scaled to the color map. This is called *pseudocolor*. The simplest way to control the scaling is by using the pseudocolor axis, i.e., **caxis**, command.

Depending on how the **caxis** function is used, it performs either a **get** or **set** on the *CLim* property of the axes object. Remember from the last chapter that *CLim* contains a 2-element vector, [cmin cmax]. The two values are used to linearly transform data values in the *CData* property of surface and patch objects to indices where each index identifies a RGB row, i.e., a color, in the *ColorMap* property of the figure. The mathematical transformation of the *CData* values to indices is described by

$$index = \begin{cases} 1 & c < c\min \\ fix \left(\dfrac{c - c\min}{c\max - c\min} \right) & c\min \leq c < c\max \\ m & c \geq c\max \end{cases} \qquad \text{Equation 8.1}$$

where c is an individual *CData* value and m is length of the color map matrix. By default, the cmin and cmax values are automatically chosen by MATLAB to correspond, respectively, to the absolute minimum and maximum *CData* values found in any of the patch or surface objects in the axes object. This allows MATLAB to use the entire range of colors in the color map over the plotted data. However, using either the function

```
caxis([cmin cmax])
```

or

```
set(axes_handle,'CLim',[cmin cmax])
```

allows you to control how your data is mapped into indices of the color map. After you set the *CLim* property with either of these methods, the *CLimMode* property of the axes will be set to "manual", and therefore, auto scaling of the color axis will no longer be done for surface and patch objects contained within that axes object. However, if at some point you would like MATLAB to determine the color limits, set the *CLimMode* back to "auto".

8.2.3 Color Maps as they Relate to Graphics Objects

To better understand how the *CData* values are translated to colors, we will look at examples for each of the three objects that are directly affected by color maps, namely; surfaces, patches, and images.

8.2.3.1   Color Maps and the Surface Object

We will start by looking at an example that illustrates how *CData* values are converted to indices for surface objects. Since the direct mapping method is straightforward and is not the default setting for a surface object, the discussion that follows regarding the determination of the color map indices assumes that the surface object's *CDataMapping* property is set to "scaled". Consider a situation in which there are three colors (red, blue, and green) in the *ColorMap*,

$$map = \begin{bmatrix} 1 & 0 & 0 \\ 0 & 1 & 0 \\ 0 & 0 & 1 \end{bmatrix} = \begin{bmatrix} red \\ blue \\ green \end{bmatrix}$$

so that $m = \textbf{size}(map,1) = 3$. If we have a 4-by-4 element *CData* matrix

$$cdata = \begin{bmatrix} -5 & -3 & 2 & 7 \\ -4 & 0 & 2 & 6 \\ 0 & 1 & 3 & 9 \\ 4 & 2 & 2 & 5 \end{bmatrix}$$

and assume that the *CLim* property contains the minimum and maximum values of the *CData* (i.e., [cmin cmax] = [-5 9]), we can readily determine the index numbers using Equation 8.1 to be

$$index = \begin{bmatrix} 1 & 1 & 2 & 3 \\ 1 & 2 & 2 & 3 \\ 2 & 2 & 2 & 3 \\ 2 & 2 & 2 & 3 \end{bmatrix}$$

Now, before you quickly create a surface plot of this data and see something different from what you might expect, think about how a surface is created with **surf**(cdata). Since, in this example, we are not supplying any x- or y-coordinate data, recognize that the x- and y-coordinates are simply the row and column indices. Therefore, the *CData* values along with the row and column indices specify 16 vertices where each neighboring set of 4 elements is connected by means of a quadrilateral. As shown below, in terms of the elements within the *CData* matrix, there will be nine quadrilaterals.

$$cdata = \begin{bmatrix} -5 & -3 & 2 & 7 \\ -4 & 0 & 2 & 6 \\ 0 & 1 & 3 & 9 \\ 4 & 2 & 2 & 5 \end{bmatrix}$$

You might wonder why we need 16 indices to the color map when there are only nine quadrilaterals. With surfaces, each vertex can be assigned a color. This allows MATLAB to perform a bilinear interpolation between the four vertex colors to determine the color at any point within the quadrilateral. If you do not want to use color interpolation, the *CData* can also be a 3-by-3 matrix in the example above. Color interpolation is only needed when the surface property *FaceColor* or *EdgeColor* is set to "interp", such as in the case when you issue the command **shading** interp. When the *FaceColor* property is set to "flat" or "faceted", the quadrilateral's color will be determined by the color index of the vertex with the smallest row and column number. Continuing with our previous example, we see that the *CData* element in the first row and column (-5) has an index value equal to one (as calculated earlier with Equation 8.1), which, in turn, indicates that the quadrilateral defined by the

$$\begin{bmatrix} -5 & -3 \\ -4 & 0 \end{bmatrix}$$

components of the matrix will be red (i.e., since the index value equals 1, the quadrilateral will use the first row in our three-color color map). Taking the same approach in determining the color of the quadrilateral defined by the component

$$\begin{bmatrix} 3 & 9 \\ 2 & 5 \end{bmatrix}$$

of the *CData* matrix, we see that it will be green.

$$cdata = \begin{bmatrix} -5 & -3 & 2 & 7 \\ -4 & 0 & 2 & 6 \\ 0 & 1 & 3 & 9 \\ 4 & 2 & 2 & 5 \end{bmatrix} \Rightarrow index = \begin{bmatrix} 1 & 1 & 2 & 3 \\ 1 & 2 & 2 & 3 \\ 2 & 2 & 3 & 3 \\ 2 & 2 & 2 & 3 \end{bmatrix} \Rightarrow \begin{matrix} red & red & green \\ red & green & green \\ green & green & green \end{matrix}$$

Proceeding with a mental image of our expectations, we can set up and plot the surface with

```
map = [1 0 0; 0 1 0; 0 0 1];
figure('colormap',map);
% colormap(map) could have also been used in line above.
cdata = [-5 -3 2 7; -4 0 2 6; 0 1 3 9; 4 2 2 5];
surface_handle = surf(cdata);
```

to obtain Figure 8.1. We see that this figure has nine quadrilaterals; three of them are red and six are green.

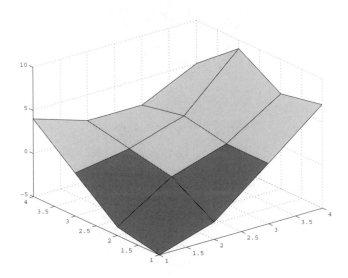

Figure 8.1 Controlling the color of a surface object.

Now consider the same surface with interpolated shading by typing

```
set(surface_handle,'facecolor','interp');
```

In Figure 8.2 we can see that the index values previously calculated are indeed used to identify the colors of the vertices and that each quadrilateral's color is bilinearly interpolated between the vertex colors.

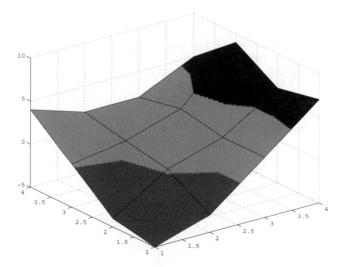

Figure 8.2 Interpolated shading.

At this point, the edges of each quadrilateral can be identified because the *EdgeColor* of the surface has been left in its default setting, black ([0 0 0]). However, in Chapter 7 you learned that edges color can be defined, too. You can specify that the edges of the quadrilaterals have a solid color by setting the *EdgeColor* either to a particular RGB value (note, the RGB vector does not have to be one of the values in the color map) or to "flat" which will use the color indices of the vertices to identify a color for the segment of the line associated with that vertex. Whenever the *FaceColor* is set to "interp", the figure will look the same when you set *EdgeColor* to "none" or "interp". This is because the "none" setting makes the edge lines invisible exposing the interpolated face colors below the edges.

You should also realize that the color of each quadrilateral or vertex does not need to relate to the height, or z coordinate, of the surface. You can also use a form such as **surf**(z,c) or **surf**(x,y,z,c). In these two forms, the color data can be whatever you want it to be as long as either

```
    size(c) = = size(z)
```
or
```
    size(c) = = size(z) - 1
```

holds true. For example, we can plot the peaks function with stripes of colors in either the y-axis direction using

```
    s = peaks(20);
    c = meshgrid(1:20);
    surf(s,c);
```

or, as shown in Figure 8.3, with stripes in the x-axis direction with

```
    surf(s,c');
    grid on;
```

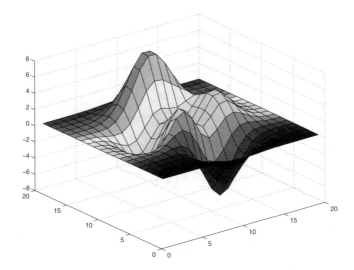

Figure 8.3 Forcing stripes across a surface.

Color stripes may not be very informative, however, color might be used to identify regions of a surface that have like curvatures, gradients, or whatever is of interest to you. For example, in Figure 8.4 color identifies the regions of the peaks function that have similar curvature.

```
s = peaks(20);
c = del2(s);
surface_handle = surf(s,c);
set(surface_handle,'FaceColor','interp');
colormap(hot(10));
```

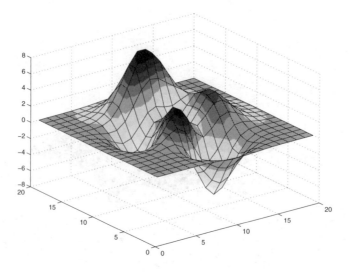

Figure 8.4 Coloring based on surface curvature.

A question that often arises is

"I have a surface that is symmetric in terms of its height; however, when I plot it with the surf command, the colors are not symmetric. What is the reason for this?"

The **surf** command, by default, will display the surface with a faceted shading (i.e., **shading** faceted). The quickest way to solve the problem is to change the shading to interpolated with **shading** interp, which varies the color in each line segment and face by interpolating the color map index, or true color value, across the line or face.

   The reason we don't get the result we would expect is that the last row and column are not used in determining the color of the individual quadrilateral faces for surface objects that are displayed in the faceted or flat shading; remember that the color value assigned to the upper left vertex of each quadrilateral, when looked at in terms of the matrix, determines the color. If you want to have faceted or flat symmetric shading, a solution is to calculate the height of the center of each quadrilateral and use this as the *CData* matrix. For example, the following code

```
[x,y] = meshgrid(-3:.5:3);
z = x.^2;
% Now plot the matrix and see that the color is not
symmetric.
surf(x,y,z);
% Calculate the CData matrix by averaging the vertex
heights.
[m,n] = size(z);
C = ( z(1:(m-1),1:(n-1)) + z(2:m,1:(n-1)) + ...
      z(1:(m-1),2:n) + z(2:m,2:n) ) / 4;
```

```
surf(x,y,z,C);
```
generates the two plots shown in Figure 8.5 and illustrates the difference in color symmetries. The left-hand side of the figure shows the non-symmetric colored surface, while the right-hand plot shows the symmetric colored surface.

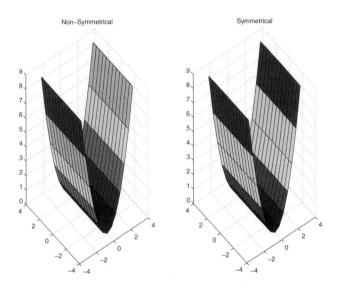

Figure 8.5 A symmetric surface with non-symmetric coloring (left) and symmetric coloring (right) achieved by determining the surface height at the center of each quadrilateral.

8.2.3.2   Patch Objects and the Color Map

Now that we have looked at surfaces, let's revisit the patch object and see what kinds of visual effects can be obtained with color. An individual patch object is created with the command **patch**(x,y,z,c) or **patch**(x,y,c), where the vectors x, y, and z define the vertex coordinates and c specifies the color data i.e., the *CData* property of the patch object. Please note that if z is not supplied, the *ZData* property of the patch object is set to the empty or null matrix, and the patch object is rendered as if the z-component of each vertex was zero. The variable, c, can be one of the built-in color names (a string), an RGB vector, a scalar, or a vector of values where there is one element for each vertex. As we saw in Chapter 7, patches can also be defined by their vertices using the form **patch**(*'Vertices'*, v, *'Faces'*, f, *'FaceVertexCData'*, fvc,...). With this form, the color data is specified by the *FaceVertexCData* property of the patch. When using *FaceVertexCData*, any corner of a face of the patch object connected to a particular vertex will have the same color associated with it. Whereas when specifying color with *CData*, corners that are shared by a patch's faces can have multiple colors assigned to them. This allows you to have complete and independent control over the colors in a particular face of a patch object.

If a simple color string or RGB vector is used for the *Cdata* or *FaceVertexCData*, the entire patch object will be one solid color (usually referred to as *flat* coloring). If the *CData* is set to a scalar, the entire patch object will also be a flat color, however in this case the color is determined by the translation of Equation 8.1 if using scaled mapping, or by the scalar to an index to the color map. If there is an element in the c variable for each vertex, Equation 8.1 can be used to identify what color will be applied to each vertex. In other words, if the *CData* property is set to a scalar or a vector (with more than 3 elements), the parent axes of the patch object will specify the color axis in the same way it does for surface objects, i.e., elements in the c vector are translated to color map indices.

In the next example, a pentagon is created and each vertex is specified to have a particular color. Five triangles are also created and used to indicate the color that is at each of the pentagon's vertices. The pentagon's coloring is interpolated across the object's face and the color map contains a 20-by-3 jet color matrix. The results of this example are shown in Figure 8.6 and Plate 9[*].

```
figure('colormap',jet(20));
axis([-1.5 1.5 -1.5 1.5])

p = patch([cos(linspace(0,360,6)*pi/180)],...
          sin(linspace(0,360,6)*pi/180),...
       [0 0 0 0 0 0], [1 1 2 5 2 1],...
          'facecolor','interp')

rotate(patch([1.1 1.3 1.3],[0 .15 -.15],[0 0 0],[1]),...
       [0 90],0);
rotate(patch([1.1 1.3 1.3],[0 .15 -.15],[0 0 0],[1]),...
       [0 90],1*72);
rotate(patch([1.1 1.3 1.3],[0 .15 -.15],[0 0 0],[2]),...
       [0 90],2*72);
rotate(patch([1.1 1.3 1.3],[0 .15 -.15],[0 0 0],[5]),...
       [0 90],3*72);
rotate(patch([1.1 1.3 1.3],[0 .15 -.15],[0 0 0],[2]),...
       [0 90],4*72);
```

---

[*] Color plates follow page 112.

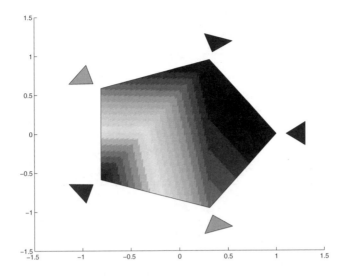

Figure 8.6 Defining the color of patch vertices.

You will learn more about the rotate graphics function in Chapter 9; however, it essentially relieves you from the responsibility of determining the coordinate transformation required to rotate an object. The syntax for rotate is

rotate(object_handle,axis_of_rotation,angle_degrees,origin_of_rotation),

where the variable *object_handle* is the handle to the object that will be rotated, *axis_of_rotation* is a 2-element ([Az El]) or 3-element ([x y z]) vector that defines the axis about which the object should be rotated. The axis of rotation passes through the point (0,0,0) unless otherwise specified with the *origin_of_rotation* variable. The number of degrees that the object will be rotated through is specified by the third argument, *angle_degrees*. (You could also use a routine such as **viewmtx** to determine a coordinate transformation matrix.)

8.2.3.3 Images and the Color Map

We have already discussed image types and their characteristics in Chapter 5. In Chapter 7 we presented the properties of image objects. The list of properties belonging to image objects are presented again here to facilitate our discussion of how the color map is influenced by these properties.

Property	Read Only	ValueType/Options	Format
General Properties of the Image Object			
CData	No	numbers	matrix or M-by-N-by-3 array
CDataMapping	No	[{direct}	scaled]	row
XData	No	[min, max] default = [1, size(CData,2)]	2-element vector
YData	No	[min max] default = [1, size(CData, 1)]	2-element vector
Properties Affecting Transparency			
AlphaData	No	default = 1 (opaque)	M-by-N matrix of double or uint8
AlphaDataMapping	No	[{none} |direct	scaled]	row

Of course these properties are in addition to the universal properties already discussed in Chapter 7. Recall that the *XData* and *YData* properties are vectors that specify the region in the xy-plane that the image object will occupy. The image is scaled to fit between the first and last element stored in each of these two properties. The elements between the first and last do not affect the image object's location in the xy-plane. For example, if the *XData* is [1 10] or [1:10] or [1 43 20 10], the image would be located in its axes object parent over the x-data values from one to ten, while the region occupied in the y-axis direction would depend on the values in the *YData* property in a similar manner.

So far, you have seen that the color of both surface and patch objects is determined by transforming the color data (*CData*) values of the object to indices in the figure's color map by means of the color limits (*CLim*) of the axes object. Image objects, on the other hand, with their default settings are unaffected by color limits of the axis. When the *CDataMapping* property is set to "direct", the color data values of images are expected to be integers between 1 and the number of rows in the color map, i.e., **size(get(gcf**,'colormap'),1). These values are integers because they are used as indices to the figure's color map without any transformation and, therefore, represent the color of a portion of the image object.

Color data values that are not integers are rounded down to the nearest integer, while those values that are less than one are assumed to be one. Those values greater than the number of rows in the color map are assumed equal to the number of rows in the color map. This is an important fact to realize and implies that if there are indexes with values outside the allowable range, those portions of the image will be "clamped" to the upper or lower color values in the color map. This image clamping translates to a loss of information and can distort or make it difficult to discern the essence of the image. Usually you will not have a problem with this if you are using any of the images that come with MATLAB since images and their color maps are usually stored in the same binary (.mat) file. However, since you might want to create your own images, you need to be aware of image clamping effects.

To illustrate, the next example shows how easily an image can be generated and demonstrates what image clamping can do. The next two figures use the following image data:

```
X = [1 1 1 1 1
     1 2 2 2 1
     1 2 3 2 1
     1 2 2 2 1
     1 2 3 2 1
     1 2 2 2 1
     1 1 1 1 1];

image(X)
colormap([.2 .2 .2; 1 1 1; .5 .5 .5]);
```

will generate the left-hand plot of Figure 8.7, while

```
image(X)
colormap([.2 .2 .2 ; 1 1 1])
```

will generate the right-hand plot.

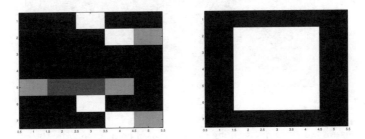

**Figure 8.7 An image clamped by a limited number of color map entries.
The desired image on the left and the clamped on the right.**

Image objects can also scale their *CData* to the color map using Equation 8.1, if you set the *CDataMapping* to "scaled". If you have some image's color data values in X, and you want to use any given sized color map to obtain the most information out of the image, the scaled setting of *CDataMapping* is usually the easiest method.

```
image_handle=image(X);
set(image_handle,'CDataMapping','scaled');
set(gca,'clim',[min(min(X))    max(max(X))]);
```

Try loading the penny.mat image and experiment with different sized color maps.

```
load penny;
figure;
i=image(P, 'CDataMapping','scaled');
colormap(copper(255));
```

will let you see the greatest amount of detail in the data matrix, P. But if you wanted to visually quantize the data into fewer levels of detail, you could change the size of the color map and MATLAB will scale the color data to it. So,

```
colormap(copper(10))
```

will produce the illustration shown in Figure 8.8.

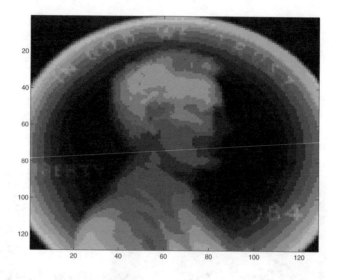

Figure 8.8 Viewing an image with a scaled color map with 10 entries.

Had you used the default direct value of the *CDataMapping* property, the data values in P greater than 10 would be clamped to the last color in the color map.

You may have noticed in the previous two figures that the y-axis labels increase in the reverse direction than is normally displayed. This is due to the fact that when an image is created, the axes object's *YDir* property is set to "reverse". This keeps the image oriented in the same manner as the data in the *CData* matrix. In addition, images created with the **image** function can be visualized only in 2-dimensional perspectives. Later in this chapter, you will learn how to visualize image data in 3-dimensional perspectives.

The *EraseMode* property of image objects can be used to control the manner in which an image object is erased and/or redrawn. This property is primarily manipulated when animating graphics objects, which we will discuss in Chapter 9. The default setting of "normal" will give the most accurate presentation of the image object. The other three modes of erasing are faster, but you will lose some accuracy. With "none", MATLAB will not update the region where the object was located before it was either deleted or moved. This is the mode that you should use, if you want to update the image's *CData* quickly but are not translating the object in the xy-plane. For instance, if you have several same-sized images and you want to flip between them, create one image object whose *EraseMode* is set to "none" and update only the *CData* with the next image's data. The "xor" setting lets you move or delete the object without affecting objects rendered behind the image; however, the image's color will be affected by objects rendered behind it. The "background" setting will make sure the image object is drawn in the correct color, but if the image is deleted or moved, an *imprint* of the erased image object will remain until you do a **refresh** or a new object that has its *EraseMode* property set to "normal" is created, moved, or deleted.

The last two properties, *AlphaData* and *AlphaDataMapping*, have to do with making part (or all) of the image transparent. These properties exist in surface and patch objects as well and are subject to an entire section later in this chapter.

8.2.4 Color Shading

To control how color is applied to surface and patch objects, you can use the graphics function named **shading**. One of the following three arguments must accompany the command: flat, faceted, or interp. The default shading applied to surfaces and patch objects is faceted. Each quadrilateral has a constant color face and edges that are highlighted with black lines. Flat shading is the same as faceted, except that there are no edge lines, while interpolated shading makes use of bilinear color interpolation between the vertices.

The command shading manipulates the *FaceColor* and *EdgeColor* properties of surface and patch graphics objects. In the following comparison table, *surface_handles* is the variable containing the handle or handles to all surface and patch objects in the current axes object whose shading you want to alter.

Using...	is the same as....
shading flat	set(surface_handles,'FaceColor','flat','EdgeColor','none')
shading faceted	set(surface_handles,'FaceColor','faceted','EdgeColor',[0 0 0])
shading interp	set(surface_handles,'FaceColor','interp','EdgeColor','none')

8.2.5 Brightening and Darkening Color Maps

The graphics function **brighten** can be used to alter a color map by either brightening or darkening the colors. Depending on the value of the argument passed to the brighten command, the intensity of the color map is either increased, i.e., brightened, or decreased, i.e., darkened. To brighten the existing figure's color map, use

```
brighten(intensity_factor);
```

where the *intensity_factor* should be a value between zero and one. To darken the existing figure's color map, use an *intensity_factor* with a value between negative one and zero.

Applying a change in sign to the intensity_factor after having already used that intensity_factor to alter the color map will lead to the original color map. In other words, the MATLAB code

```
brighten(intensity_factor);
brighten(-intensity_factor);
```

would yield the color map that existed before the two statements were executed.

You can also create a new color map matrix without affecting the current figure's *ColorMap* property by using an output argument such as

```
new_colormap = brighten(map,intensity_factor);
```

or

```
new_colormap = brighten(intensity_factor);
```

where the first syntax form returns an altered color intensity of the map you passed to the function and the second form returns an intensity-altered map of the current figure.

We can look at the RGB plots (see following figures) for the hot color map in its normal, brightened, and darkened mode using

```
plot(hot);axis([1 64 0 1])
title('hot color map');
plot(brighten(hot,.75));axis([1 64 0 1])
title('brighten(hot,.75)');
plot(brighten(hot,-.75));axis([1 64 0 1])
title('brighten(hot,-.75)');
```

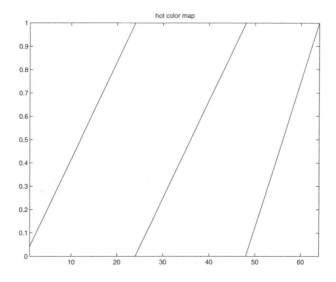

Figure 8.9 Original red, green, and blue components of the hot color map.

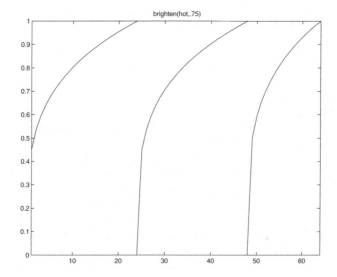

Figure 8.10 Plotting the red, green, and blue components of a brightened version of the hot color map.

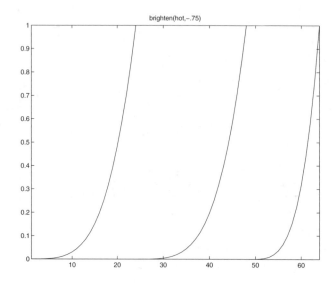

Figure 8.11 Plot of the red, green, and blue components of a darkened version of the hot color map.

8.2.6   Spinning the Color Map

An interesting visual effect can be created by *spinning* the color map. This essentially entails shifting the color map colors up or down in terms of row index numbers. Those colors that either would be shifted below the first index or above the last indexed row in the color map matrix are wrapped around to the end or the beginning depending on how the rows are shifted. After the color map matrix has been redefined, the map is quickly applied to the current figure. Now, if you do enough of these shifts quickly, the visual effect is as if the colors were moving across the surface in the figure. The **spinmap**(*time,shift_increment*) function makes this easy to accomplish. The variable, time, is the period in seconds (whether or not it truly represents seconds depends upon the speed of the platform on which MATLAB is running) over which the color map should spin (default is five seconds) and *shift_increment* is the number of rows by which each color in the current color map should be shifted down. An upward shift can be achieved by providing a negative integer.   Larger values, in terms of absolute value, of the *shift_increment* argument lead to a faster rotation through the color map indices. Be aware however, that because of differences in the many graphics cards in modern computers, spinning the color map might produce unexpected results; consequently you might have to change the color mode of your environment. If you try to use **spinmap** and get a message like,

```
Warning: Colormap animation is only possible for 256
color screens.
```

try changing the number of colors you are using for your display. You will also have to restart MATLAB after changing colors so that system properties are reported properly to MATLAB.

To see an example of the spinmap function, try the following:

```
figure('ShareColors','off');
peaks(20);
shading interp;
spinmap(10,1);
```

If you want to slow down the rate by more than a single row index shift, you can increase the size of the color matrix. For example,

```
colormap(hsv(128));
spinmap(10,1)
```

will be a smoother and longer rotation through the color map colors than with the previous example. To have a little fun, try this:

```
figure('ShareColors','off');
patch([0 0 10 10 0 1 9 9 1 1 0],...
      [0 10 10 0 0 1 1 9 9 1 0],...
      zeros(1,11),[1 2 3 4 5 5 4 3 2 1 1],...
      'EdgeColor','none');
colormap(flag(128));
spinmap(5,1);
```

8.2.7 Making Use of the Invisible Color with NaN

MATLAB's Not-a-Number (NaN) representation is a convenient way of making portions of a surface invisible. For example, if you zoom into a region of some 3-dimensional plot, you may not like the way that MATLAB clips the regions outside of the x-, y-, and z-axis limits. This would be a prime example of when the use of NaNs will help achieve your desired results. Another case might involve a situation in which you have altered the color axis (*CLim*) property to force a color variation over a particular portion of a surface and you want those regions outside of the limits to be invisible instead of clamped to the first or last color in the color map.

Power!

There are several techniques that can be used to achieve these and other similar types of results. Typically, they involve setting the elements of the color data, i.e., the *CData* property, to NaNs. Let's first look at an example in which we want to zoom in on a particular region of the 3-D plot shown in Figure 8.12.

```
[x,y] = meshgrid(-3:0.1:3);
z = sin(sqrt(x.^2+y.^2)).*exp(-(sqrt(x.^2+y.^2)));
surface_handle = surf(x,y,z);
```

```
shading flat
axis([-3 3 -3 3 min(min(z)) max(max(z))])
```

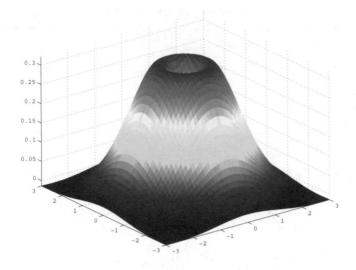

Figure 8.12 A flat shaded surface plot.

Zooming in on a region of the surface with

```
axis([0 3 0 3 min(min(z)) max(max(z))])
```

will produce Figure 8.13, which, as you can see, obscures the x- and y-axis tick mark labels.

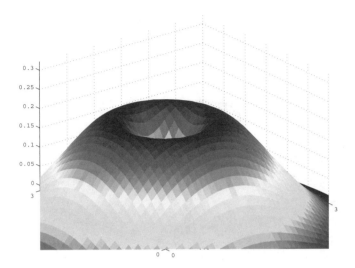

**Figure 8.13 Zooming in on a region with the axis command can obscure
axis tick mark and label information.**

However, with the following code we can make those regions of the
surface that fall outside desired limits invisible as shown in Figure 8.14.

```
indexs = find(~(x <= max(get(gca,'xlim')) & ...
               x >= min(get(gca,'xlim')) & ...
               y <= max(get(gca,'ylim')) & ...
               y >= min(get(gca,'ylim')) & ...
               z <= max(get(gca,'zlim')) & ...
               z >= min(get(gca,'zlim')) ) );
c = get(surface_handle,'cdata');
c(indexs) = NaN*c(indexs);
set(surface_handle,'cdata',c);
```

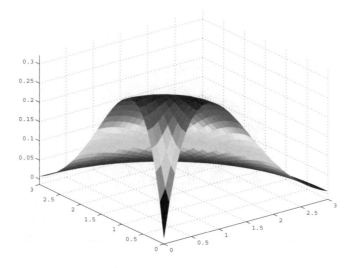

Figure 8.14 Using NaNs to remove unwanted portions.

Now, looking at this example, you might be thinking, and correctly so, that the same results can be achieved by plotting the quadrant that we are interested in by redefining x, y, and z. In addition to illustrating a technique, this previous solution may be useful in situations where perhaps you need to clip parts of the surface that extend beyond some set of limits along the z-axis.

In the next example we plot two spheres, where one sphere is inside the other as shown in Figure 8.15. Here is the code:

```
figure
view(3);
[x,y,z] = sphere(20);
% Create the outer sphere.
z1 = z;
z1(:,1:4) = NaN*z1(:,1:4);
c1 = ones(size(z1));
s1 = surface(2*x,2*y,2*z1,c1);
% Create the inner sphere.
z2 = z;
c2 = 2*ones(size(z2));
c2(:,1:4) = 3*ones(size(c2(:,1:4)));
s2 = surface(1.5*x,1.5*y,1.5*z2,c2);
colormap([0 1 0;.5 0 0; 1 0 0]);
grid;
set(gca,'box','on');
```

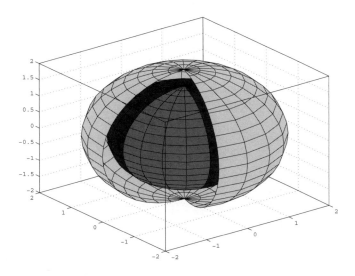

Figure 8.15 Cutting a hole in a surface with NaNs to make the surfaces behind visible.

Three colors are used in the color map. Green is used for the outer sphere, while two shades of red are used for the inner sphere. The darker red helps it look as if the outer sphere is casting a shadow on the inner one - a clever trick, but as you will learn in the section on lighting, there are commands that make it easy for you to apply various lighting models and color shading to your graphics.

As a final note to this section, even though NaNs were applied to either the *CData* or *ZData* surface properties in the two examples above, you should be aware that the NaNs could have been applied to either the *XData* or *YData* properties just as well. In addition, using NaNs to make portions of an object invisible is also applicable to line objects. An NaN in the *XData*, *YData*, or *ZData* properties of a line will make invisible segments about the coordinate with NaN. It is almost as if the "pen" that draws the line is lifted off the screen when MATLAB sees that the next coordinate contains an NaN. The pen is set back on the screen at the next coordinate that does not contain an NaN. To illustrate this very useful ability, the following code will clip off the top and bottom portions of a sine wave as shown in Figure 8.16.

```
x = [0:pi/16:4*pi];
y = sin(x);
index = find(abs(y) > .5);
x(index) = NaN*x(index);
plot(x,y);
```

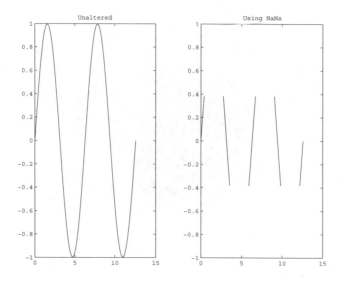

Figure 8.16 Using NaNs to "lift the pen."

8.2.8   Creating Simple Color Bars

It is often useful to visualize data in 3-dimensional perspectives so that a lot of data can be seen in one figure. However, it is often difficult to extract specific information from these types of plots. For example, 3-dimensional surface plots may show you the general regions of minima and maxima, but even with a grid, it is difficult to determine the height of the surface at any location on the surface. You have seen that color can aid this process considerably. Furthermore, a color bar can be used to make it even easier for the observer to associate colors with the surface values. MATLAB provides a graphics command called **colorbar** to make it very easy to generate a color-to-value association bar.

By default, typing **colorbar** after having created a surface plot creates a vertical color bar to the right of the axes with the 3-dimensional view. The following code will produce the plot of the peaks function shown in Figure 8.17 with an associated color bar.

```
surf(peaks(30));
colormap(hot));
colorbar
```

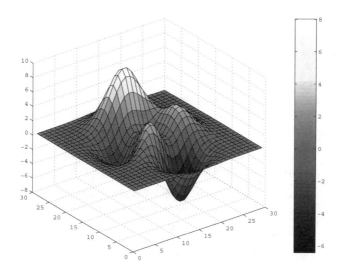

Figure 8.17 A color bar allows easier association of color to value.

You can also tell MATLAB to generate the color bar below the figure of interest by using

```
colorbar('horizontal')
```

If you require even more flexibility in the placement of the color bar, create and store the handle of an axes object in the desired position for the color bar. Then pass the graphics handle of that axes object to the colorbar function with

```
colorbar(axes_object_handle);
```

where *axes_object_handle* is the axes object that will contain the color bar. If the width is less than the height of the axes object, the bar will be labeled vertically; otherwise, the colorbar will be labeled horizontally.

8.2.9 The Pseudocolor Plot

A pseudocolor plot (sometimes referred to as a checkerboard plot) can be created with the **pcolor** function. This graphics function creates a surface object in which the *ZData* elements are set to zero and displays the plot in a perspective which makes it appear as if you are looking directly down on the surface (i.e., **view**([0 90])). By default, the *FaceColor* of this surface object is faceted. This function's syntax is very similar to the syntax used by **surf**, except that you do not supply the z-axis coordinates of the vertices. You need only specify the vertice colors such as with **pcolor**(C). However, you may specify the x- and y-axis components of the vertex coordinates with

pcolor(x,y,C) or **pcolor**(X,Y,C). If vectors x and y are supplied, the length of x must correspond to the number of columns in C and the length of y must correspond to the number of rows. The actual color of each quadrilateral, as with all surface objects, is determined by scaling the *CData* values with the color axis limits to an index to the color map. When the surface is being displayed in the default "faceted" shading, the color in the ith row and jth column is determined by the element C(i,j). However, when the shading is interpolated (i.e., the *FaceColor* property is set to "interp"), a bilinear color interpolation between the four vertices of each quadrilateral is performed.

It is useful to identify the similarities between the **pcolor** and **image** graphics functions by comparing the graphics objects that they create.

Speed!

The surface object created by **pcolor**(C) will have the same number of vertices as the number of color cells that an image object created with **image**(C). Unless you need to generate parametric grids or need to have control over the spacing of the color cells, it is often advantageous, in terms of rendering speed, to use **image** instead of pcolor.

The pcolor function is a useful means of visualizing the contents of the colormap. For example, if you want to display a 32-element hsv color map, type

```
M = 32;
figure
pcolor([1:M;1:M]');
colormap(hsv(M));
set(gca,'Position',[.4 .1 .2 .8])
title('hsv(32) colormap')
```

The result is shown in Figure 8.18.

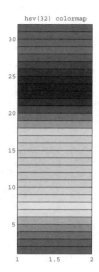

Figure 8.18 Pseudocolor plot of the hsv color map.

Now that we have a simple way of displaying color maps, try to answer the next frequently asked question, without looking at the answer, by using your knowledge of specifying color axis scaling and creating color maps.

Since the figure object contains the color map and axes objects are children of figures, how can I have multiple axes objects using the colors from different color map generation functions in the same figure as shown in Figure 8.19?

You might have noticed, for example in using a function like **subplot**, that using the **colormap** command affected all subplots – not always what you want to do!

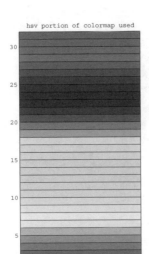
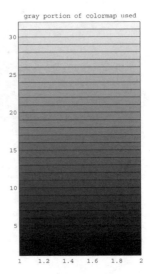

Figure 8.19 Using multiple color maps in the same figure.

This leads to the following question,
"If I want to use several concatenated color maps so that I can have multiple axes objects using different portions of the figure's color map as discussed, how do I determine the minimum and maximum color values?"

The following two equations can be used to determine the required color axis values:

$$\text{cmin} = \text{mincdata} - \frac{\text{minmapindex} - 1}{\text{maxmapindex} - \text{minmapindex} + 1}(\text{maxcdata} - \text{mincdata})$$

$$\text{cmax} = \text{mincdata} + \frac{N - \text{minmapindex} + 1}{\text{maxmapindex} - \text{minmapindex} + 1}(\text{maxcdata} - \text{mincdata})$$

where cmin and cmax are the minimum and maximum color axis limits specified in *CLim* (i.e., caxis([cmin cmax])). The mincdata and maxcdata are the minimum and maximum color data values that are to be plotted in the axes object. The variable minmapindex and maxmapindex are the minimum and maximum index numbers to the color map that contains N colors (i.e., the figure's *ColorMap* property is an N-by-3 RGB matrix). For example, consider a situation in which the data in one of the axes objects varied between -1 and 1 and the figure's colormap property had 64 rows. If you want the color of the contents of this axes object to be scaled to the first 32 rows of the color map, first determine the value of the variables needed in the previous equations:

```
mincdata = -1;
maxcdata = 1;
N = 64;
```

```
minmapindex = 1;
maxmapindex = 32;
```

Next, plug these values into the equations to obtain

```
cmin = -1;
cmax = 3;
```

Finally, set the axes object's *CLim* property to [-1 3] by typing

```
caxis([-1 3]);
```

Plate 10 was created with this technique with the code shown below.

M-File

```
% Define color map.
colormap([hsv(32);hot(32);cool(32);flag(32)]);

% Create first subplot using first quarter of color map.
subplot(221)
x = 0:.02:5*pi;
y = sin(x);
surface([x;x],[y;y],0*[x;x],[y;y],...

'facecolor','none','edgecolor','flat','linewidth',3)
set(gca,'box','on');
axis([min(x) max(x) [min(y) max(y)]*1.1 ])
% Use equations found at beginning of this section.
cmin = min(y) - (1 - 1)*(max(y)-min(y))/(32 - 1 + 1);
cmax = min(y) + (128 - 1 + 1)*(max(y)-min(y))/(32 - 1 +
1);
caxis([cmin cmax]);

% Create second subplot using middle half of color map.
subplot(222)
u = 0:.02:10*pi;
x = exp(-.05*u).*cos(u);
y = exp(-.05*u).*sin(u);
z = .05*u;
surface([x;x],[y;y],[z;z],[z;z],...

'facecolor','none','edgecolor','flat','linewidth',2)
view(3);axis([-1 1 -1 1 0 1.5]);grid;
set(gca,'ztick',[0 .5 1 1.5])
% Use equations found at beginning of this section.
cmin = min(z)- (33 - 1)*(max(z)-min(z))/(96 - 33 + 1);
cmax = min(z)+ (128 - 33 + 1)*(max(z)-min(z))/(96 - 33 +
1);
caxis([cmin cmax]);

% Create third subplot using last quarter of color map.
subplot(223)
x = 0:.2:5*pi;
y = sin(x);
surface([x;x],[y;y],0*[x;x],[y;y],...

'facecolor','none','edgecolor','flat','linewidth',.5)
set(gca,'box','on');
```

```
axis([0 5*pi -1.1 1.1])
% Use equations found at beginning of this section.
cmin = min(y)- (97 - 1)*(max(y)-min(y))/(128 - 97 + 1);
cmax = min(y)+ (128 - 97 + 1)*(max(y)-min(y))/(128 - 97 +
1);
caxis([cmin cmax])
caxis([-7 1])

% Create fourth subplot using middle half of color map..
subplot(224)
u = 0:.02:10*pi;
x = exp(-.05*u).*cos(u);
y = exp(-.05*u).*sin(u);
z = .05*u;
surface([x;x],[y;y],[z;z],[z;z],...

'facecolor','none','edgecolor','flat','linewidth',2)
% Use equations found at beginning of this section.
cmin = min(z) - (33 - 1)*(max(z)-min(z))/(96 - 33 + 1);
cmax = min(z) + (128 - 33 + 1)*(max(z)-min(z))/(96 - 33 +
1);
caxis([cmin cmax])
set(gca,'box','on');
```

8.2.10 Texture Mapping

So far we have looked only at defining the vertex colors for surfaces which could then be used to color a particular quadrilateral with a flat or bilinearly interpolated color. In addition, we have also explored image objects and seen that we are limited to viewing them in the xy-plane and only in 2-dimensional perspectives. Now wouldn't it be great if we could wrap an image over any surface of our choosing? Well, MATLAB does provide a method for doing this, and it is called texture mapping. Reviewing the surface object properties, recall that the *FaceColor* property has the following five choices:

[none | {flat} | interp | texturemap] - or - a ColorSpec

When the *FaceColor* property is set to "flat", the size of the *CData* matrix must either be the same as the matrix stored in the *ZData* property or have one less row and column. If the *FaceColor* property is set to "interp", the *CData* matrix must be of the same dimensions as the *ZData* matrix of the surface object. Only when the *FaceColor* property is set to "texturemap" are you unlimited as to the size of the *CData* matrix.

Speed!

If the FaceColor Property is set to...	then the CData matrix size is...
none	unrestricted - (makes no difference)
flat	size(CData) = size(ZData) or
	size(CData) = size(ZData) - 1
interp	size(CData) = size(ZData)
texturemap	unrestricted
ColorSpec	unrestricted - (makes no difference)

We shall now demonstrate how to take a surface and an image and combine them so that the image is mapped to the surface. The surface we will use is the portion of a cylinder produced by,

```
[x,y,z] = cylinder(1,30);
surface_handle = surf(x(1:2,15:30),...
                      y(1:2,15:30),...
                      z(1:2,15:30));
```

and the image will be the MATLAB clown of Chapter 5 notoriety. Our subjects are shown in Figure 8.20.

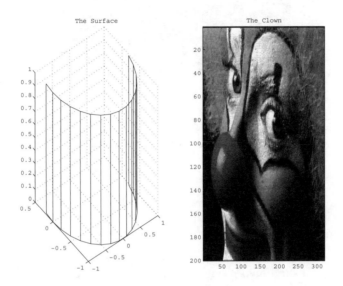

Figure 8.20 We will map the image to the surface.

The procedure here is to first load the clown image and then place it into the *CData* property of the cylindrical type surface. Here is the code that will do just that.

```
load clown
[x,y,z] = cylinder(1,30);
figure('Colormap',map);
surface_handle = surf(x(1:2,15:30),...
                      y(1:2,15:30),...
                      z(1:2,15:30));
set(surface_handle,'FaceColor','texturemap','cdata',flipu
d(X));
set(gca,'box','on');
```

The result is shown in Figure 8.21.

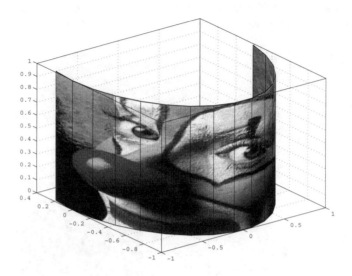

Figure 8.21 A texture mapped surface.

If at a later time you want to alter the *FaceColor* property from "texturemap" to one of the other possible settings, make sure that the *CData* matrix is appropriately resized so that rendering errors do not occur.

Warning!

In the next example we will load in some topographic data that comes with MATLAB and texture map it onto a wavy map-like surface. The difference in this case is that the data was not stored in an image format (i.e., there are elements of the data matrix that are non-integer or negative). Therefore, we see that any data set you have can be texture mapped to a surface object.

Here is the code that will create the surface as shown in Figure 8.22.

```
[x,y] = meshgrid(1:20);
z = (x-10).^3+(y-10).^3;
s1 =surf(z)
axis('off')
```

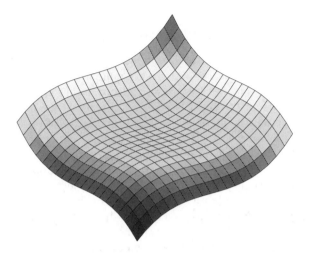

Figure 8.22 The surface object that will be texture mapped.

We then load and apply the texture map data with the following code and produce Figure 8.23 which can also be seen in color in Plate 11.

```
load topo
set(s1,'facecolor','texturemap','cdata',topo)
colormap(topomap1);
```

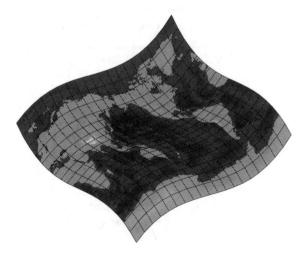

Figure 8.23 A non-image data set mapped to a surface object.

8.3   Modeling Object Lighting

In Chapter 4, before we introduced handle graphics, we looked at 3-D plots and learned that in most cases we can get quite impressive results using high-level 3-D plotting commands like **mesh**, and **surf**. In this chapter, having armed you with the concept of object handles in Chapter 7, we have already taken you into the deep areas of color. The next property we will now study in some detail is object lighting. Light, unlike color, is not a direct property, but rather what we might call an indirect property. That is, we can't point to an object in our figures and say, "that's a light object," but rather the appearance of surface and patch objects is affected by the unseen light objects present in our figure. Where the light is, its intensity, color, different lighting models etc., is the subject of this section. First we will look at the properties of the light object, then we will exercise and demonstrate those properties with functions like **camlight, lightangle, surfl, surfnorm, material, diffuse**, and **specular**.

8.3.1   Light Properties

In order to reveal detail and accentuate important information in a 3-dimensional scene, MATLAB provides the capability of adding lighting effects with the graphics object light. This object can be placed in relation to or directed toward other graphics objects in the scene. The visual attributes of patch and surface objects are affected by light objects. To create a light object, you use the function **light**. The properties of this object were only briefly mentioned in Chapter 7, but are presented here and summarized in the following table.

Property	Read Only	ValueType/Options	Format
Properties Defining the Light Object			
Color	No	[R G B] or color string	3-element vector or row
Position	No	x-, y-, z-coordinates in axes units Default: [1 0 1]	3-element row
Style	No	[{infinite} | local]	string

A light by default will be a white light, but you may alter it to whatever color you want with its *Color* property. This property may be set to a three-element RGB value, such as [1 0 0] (which would make a red light), or a string specifying the color you want, like 'red'.

The meaning of the *Position* property will depend on whether you have set the *Style* property to "infinite" or "local". The default *Style* property value is "infinite", which means that the light source is placed at infinity and the rays radiating from the source are pointing in the direction specified by the *Position* property. If the Style property is set to "local", the light source will be located at the point specified by the *Position* property, and will radiate in all directions from that point.

8.3.2 Functions that Make Use of Light

Lighting models are used to create highlights to curvatures and faces of a surface graphics object. Lighting models determine the amount of reflectance that occurs from a light source at a specific location with respect to the surface. The reflectance is then a measure that can be scaled and transformed into indices pointing to particular rows of a color map.

The first function that we will look at is **surfl**. This function creates a 3-dimensional surface plot where the shading is based on a mixture of *diffuse*, *specular*, and *ambient* lighting models. Using **surfl** is practically automatic and so requires the least amount of specification by the user. There are several ways that **surfl** can be used. If you supply only the height information, Z, to the surface with **surfl**(Z, 'light'), the lighting will, by default, be 45 degrees counterclockwise in azimuth from the current view point, i.e., if [az,el] = **view.** With **surfl**, in addition to the surface object, there will be created a white light source placed at infinity with its *Position* property set to [0 -0.707 0.707]. You can use **surfl** without 'light', but it will not create a light object; instead, it will alter the color map of the surface object to make it *look* like there is a light object in the specified direction.

Just as with **surf**, you may also specify the x- and y-coordinates of the vertices with **surfl**(X,Y,Z). If you need to specify a different light source direction, you can use **surfl**(Z,*s*) or **surfl**(X,Y,Z,*s*), where *s* is either a 3-element vector, [Sx Sy Sz], or a 2-element vector, [az el], defining the direction from the object to the light source. Consider Figure 8.24 which is generated with the following code:

```
[X,Y] = meshgrid(-3:.1:3);
Z = sin(2*X).*sin(Y).*sqrt(X.^2 + Y.^2);
surface_handle = surfl(X,Y,Z,[0 30]);
shading flat;
colormap(gray);
set(gca,'box','on');
```

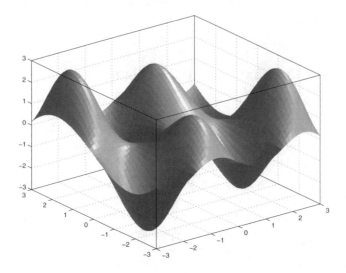

Figure 8.24 Automatic lighting provided by surfl.

Typically, the best results are obtained with flat or interpolated shading applied to the surface and that the surface be defined on a fine grid. In addition, simple color maps (i.e., maps that are made up of many shades of the same color) should be used. The *gray*, *bone*, and *pink* color maps are usually ideal for these types of plots.

To better explain how the light direction vector, *s*, is used, the following table contains example values of *s* and their interpretation.

If the light source vector s is...	*then the source of the light is...*
[0 0 1] or [0 90]	directly above surface
[0 0 ⁻1] or [0 ⁻90]	directly below surface
[1 0 0] or [90 0]	pointed down the x-axis
[0 ⁻1 1] or [0 45]	pointed at a slight angle down on the surface

In the code that produced Figure 8.24, we could have also generated a light object, by using

```
surface_handle = surfl(X,Y,Z,[0 30],'light');
```

instead of

```
surface_handle = surfl(X,Y,Z,[0 30]);
```

Try the previous example and create a light object. Then to find the graphics handle of the light object, use

```
h_light(1)=findobj('type','light');
```

Now we could make this light object radiate blue light by using the command

```
set(h_light(1),'color','blue');
```

We could also create another light object off to the right using the command

```
h_light(2) = light('color','green','style','local');
```

and generate the following figure which can also be seen in Plate 12.

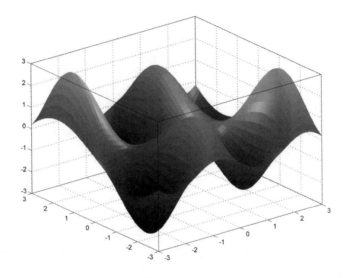

Figure 8.25 Surface with one green light and one blue light.

If you want more control over the lighting model, you can also use **surfl**(X,Y,Z,s,k), where k is a 4-element vector that defines the ambient, diffuse, specular, and spread coefficients ([k_a k_d k_s spread] that normally default to the values [0.55 0.6 0.4 10]). These coefficients are used to apply a weighting to reflectance values that are returned from the various light model functions.

The functions that generate the reflectance values all utilize the normals to the surface at the vertex locations. The command **surfnorm** (which we first saw in an example in Chapter 4) has been created to compute, and if desired display, the 3-dimensional surface normals. If you want to display the normals of a matrix, Z, that represent the height of the surface at the vertex locations, use **surfnorm**(Z).

Each patch and surface object already contains the normals calculated at the vertices of the object. These are stored in the *VertexNormals* property. As long as the object's *NormalMode* is set to *auto*, MATLAB will recalculate them any time you make changes to the object that would affect the normals.

The x- and y-axis locations of the vertices are assumed to be the row and column numbers of the Z matrix elements. If the x- and y-coordinates of the surface are known, you can use **surfnorm**(X,Y,Z). For example, if we wanted to display the normals of a sphere, we could type

```
[x,y,z] = sphere(10);
surfnorm(x,y,z);
grid on;
```

Which will produce Figure 8.26.

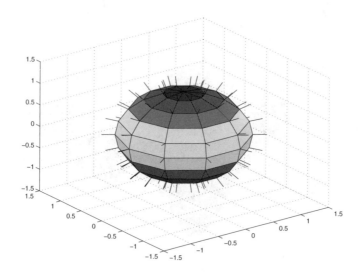

Figure 8.26 Surface normals are used to determine how an object reflects light.

To store the components of the normals, use

```
[nx,ny,nz] = surfnorm(z);
```

or

```
[nx,ny,nz] = surfnorm(x,y,z);
```

After storing the normal components, if you are generating your own normals for the surface and patch objects, you can make use of the functions **diffuse** or **specular** depending on your lighting model preference. Both models generate reflectance values in the range zero to one, where a zero means that none of the light is reflected and a one means that all of the light is reflected. In modern versions of MATLAB, since the normals for each patch and surface object are stored in the object's *VertexNormals* property, you do not need to calculate them. Also, it is important to remember that the edge (*EdgeLighting*) and face lighting (*FaceLighting*) that you have specified for your objects has a very significant visual effect. More often than not, you will find it convenient to use **lighting none**, **lighting flat**, **lighting gouraud**, or **lighting phong** to specify the edge and face lighting properties. The *flat* lighting method forces the entire face of a surface or patch object to have the same color at each pixel. The *gouraud* lighting method determines the colors at the vertices of the faces using the normals and then interpolates these colors across the face, while *phong* interpolates the normals across the face and calculates the color at each pixel.

8.3.2.1 Lighting Commands

We need to mention here, that in addition to setting values in patch and surface objects lighting properties, there are several MATLAB commands (such as **lighting** that you have already seen) that can not only let you create a light object, but quickly set some interesting lighting effects. The following table lists the MATLAB lighting commands.

Command	Description	Arguments
camlight	sets the position of a light, creates one if it doesn't exist	headlight, right, left, [az,el]
light	creates a light object	'Property1','Value1'',...
lightangle	positions a light in spherical coordinates, creates a light if it doesn't exist	az, el
lighting	select a lighting method	flat, gourard, phong, none
material	sets the reflectance property	shiny, dull, metal, or [ka kd ks n sc] (see discussion)

These commands are best illustrated with an example. The following code will quickly generate the surface object shown in Figure 8.27.

```
ezsurf('sin(sqrt(x^2+y^2))/sqrt(x^2+y^2)',...
    [-6*pi,6*pi]);
```

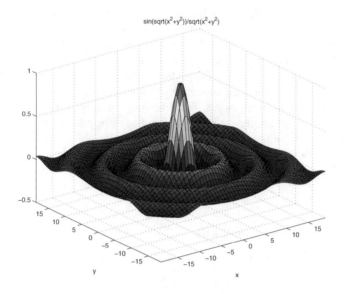

Figure 8.27 A surface we will use.

Now try the following and see what happens.

```
shading interp
lightangle(-45,30)
material dull
```

Play around with these commands to see what they can do.

8.3.3   Lighting Models

In MATLAB you can achieve the effects of lighting in two ways: one is to create a light object with one of the light object creating commands (a light object can also be created simultaneously with **surf**, **mesh**, **pcolor**, **fill**, **fill3**, **surface**, and **patch**); the other is to use a *lighting model*. Lighting models do not rely on the presence of a light object, instead the appearance of light is created by altering the *CData* values of an object. Three different functions apply these models to the surface or patch object. They are **diffuse**, **ambient**, and **specular**. Remember, instead of applying these models, you could achieve similar results by creating light objects and specifying the appropriate properties of the patch or surface objects. We will point this out in the following discussions.

8.3.3.1   The Diffuse Lighting Model

The **diffuse** function uses an algorithm that generates reflectance values based upon Lambert's Law for diffuse surfaces. This function calculates the reflectance as a function of the angle between the surface normals and the direction of the light source (reflectance = cos(θ), where θ is the angle). When the normal and light source directions are the same, the reflectance will be the largest.

```
diffuse_reflection  = diffuse(nx,ny,nz,s);
```

where nx, ny, and nz are normal components calculated by surfnorm, and s is the direction of the light source with respect to the surface. The light source direction can be provided as either a 3-element (x,y,z) or 2-element (az,el) vector. The following code will create a sphere surface object with diffuse lighting characteristics as shown in Figure 8.28.

```
[x,y,z] = sphere(20);
[nx,ny,nz] = surfnorm(x,y,z);
diffuse_refl = diffuse(nx,ny,nz,[0.5 -1 1]);
surf(x,y,z,diffuse_refl);
shading interp;
colormap(gray);
```

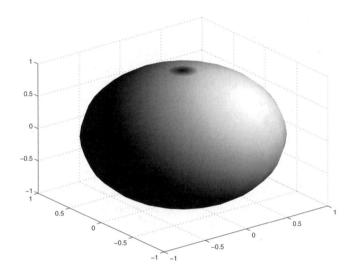

Figure 8.28 Applying a diffuse model – no light object created.

The method just shown achieves its results by altering the color data values (*CData*) of an object and therefore does not require a light object. You can verify that no light object was created by typing

```
findobj('type','light')
```

which will return

```
ans =

   Empty matrix: 0-by-1
```

You could control the diffuse reflection using the patch or surface object's *DiffuseStrength* property. However, for this property to have an effect on the object, there must be a light source (i.e., you must create a light object with one of the light creating commands). The only restriction on the value you assign to this property is that it must be greater than or equal to zero. One way to describe diffuse reflection is to think of it as the spread or creeping of the light across the object. By increasing the value, you increase the intensity or the spread of the diffuse reflection.

8.3.3.2   The Ambient Lighting Model

Ambient light shines across all surface and patch objects in a uniform manner. The color of the ambient light is by default white (i.e., [1 1 1]) and can be changed only on a per axes object basis, since the axes object stores the color in its *AmbientLightColor* property. The ambient light's relative effect on the objects for a given axes object can be controlled by setting the patch and surface *AmbientStrength* property with a value greater than or equal to zero. Regardless of the value set to the *AmbientStrength*, a light object must be present in the scene if you want to see the ambient light.

Assuming a light object is present, a zero *AmbientStrength* setting means that the ambient light has no effect on the object. Pixels of the object that have non-zero color components (RGB) corresponding to non-zero components of the ambient light color will be affected by the ambient light. You can think of it as a multiplicative effect, whereby the pixel's red, green, and blue components, [Rp Gp Bp], and the ambient light's components [Ra Ga Ba], and the *AbientStrength*, A, are multiplied to determine the pixels final color intensity,

```
Pixel Color Intensity = [Rp Gp Bp] .* [Ra Ga Ba] * A
```

For example, if an object is green, then none of the ambient light's red and blue components will contribute to the light on that object. However, the object's green components will get brighter. Experiment with the following code which will create Figure 8.29:

```
z=ones(2,4);
c(:,:,1) = [0 0 0; 0 0 0]; % Red component of each face
c(:,:,2) = [1 .6 .3; 1 .6 .3]; % Green component
c(:,:,3) = [0 0 0; 0 0 0];     % Blue component
s=surf(z,c);
set(s,'diffusestrength',0,...
       'specularstrength',0,...
       'ambientstrength',1);
l=light;
axis equal
```

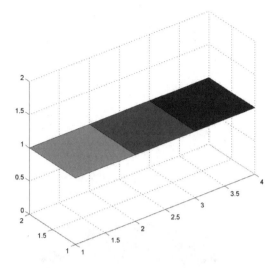

Figure 8.29 Test surface for ambient color effects.

You will notice that if you change only the red and blue components of the axes object's *AmbientColor* using,

```
set(gca,'AmbientColor',[.5 1 0]);
```

the three shades of green will not change. However, if you make the green component less than 1, the shades of green will get darker.

Try increasing the surface's *AmbientStrength*, and you will notice that the shades of green will all be the same brightness when using values greater than 3.33 (I = 1/(Rg*Ra) = 1/(1*0.3)), since you have pushed the brightness of all three faces to their maximum value of 1.

8.3.3.3   The Specular Lighting Model

The **specular** function's algorithm generates the largest reflectance values when the normals are in the direction halfway between the light source and the viewer. To use this function, you can use either

```
specular_reflectance = specular(nx,ny,nz,s,v);
```

or

```
specular_reflectance = specular(nx,ny,nz,s,v,spread);
```

where nx, ny, and nz are determined with surfnorm, s is the direction of the light source from the surface, v is the direction of the viewer from the surface, and spread is a measure of how quickly the reflectance falls from the peak reflectance value. The s and v variables must be defined as either a 3-element ([x y z]) or 2-element ([azimuth elevation]) directional vector. The spread variable defaults to 10 if not supplied. Spread values larger than 10 force the reflectance to fall more quickly. The following code demonstrates using **specular** with the results shown in Figure 8.30.

```
[x,y,z] = sphere(20);
[nx,ny,nz] = surfnorm(x,y,z);
specular_refl = specular(nx,ny,nz,[0.5 -1 1],[-37.5
30],1);
surf(x,y,z,specular_refl);
shading interp;
colormap(gray);
```

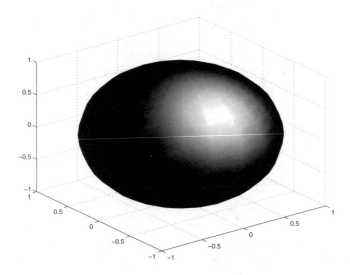

Figure 8.30 A sphere enhanced by specular lighting – again, no light object created.

Again, we point out that since no light object creating command was issued, there is no light object in this figure. The **specular** function changed the *CData* values to give the appearance of lighting. If we had a light object, we could create diffuse reflection using the patch or surface object's *SpecularStrength*, *SpecularExponent*, and *SpecularColorReflectance* properties.   For these properties to have an effect on the object, there must be at least one light source (i.e., you must create a light object with the light function). By increasing the *SpecularStrength* (any finite value greater than or equal to zero), you increase the intensity of the the specular reflection.   By increasing the *SpecularExponent* (a value greater than zero), you increase the size of the "hot spot".   The *SpecularColorReflectance* property lets you decide the fraction (using values between 0 and 1) of the color of the specular reflectance. Values closer to 0 use more of the object color in the reflectance, while values closer to one use more of the light color as a percentage of the reflected color.

8.3.3.4   Combining Lighting Models

The best way to learn about the different lighting models is to experiment. You should also remember that nothing prevents you from adding together the reflectance values generated from the **specular** or **diffuse** functions. In fact, very interesting and visually pleasing 3-dimensional plots can be created with combinations of multiple light sources and different reflectance models. For example, Figure 8.31 (see Plate 13) was created with the following code.

```
n = 20;
t = (0:n)'*2*pi/n;
x = [cos(0:.1:(2*pi)) ones(1,10) -1 -2 -3]+3;
y = [fliplr(1:(length(x)-3)) 1 1 1 ];
t = (0:20)'*2*pi/20;
xx = cos(t)*x;
yy = sin(t)*x;
zz = ones(n+1,1)*y;
[nx,ny,nz] = surfnorm(xx,yy,zz);
reflectance = specular(nx,ny,nz,[-80.5 30],...
[-70 -30],5) + diffuse(nx,ny,nz,[230 40]);
figure('colormap',hot);
surface_handle = surf(xx,yy,zz,reflectance);
shading interp
axis('off');
```

Figure 8.31 Mixing specular and diffuse reflectance models.

8.3.3.5   A Final Word on Light Objects

Now that you have seen the properties of light objects, and have seen the results of light models, you probably have a good feel for the interaction between a patch or surface object and light – whether modeled light or a light object. Here is an example that introduces a light object into a figure, and by setting various surface properties creates a very natural-looking result. Consider the following code, recalling Figure 8.27.

```
ezsurf('sin(sqrt(x^2+y^2))/sqrt(x^2+y^2)',...
    [-6*pi,6*pi]);
view(0,75);
shading interp
%create a light object
h_light=lightangle(-45,30);
%use findobj to get the surface handle
h_surf=findobj('type','surface');
%now change the surface properties that
%are affected by light
set(h_surf,'FaceLighting','phong',...
    'AmbientStrength',0.3,...
    'DiffuseStrength',0.6,...
    'SpecularStrength',0.7,...
    'SpecularExponent',0.25,...
    'BackFaceLighting','unlit')
```

Try altering the properties of h_surf and h_light to see what you can do.

Remember, when you are using the light models, you are changing the patch or surface object's *CData*. When you use a light object, your patch or surface object isn't being inherently changed. The choice as to which method to use is dependent on your intended purpose of your patch and surface objects.

Warning!

8.3.4 Creating Color Varying Lines with Surface Objects

In Section 8.2.9 we explored the pseudocolor plot. You probably have been hoping that there is some way you can have lines change color to reflect different values just like with surface objects.

Even though you cannot use color maps with line objects, this does not mean that you cannot create graphics that look like a line with varying colors. The next couple of examples in this section show how you can create lines whose colors are specified by mathematical expressions.

Power!

The interesting point is that we will not use the line object; rather, we will use a thin surface object and create what will be called a *virtual line*. Since surface objects can be defined by x-, y-, and z-axis data, we can create virtual lines that are in either the 2- or 3-dimensional plotted perspectives. The following example shows how to create a virtual line in which the color is a function of the y-coordinate data values.

```
% Define the coordinates of the virtual line
x = 0:.02:5*pi;
y = sin(x);
z = 0*x;
```

```
% Define the color values of each coordinate of the line
c = y;
% Generate the plot.
figure;
surface([x;x],[y;y],[z;z],[c;c],...
    'facecolor','none',...
    'edgecolor','flat',...
    'linewidth',3);
set(gca,'box','on','xtick',[0:pi:5*pi],...
        'xticklabels','0|pi|2pi|3pi|4pi|5pi');
axis([0 5*pi -1.1 1.1])
```

From the code you see that the surface object's *FaceColor* is set to "none" and the *EdgeColor* is "flat". You can just as easily set the *EdgeColor* to "interp"; however, it will take longer for the line object to render and with some versions of MATLAB, you will not be able to control the *LineWidth*. Figure 8.32 shows the result, although you might want to look at Plate 14 to better appreciate it.

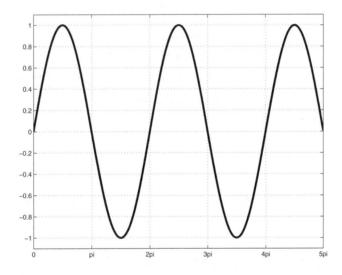

Figure 8.32 Making a virtual line with surface to create a line where the color changes as a function of the y-coordinate.

We should point out that it is not necessary to set the *FaceColor* to "none", unless you want to create several lines with the same surface object. You may have realized that you can just as easily create multiple lines in which each line's color varies as a function of the x, y, or z data. Each row or column of the matrices in the *XData*, *YData*, and *ZData* can be used to represent a line by setting the *MeshStyle*, respectively, to row or column instead of its default value of both.

For example, we can create the plot shown in Figure 8.33 (see Plate 15 for the color representation) with the following code.

```
u = 0:.2:4*pi;
x = cos(u);
y = sin(u);
z = u;
figure('colormap',cool(64));
h_surface = surface([0*x;x],[0*y;y],[z;z],...
    'facecolor','none',...
    'edgecolor','flat',...
    'meshstyle','row',...
    'linewidth',3);
view([-40 40]);
grid on;
```

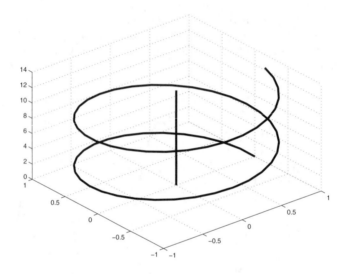

Figure 8.33 Creating multiple color lines with one surface object.

8.4   Object Transparency

Transparency is a powerful visualization technique that allows you to see an object while at the same time see information that would otherwise be obscured if the object was fully opaque. In MATLAB, you can create varying degrees of transparency, based on your needs, in image, patch, and surface objects. Transparency is useful not only in seeing what information lies behind or within some other (as in volume plots), but also can be used as another dimension for data.

8.4.1 Alpha Properties

The following table lists the object properties that affect transparency.

Property	Read Only	ValueType/Options	Format
AlphaData	No	m-by-n matrix of transparency data for image and surface objects	matrix
AlphaDataMapping	No	none | direct | scaled none = default for images scaled = default for patches	row
FaceAlpha	No	Transparency for faces	row or scalar
EdgeAlpha	No	Transparency for edges	row or scalar
FaceVertexAlphaData	No	Alpha data property for patches	row or scalar
ALim	No	Alpha axis limits	vector
ALimMode	No	Alpha axis limits mode	row
Alphamap	No	Figure Alphamap	matrix

8.4.1.1   AlphaData

Just like *CData* contains color data for surfaces, each element of the alpha data contained in *AlphaData* is mapped to a transparency value in the *Alphamap*. This property applies to surface and image objects (*FaceVertexAlphaData* is its counterpart for patches).

8.4.1.2   Alphamap

This is the set of alpha values (numbers between 0 and 1) that MATLAB uses to determine transparency for an object, i.e., the alpha map. The data stored in *Alphamap* is an m-by-1 array where the first row is the first alpha value and m is the last. The default alpha map has 64 values linearly progressing from 0 to 1. *Alphamap* applies to surface, patch, and image objects.

8.4.1.3   ALim

ALim is an axes property that applies to any axes children using alpha data. It is a two-element vector, stated as [*amin amax*], that specifies how the alpha map (in *AlphaData*) is mapped. The value of *amin* is mapped to the first alpha map value, and *amax* is the value mapped to the last alpha map value. Data values between are linearly interpolated across the alpha map, while values beyond the limits are "clamped" to the limits. Setting *ALim* will set *ALimMode* to "manual".

8.4.1.4   ALimMode

Working with *ALim*, *ALimMode*, can take the values of "auto" or "manual". In the default "auto" mode, the *ALim* property is automatically set to span the range of all objects' *AlphaData* (for surface objects) or *FaceVertexAlphaData* (for patch objects). In the "manual" mode, the value of *ALim* is not changed when the *AlphaData* limits of axes children change.

8.4.1.5   AlphaDataMapping

Hand-in-hand with *AlphaData* is the property *AlphaDataMapping* which is used to determine how the alpha data is to be interpreted. The three choices are:

none – transparency values are clamped to be between 0 and 1, the default for images.

scaled - forces the *AlphaData* to span the range of *ALim*, the default for patches.

direct – uses the *AlphaData* as indices directly into the alpha map.

8.4.1.6   FaceAlpha

FaceAlpha specifies the type of transparency to be used for a patch or surface face. Since this applies only to patch and surface objects, it uses the data stored in *FaceVertexAlphaData*. It can be any one of the three following values:

A scalar value – a number between 0 and 1 where 0 is fully transparent, i.e., invisible, and 1 is completely opaque (the default).

flat – the values stored in *FaceVertexAlphaData* determine the transparency at each face.

interp – binary interpolation of the alpha data in *FaceVertexAlphaData* at each vertex determines the transparency of each face.

texturemap – for surface objects only, uses transparency for the texture map.

8.4.1.7   EdgeAlpha

Similar to *FaceAlpha*, *EdgeAlpha* lets you control the transparency of the edges of patch faces and surfaces. The possible values for patch objects are:

scalar - a single scalar value between 0 and 1 where 1 (the default) is fully opaque and 0 is invisible.

flat - alpha data, i.e., the contents of *FaceVertexAlphaData*, of each vertex controls the transparency of the edge that follows it.

interp - linear interpolation of the alpha data (*FaceVertexAlphaData*) at each vertex determines the transparency of the edge.

The only difference with surface objects is that instead of *FaceVertexAlphaData*, the "flat" and "interp" options apply to *AlphaData*.

8.4.1.8   FaceVertexAlphaData

FaceVertexAlphaData is a m-by-1 matrix that specifies the face and vertex transparency for patch objects (as defined by *Faces* and *Vertices* properties), the interpretation of which depends on the dimensions of the data. The contents of *FaceVertexAlphaData* can be:

scalar – a single scalar value that will be applied to each patch.

matrix – m-by-1 matrix specifying one transparency value per face, where m is the number of rows in the *Faces* property or the number or rows in the *Vertices* property.

8.4.2   Alpha Functions

Alpha functions are those functions that will create or affect transparency effects in surface, patch, or image objects. The following subsections present a summary of the three alpha functions, namely **alpha**, **alphamap**, and **alim.**

8.4.2.1   alpha

An object's alpha data, i.e., the value stored in *AlphaData*, can be specified with the **alpha** function.. The possible inputs to **alpha** are given in the following table:

Usage	Interpretation
Specifying a single alpha value for the entire object.	
alpha(scalar)	Sets the face alpha to be the value of scalar where 0 = invisible and 1= opaque
alpha('flat')	face alpha set to 'flat'
alpha('interp')	face alpha set to 'interp'
alpha('texture')	face alpha set to a 'texture'
alpha('opaque')	same as alpha(1)
alpha('clear')	same as alpha(0)
Specifying a different alpha value for each element in an object's data.	
alpha(matrix)	alpha data set to matrix
alpha('x')	alpha data set to x data
alpha('y')	alpha data set to y data
alpha('z')	alpha data set to z data
alpha('color')	alpha data set to the same as the color data
alpha('rand')	alpha data set to random values
Specifying the *AlphaDataMappingMethod* property.	
alpha('scaled')	sets *AlphaDataMappingMethod* to 'scaled'
alpha('direct')	sets *AlphaDataMappingMethod* to 'direct'
alpha('none')	sets *AlphaDataMappingMethod* to 'none'

8.4.2.2   alphamap

The function **alphamap** is provided to let you set an object's *Alphamap* property. The following table shows the different usage specifications for **alphamap**.

Usage	Interpretation
Forms that create a new alpha map.	
alphamap('default')	sets *Alphamap* to default values.
alphamap('rampup')	creates a linear alpha map with increasing opacity
alphamap('rampdown')	creates a linear alpha map with decreasing opacity
alphamap('vup')	creates an alpha map that is transparent in the center, and linearly increasing to the beginning and end
alphamap('vdown')	creates an alpha map that is opaque in the center, and linearly decreasing to the beginning and end
alphamap(*matrix*)	creates a new alpha map with the values of *matrix*.
Forms that modify the existing alpha map.	
alphamap('increase')	makes the alpha map more opaque
alphamap('decrease')	makes the alpha map more transparent
alphamap('spin')	rotates the alpha map
alphamap('')	creates an alpha map that is transparent in the center, and linearly increasing to the beginning and end
alphamap('vdown')	creates an alpha map that is opaque in the center, and linearly decreasing to the beginning and end
alphamap(param, length)	affects parameters that create new alpha maps making them *length* long
alphamap(change, delta)	changes alpha map parameters by *delta*
alphamap(figure, param, length | change| change, delta)	sets a figure's alpha map "param"
Forms that return information.	
amap = alphamap	returns the current alpha map
amap= alphamap(figure)	returns the current alpha map from the handle *figure*
amap = alphamap(param)	returns the alpha map based on *param* without setting the property

8.4.2.3   alim

The function **alim** can be used to set the value of the *ALim* and *ALimMode* properties. The general form of use is **alim**([*amin amax*]) which will set the alpha limits. You can also use it as **alim**(*mode*) where *mode* is one of the valid *ALimMode* strings ("auto" or "manual").

The **alim** function can also be used to return the contents of the *ALim* property or the setting in *ALimMode*. Typing al=**alim** will return the alpha limits of the current axis, i.e., the data stored in the *ALim* property.

8.4.3 Setting a Single Transparency Value

As you can see, you can use **alpha** to specify the contents of *AlphaData* and to set *AlphaDataMappingMethod*. The **alpha** function can be very convenient to use whenever you want a quick transparency of equal value across an object as is demonstrated with the following code, which harkens back to the isosurface plot of Figure 4.42 in Chapter 4. The result is shown in Figure 8.34 and in color in Plate 16.

```
[x y z v] = flow;
h_p=patch(isosurface(x, y, z, v, -3));
daspect([1 1 1]);
set(h_p, 'FaceColor','green','EdgeColor','none');
view(3)
axis tight
grid on
camlight; lighting phong
alpha(.5) %set alpha for all
```

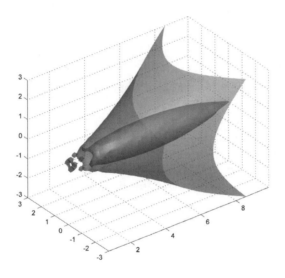

Figure 8.34 Setting a single transparency value with alpha.

8.4.4 Mapping Data to Transparency

We have already seen how to use our plot data by mapping it to *CData* so that the color is a function of the data. We can do the same sort of thing with *AlphaData,* essentially making the degree of transparency a function of some data. Consider a surface similar to mesh plot of Figure 4.3 created by

```
[X,Y] = meshgrid(linspace(0,2*pi,50),linspace(0,pi,50));
Z = sin(X).*cos(Y);
hsurf=surf(X,Y,Z);
```

```
set(hsurf,'CData',gradient(Z));
set(hsurf,'AlphaData',gradient(Z));
set(hsurf,'FaceAlpha','flat');
set(hsurf,'EdgeColor','none');
```

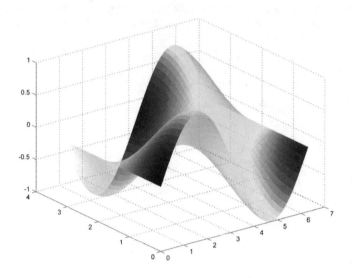

Figure 8.35 Using data to specify transparency.

In this example we used the **gradient** function to indicate by both color and transparency where the slopes of the curves are equal. Try setting the *FaceColor* to a constant, e.g., 'blue', and set *EdgeColor* to [0.8 0.8 0.8]. Can you change the *AlphaData* to another function of x, y, or z?

8.5   Illustrative Problems

1. Experiment with the different *EdgeColor* and *FaceColor* settings so that you become more familiar with their effects. Also look at altering the color axis limits by setting the *CLim* property. For example, try

```
set(gca,'clim',[-5 5])
```

   or

```
set(gca,'clim',[-20 10])
```

   If the results don't make sense, go back to Equation 8.1 and calculate the index to the color map using the new *cmin* and *cmax* values. You may also want to try adding more colors to the color map with

```
set(gcf, 'colormap',jet(20); set(gca,'climmode','auto')).
```

2. Create a surface plot and apply various intensity levels to it using the **brighten** function and some of your favorite color maps.

9 ANIMATION

IN THIS CHAPTER...
9.1 FRAME-BY-FRAME CAPTURE AND PLAYBACK 361
9.2 ON-THE-FLY GRAPHICS OBJECT MANIPULATION 372
9.3 CHOOSING THE RIGHT TECHNIQUE..383

With our study of MATLAB graphics we have concerned ourselves with the visualization of information. As such we have explored 2-D and 3-D presentations of data, using color and transparency to add more dimensions or to emphasize aspects of our plots, and now we add yet another dimension – motion. As you are about to learn, animating a graphic can be both enjoyable to watch and provide great insight into the nature of data. Graphical representations of a physical object or interactions between objects that a simulation is attempting to model allow someone who did not have a role in the design of the simulation to follow along and gain an intuitive feel for the results that are being generated. Additionally, we are becoming an increasingingly graphically oriented world, and we take great comfort in "seeing" information. Although this sense of comfort is purely psychological and really shouldn't play a role in the amount of confidence that is put into a program, in many cases it does carry significant weight. In order to fully comprehend and appreciate all of the animation capabilities that the MATLAB graphics engine provides, it will be even more important for you to implement the MATLAB code found in this chapter. At the very least, run the programs as you are reading along and you will gain a great appreciation for the strength animation can add to your visualizations.

In this chapter we will focus on two different methods of creating an animation. The first is a frame-by-frame capture and playback technique, like frames in a movie, and the second is a true on-the-fly graphics coordinate location manipulation. Both have their advantages and disadvantages and are geared for animating different types of graphics objects in various types of circumstances. You will learn how to create animations with both of these techniques and will understand when it is advantageous to use one or the other. You will also learn how to translate your MATLAB movies into standard AVI format, which will allow you to make your movies easily playable on any computer and conveniently insertable to presentation applications like Microsoft's PowerPoint.

9.1   Frame-by-Frame Capture and Playback

To create a movie, there are only several MATLAB commands that you will need to use. There is the function **moviein**, which preallocates enough memory to capture a specified number of frames of a movie, the function **getframe** for capturing the individual frames, and the function **movie** for playing back a series of captured frames. The manner in which the two commands are used is fairly straightforward; however, to get the results you want, you need to learn about the nuances of each command and understand what each command is doing.

Although it is not necessary to preallocate memory for the data matrix that stores the frames of the movie, it is recommended because the amounts of data that must be stored are usually large and because a slight speed improvement with regard to the time it takes to add a movie's frame to the movie data matrix can be realized. The simplest way to preallocate memory for a movie is by using

```
M = moviein(N);
```

where N is the number of frames that you intend to record. If no axes object currently exists, this command will create one (and, if necessary, a figure object parent for that axes object will be created).

You are now ready to create the frames of your movie. However, we will point out that the general form of the **movie** function is **movie**(*H,M,N,FPS,LOC*), where *H* is the handle to a figure or an axis (this means you can have a movie that is in one of these objects), *M* is structure array that contains the frames for the movie, *N* is the number of times the movie will be played, *FPS* is the rate the movie is to be played in frames-per-second (default is 12 fps), and *LOC* is the location where the movie is to be played relative to the lower left-hand corner of object *H* (*LOC* is always in pixels, regardless of *H*'s *Units* property).

Another word about *LOC*; unlike the usual position defining vector we have seen, *LOC* does not specify the size of the movie – only the position. The width and height of the movie is established when it is recorded. Nevertheless, you still have to give *LOC* four elements.

Warning!

The approach to the animation you are about to create relies on taking an individual snapshot of each frame for the movie, then quickly flashing them back at a fixed rate, consequently, to make a smooth and fluid movie, you will want each individual frame or snapshot to be only slightly different from the previous. Therefore, in addition to the changes between sequential frames, the smoothness of the movie will also depend on the speed of your machine. In this example we will use the bessel function to create a visually interesting set of frames of a 3-dimensional surface.

```
% Create a figure that is a little smaller than
% standard to save
% memory since we will be storing 20 frames.
%If your machine has a lot of memory you do
% not have to define the position property of the
%figure.
movie_figure = figure('position',[100 250 300 200]);
M = moviein(15);
[x,y] = meshgrid([-10:0.5:10]);
for frame_number = 1:15
    z = bessel(0,( frame_number-1)*.2 + sqrt(x.^2 +
y.^2));
```

```
   surf(x,y,z);
   axis([-10 10 -10 10 -.5 1]);
   % Bring the figure to the front before taking a
snapshot.
   figure(movie_figure);
   M(:,frame_number) = getframe;
end
```

Now that we have created and stored all of the frames in our movie, we can play the movie back with

```
movie(M);
```

If you have a fast machine and the movie played through so fast that you didn't even get to enjoy it, try playing forward and backward with

```
frame_order = [1:15 14:-1:1];
number_repeats = 5;
movie(M,[number_repeats frame_order]);
```

Now that you have had a quick introduction, let's look at the different command syntaxes and develop a true understanding of what each command is doing so that you can develop a world-class movie. We will start with **getframe**, because the data that is returned with this function has implications for **movie** and **moviein**.

9.1.1 Taking a Snapshot

To take a snapshot of the current figure or axes object for use in a movie, we use the command **getframe**. This function can return a vector that is stored in a special MATLAB format called pixmap that is used for movies. Although now an obsolete usage, **getframe** can also be used to return two matrices for creating an image object (the first matrix is the image's color data matrix and the second is the associated color map). But as we said, this is now an obsolete use and instead MATLAB provides the function **frame2im** that will convert an individual movie frame to an indexed image.

In this chapter we discuss only the forms of **getframe** used for recording movie frames. The first two elements of the returned pixmap vector identify the size of the frame in pixels ([width height]), and the remaining elements of the vector represent the actual pixels of the frame that is stored. There are several forms of the command that can be used for movies. The forms are all very similar, the only difference being how you specify the region over which a bit-mapped snapshot should be taken. Since you are essentially specifying only a region on the screen, you must be aware that should another element lie on top of and obscure part of that region, for example, another Figure Window or even the Command Window, the bit-mapped image will contain the pixel representation of those elements. The first form of **getframe** is simply

```
M = getframe;
```

This form will use the current axes object to define the frames region. The boundaries of the axes object, as defined by its *Position* property, specify the region in which the snapshot is made. If you are unsure where the boundaries of the axes object are, you can always set the axes *Box* property to "on" and the *View* property to [0 90] (i.e., the 2-dimensional view), remembering that from this perspective the box and everything within the box are the region defined by the axes *Position* property. By creating a movie frame with this form, you may lose some or all of the tick mark labels, the x-, y-, and z-axis labels, and the title associated with that axes object. How much of this information you lose depends on the rendering perspective (i.e., a 2- or 3-dimensional view) of the axes object. In 2-dimensional views, all labels and titles will be lost; while in 3-dimensional views, part of the labels and titles will be lost or cut off.

You also have the option of recording a frame from any axes or figure object by using

```
M = getframe(object_handle);
```

where *object_handle* is the graphics handle of an axes or figure object. When an object handle is not supplied, it defaults to the current axes object. For example, using M = **getframe**(gcf) allows you to take a snapshot of everything within the current figure (i.e., multiple axes objects and their respective labels).

Finally, you are not limited to specifying the frame's position with the position of an axes or figure object. An arbitrary region, with respect to a figure's lower left-hand corner, can be specified using

```
M = getframe(object_handle,rectangle_vector);
```

where *rectangle_vector* is similar to the usual four-element position-defining vector, [left bottom width height]. The units of the *rectangle_vector* variable are the same as the *Units* property of the object with graphics handle *object_handle* (remember that a figure's factory default *Units* are in pixels). The *rectangle_vector*, when not supplied, defaults to the *Position* property of the figure, but with the first two elements set to [0 0] since the *rectangle_vector* position is defined with respect to the lower left corner of the figure.

For example, let's say you wanted to make a movie using only the upper right subplot in a figure that has four subplots. The third form of **getframe** would be most applicable. To demonstrate this, let's create the set of subplots shown in Figure 9.1 with

```
figure;
subplot(221);
plot(1:10);
subplot(222);
x = 0:0.1:(2*pi);
plot(x,sin(x))
title('sin wave');xlabel('x');ylabel('y');
subplot(223);
```

```
sphere(15);
subplot(224);
cylinder([1 .5 1]);
```

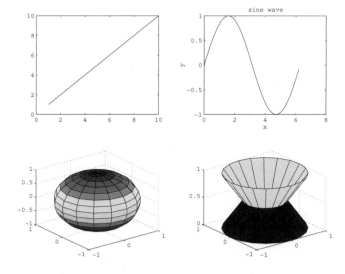

Figure 9.1   Sample subplots used to generate a movie.

Now create the vector that defines the upper right-hand quadrant of the figure using

```
figure_position = get(gcf,'position');
rectangle_vector = [figure_position(3:4)/2 ...
                    figure_position(3:4)/2];
```

Then plot and take a snapshot of the different versions of the sine wave in the appropriate axes with

```
subplot(222)
for loop = 0:20
   plot(x,sin(x+2*pi*loop/21));
   title('sin wave');xlabel('x');ylabel('y');
   M(:,loop+1) = getframe(gcf,rectangle_vector );
end
```

Finally, we can play the movie in a new figure by typing

```
figure
movie(M);
```

You may have noticed that the new figure has axes with x and y limits ranging from zero to one and that the movie is being played on top of these axes and takes up only a small portion of the figure. This is quite likely not what you would really like to see. The following set of commands might give you the type of movie you are looking for

```
figure('pos',rectangle_vector );
movie(gcf,M,5);
```

Do not change the size of the frame that is being recorded from snapshot to snapshot. Changing the size of the frame either by resizing a figure or redefining the rectangle_vector will result in a change in the size of the vector that is returned from the **getframe** function. This, in turn, will give you errors if you are storing these vectors in a movie matrix.

Warning!

9.1.2 Playing a Movie

We are now in a position to learn more about the command **movie**. As with most MATLAB functions, and as we mentioned at the beginning of this chapter, there are several ways that this function can be used. The form

```
movie(M)
```

plays the frames that are stored by columns in matrix M in the current axes object (if an axes object does not exist, a new one will be created). Since the movie is played in an axes object, the x- and y-axis lines and tick mark lines of the axes object will be visible, unless you have made them invisible with a command such as **set**(**gca**,'visible','off') or **axis**('off').

Two other forms are

```
movie(M,N)
```

and

```
movie(M,N,FramesPerSecond)
```

where N is either a scalar or vector defining the number of times and the order in which the frames are to be played, while FramesPerSecond is a scalar that specifies the rate at which MATLAB should try and play a movie. By default, MATLAB tries to play a movie at 12 frames per second, but its success at achieving this rate depends on your machine's speed. The first element of the variable N will identify the number of times that the frames should be played. If it is a negative number, the movie will be played forward and backward that many times. For example, if N equals negative three, the movie will be played forward and backward three times. If there is more than one element in N, the remaining elements specify which frames and the order in which the frames will be played. For example if there are five frames in the movie matrix M,

```
NumberOfPlays = 1;
FrameOrder = [1 2 3 4 5 4 3 2 1];
movie(M,[NumberOfPlays FrameOrder])
```

will play the movie one time forward and backward without repeating the fifth frame. The following code

```
NumberOfPlays = 2;
```

```
FrameOrder = [1:5 4:-1:2];
movie(M,[NumberOfPlays FrameOrder])
```

will play the frames in the following sequence:

1, 2, 3, 4, 5, 4, 3, 2, 1, 2, 3, 4, 5, 4, 3, 2.

This is different from using **movie**(M,-2) which would play the frames in the following sequence:

1, 2, 3, 4, 5, 5, 4, 3, 2, 1, 1, 2, 3, 4, 5, 5, 4, 3, 2, 1.

Power!

The freedom of defining the order in which the frames are played and the number of times that the order is sequenced through (as in the previous example), gives you the capability, in some circumstances, to save memory by reusing frames instead of storing identical frames in the movie matrix.

The final two forms of movie serve to provide you with the freedom of specifying the location at which the movie will be played. Using

```
movie(object_handle,M,N,FramesPerSecond)
```

plays the movie in the graphics object with the handle object_handle. The N and FramesPerSecond variables are optional with this form if their default values of one and twelve are satisfactory. The last form that will be discussed is **movie**(object_handle,M,N,FramesPerSecond,rectangle_vector). This form will play the movie at the location specified by rectangle_vector. This is also a 4-element position vector [left bottom width height], in which the left and bottom elements are defined with respect to the lower left-hand corner of the axes or figure object identified by the handle object_handle. The width and height do not really matter since the movie matrix defines these two values.

After a movie has been played, any command or action that forces the window to be refreshed will result in removal of any remnants of the movie that were in that window. For example, resizing the figure or typing refresh will remove any trace of the movie from the Figure Window. This is because the movie is only a bit-mapped representation that is flashed onto the Figure Window at a specified location. A movie is not a graphics object (i.e., it has no handle and is not a child of a figure or axes object).

9.1.3 Preallocating Memory

As stated earlier, it is not necessary to preallocate memory for the movie data matrix. However, in the name of speed and reduction of wasted effort, it is recommended that you do preallocate. The savings in terms of speed is achieved because MATLAB does not need to reallocate memory every time a new frame is added to the matrix. The probability that your efforts are wasted or made in vain are less likely because you will find out at the time you try to preallocate the memory whether or not you have enough memory. If you force MATLAB to reallocate on the fly you may get an "Out of memory" error when it attempts to store a new frame, and you will lose the time it took to reach that point.

There are different ways to preallocate the memory. The most basic,

```
M = moviein(N)
```

is used if you are recording the contents of an axes object. The variable N specifies the number of frames that will be stored. This ends up being the number of columns needed in matrix M. The number of rows is determined by the function as it calculates what the length of the pixmap vector will be to store the data in the current axes object.

If you are recording the contents of an entire figure, you can use

```
M= moviein(N,figure_handle);
```

where figure_handle is the figure in which the frames will be recorded. In both of these forms of **moviein**, do not resize the figure or axes object as this will force a change in the number of rows required to store the pixmap data.

Finally, if you are recording the contents over some arbitrary region of a figure with handle figure_handle, use

```
M = zeros(length(...
            getframe(figure_handle,rectangle_vector)),N);
```

where rectangle_vector is the 4-element vector, [left bottom width height] (defined in the units of the figure and with respect to the figure's lower left-hand corner), that defines the recorded region.

9.1.4   Practically Speaking

Although there are many different circumstances and ways in which you will want to create your movie, it is most likely that you will create a movie of all the events in a figure or portion of the figure. For this reason, it is recommended that you create the frames with respect to the figure.

9.1.4.1   Recording the Entire Figure

If you are recording all events in a figure, use

```
M = getframe(figure_handle);
```

to record the frames and play them back in the same or new figure with the form

```
movie(figure_handle,M,N)
```

If the figure in which you are playing the movie is a new Figure Window, make sure that the width and height of the new figure are the same as the frames that were recorded. If you didn't keep this information, you can always get it directly from the movie matrix, M, by using something like

```
WidthHeight = M(1:2,1)';
NewFigure = figure;
```

```
NewFigurePos = get(NewFigure,'position');
set(NewFigure,'position',[NewFigurePos(1:2)
WidthHeight]);
```

and then play the movie in this new figure with

```
movie(NewFigure,M);
```

As an example, let's create a movie that has two axes objects and animation simultaneously occurring in both.

```
figpos=[100 200 150 125];
h_fig = figure('Position',figpos);

for framenumber = 1:20;
    subplot(121);
    plot(sin(0:0.1:(2*framenumber*pi/20)),...
            0:0.1:(2*framenumber*pi/20),'--r');
    axis([-1 1 0 2*pi]);
    subplot(122);
    plot(0,exp(-(framenumber-1)/3),'*g');
    axis([-1 1 0 1]);axis('off');
    drawnow;
    M(framenumber)=getframe(h_fig)

end

delete(h_fig)
h_newfig=figure('Position',figpos)
axis off
movie(h_newfig,M,-5)
```

In this example, we simply take the entire figure as the frame with **getframe**(h_fig), store those frames in M, then play M in a new figure that is the same size as the original figure.

9.1.4.2   Animating a Portion of the Figure

You might be thinking at this point,

"What if I want to animate only a portion of a figure?"

Which is a very good question. This can be achieved quite handily by using the form

```
M = getframe(figure_handle,rectangle_vector);
```

Then play the movie back in the same figure in the same region in which it was recorded with

```
movie(figure_handle,M,N,FramesPerSecond,...
rectangle_vector);
```

For example, if you have

```
h_fig = figure;
subplot(2,1,1);
plot(abs((0-5)));axis([0 11 0 5])
subplot(2,1,2)
plot(sin(0:4*pi));axis([0 4*pi -1 1])
```

you can animate and record the top axes with

```
subplot(2,1,1);
%Define the position of the region that
%is to be recorded.
figposition = get(fig,'position');
rectangle_vector = [0 figposition(4)/2 figposition(3)
figposition(4)/2];
for framenumber = 1:10;
    plot(abs((0:(framenumber -1))-5));axis([0 11 0 5]);
    M(framenumber) = getframe(h_fig,rectangle_vector );
end
```

and then play back the movie with

```
movie(fig,M,-3,12,rectangle_vector);
```

To play the movie in a new figure with a width and height equal to the width and height of the recorded frames, use something similar to

```
NewFigure = figure('position',[left bottom ...
                    rectangle_vector(3:4)]);
movie(NewFigure,M,N,FramesPerSecond);
```

where left and bottom in the first line are used to position the Figure Window at some arbitrary location on the screen. The first line creates a figure with the correct proportions and the second plays the movie. To continue with our previous example, try

```
left = 100;
bottom = 150;
NewFigure = figure('position',...
                    [left bottom rectangle_vector(3:4)]);
movie(NewFigure,M,-3);
```

To summarize, carefully keep track of the width and height (in terms of pixels) of the region that is being recorded. In addition, be careful and account for these same pixels during movie playback. This will help you avoid problems.

   Another question you might ask is,

"Am I limited to recording only the contents of a particular figure or axes object?"

FAQ

The answer is, "No." In certain situations, you may want to record regions outside a single Figure Window. For example, you may want to record the events occurring in multiple Figure Windows or perhaps you just want to include the Figure Window borders. There is no requirement that limits you to specifying frame region within the figure boundaries. They can arbitrarily be defined with respect to the lower left-hand corner of the figure. For instance, we could include a 20-pixel border around the current Figure Window with

```
figure_position = get(gcf,'position');
rectangle_vector = [0 0 figure_position(3:4)] +  ...
                   [-20 -20 40 40];
```

and then use the **getframe** function form

```
M = getframe(gcf,rectangle_vector);
```

Please note that there may be discrepancies between the actual colors and those used in the movie for the recorded regions that lie outside of the figure with respect to where the region is defined.

9.1.5 Making an AVI Movie

The Audio-Video Interleaved file format, or AVI file, is a standard movie file format that is ubiquitous in the computer world. Once you have created movie frames in MATLAB, you can use the **movie2avi** function to create an AVI file that you can share with others. The easiest form of the **movi2avi** function is **movie2avi**(*mov,filename*) where *mov* is the movie created with **getframe**, i.e., what you would use with **movie**, and *filename* is a string giving the name of the AVI file you want to create. You don't need to include ".avi" in the filename; that is done automatically. There are a number of parameters and values you might want to specify to improve your AVI files. When you use **movie2avi** the default frame rate is 15 fps, which may or may not be desired. The parameter "fps" can be used to specify a different frame rate. Another handy parameter is "videoname", which allows you to give a name to the video stream up to 64 characters. As an example, consider again the first animation example (the Bessel function). Once you have M, you can create an AVI file named "bessel" at 12 frames per second and with a video name of "3-D Bessel Function Animation" with the following,

```
movie2avi(M,'bessel','fps',12,...
'videoname','3-D Bessel Function Animation')
```

You can also specify different AVI standard compression modes to be used when you make an AVI with the "compression" parameter. It can take the strings "Indeo3", "Indeo5", "Cinepak", "MSVC", or "None". For Windows computers, the default is "Indeo3". You can affect the quality of the movie by setting the "quality" parameter, which can take any value between 0 and 100 where higher numbers are higher quality. Higher quality comes at the price of files size. The default quality is 75.

9.2   On-the-Fly Graphics Object Manipulation

Manipulating a graphics object's coordinates with small incremental changes is usually what most people envision when thinking about an animation process. Creating several snapshots in advance and rapidly playing them back with a computer is fine if you have lots of memory or if the length and size of the movies is relatively small. However, if you want to create long, animated sequences and you have a processor and software package that can perform the needed mathematical calculations and display the graphics quickly and smoothly, making the incremental changes on the fly can be advantageous. The key point is that the changes must be made fast enough so that the motion looks continuous to the user.

The MATLAB graphics environment, when programmed correctly and appropriately, can be used to provide a user with the perception that fluid animated sequences are occurring. Using a method that does not play back a sequence of static prestored snapshots also provides you with the freedom of adaptively animating graphics in response to a user's actions, such as the mouse pointer locations. In this section of the chapter we will examine several ways to program MATLAB to give you these types of capabilities. As you will see, these techniques do not mean that the frame playback method is useless; there is a time and place for each one, and it will become your responsibility to learn when to use one over the other. It is one of the goals of this book to make you aware of some of the questions that you will need to ask so that you can make a well-informed decision.

9.2.1   Simple Animation Functions

Perhaps the simplest on-the-fly animation you can do in MATLAB is achieved with two functions that come with MATLAB that allow you to create a 2- and 3-dimensional curve tracing animation. The 2-D form is called **comet** while the 3-D form is **comet3**. The basic use of these functions involves simply determining the path that you would like to have traced (i.e., the x- and y-coordinates for the 2-D form, and the x-, y-, and z-coordinates for the 3-D form) and passing them to the functions. You should note that since we are predetermining the coordinate values, this method does not lend itself toward adaptively changing the path to some arbitrary event or stimulus such as that provided by a user's actions. (However, if you look at the file comet.m, you might get some clever ideas!) Regardless of that point, these functions can be informative in the sense of watching the progression of a line.

The line trace animation is started with either **comet**(x,y) or **comet**(x,y,p). The x and y identify the coordinates of the trace. The variable p, when supplied, determines the distance by which the comet's tail should follow the front of the trace. By default, it is set to 10% (i.e., $p = 0.10$) of the length of vector x; however, you may specify that it have a value in the range $0 \le p < 1$. The 3-D trace is the same as the 2-D trace except that a z-coordinate is supplied using **comet3**(x,y,z) or **comet3**(x,y,z,p). In both cases, the comet trace is created from the first three colors from the color order, and with the default is essentially a blue circle with a dark green tail that turns to a red line. These are incrementally drawn from coordinate point to coordinate point with the red line traced on top of the green after a specified delay. The axis limits are predetermined so that all portions of the line trace will be displayed. As an example, consider the 3-dimensional comet created with the following code:

```
t = 0:0.01:10*pi;
x = t.*sin(t);y=t.*cos(t);
comet3(x,y,t);
```

This will look like Figure 9.2 after the animation is completed. If you have a really fast computer and the animation happens too fast for you, try reducing the increment of *t* to something like 0.001.

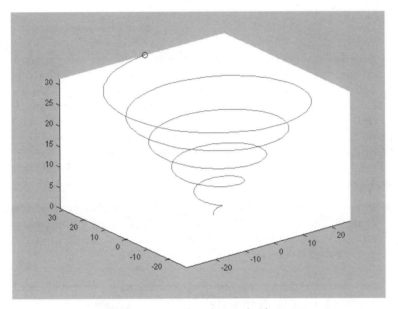

Figure 9.2   **The aftermath of comet3.**

Much of the figure's contents are lost if you resize or refresh the Figure Window after the animation has finished. This is because each component within the animation was programmed to be drawn and erased in a particular fashion. This means that if you try to do something to such a figure, like print it, you will not see the "erased" plot. In the next section, you will learn about the properties of graphics objects that are manipulated to produce animations, and what must be done in order to output an image produced by these methods.

9.2.2 The Wrong and Right Way to Animate Graphics

Before we learn how to animate graphics, it is often helpful to see and learn from the wrong way of animating graphics. You will find that using the **plot** command over and over is very inefficient and yields poor results. To provide a level of comparison, we will try and animate a blue circle along the same path that was used in the **comet3** example. First we can recreate the data with

```
t = 0:0.01:10*pi;
x = t.*sin(t);
y=t.*cos(t);
axislimits = [min(x) max(x) min(y) max(y) min(t) max(t)];
figure
for indexnumber = 1:length(x)
   plot3(x(indexnumber ),y(indexnumber ),...
         t(indexnumber ),'bo');
   axis(axislimits);
   drawnow;
end
```

We see that the circle does indeed appear to be moving along the trajectory specified by the (x,y,t) coordinates. However, it is moving very slowly and the axes flicker every time the circle is plotted. This happens because every time the **plot3** command is issued, a new graphics object (i.e., the circle) is created and the axes object is refreshed. The **drawnow** command forces MATLAB to flush the event queue and to render the newly created graphics. If this command was not there, you would not see the circle appear until its final position has been drawn in the last iteration of the **for...end** loop.

In addition to the **drawnow**, there are only three other events that force the screen to be updated. These commands are summarized in the table below. With respect to animations, the **drawnow** function is the most useful; however, the **pause** command can also come in handy.

Power!

Events that force MATLAB to flush the event queue	Examples
execution of the drawnow comand	drawnow;
issuing a figure command	figure; figure(1);
execution of the pause command	pause; pause(1);
using getframe	M(k)=getframe
execution of waitfor	waitfor(h,'PropertyName')
a return to the command prompt	keyboard; value = input('Enter a value')

Getting back to the example, you might be thinking that instead of plotting the point with **plot3** we could use the lower level command called **line**. This way we would have to set up the axes object only once and then create a new line object at the coordinates along the trajectory while at the same time deleting the old line object. For instance, using the same data as above, we can try:

```
figure
axis(axislimits);

line_handle = line(x(1 ),y(1 ),t(1 ),...
        'color','c',...
        'linestyle','o');
for indexnumber = 2:length(x)
    delete(line_handle );
    line_handle = line(x(indexnumber ),...
                        y(indexnumber ),...
                        t(indexnumber ),...
                        'color','b',...
                        'linestyle','o');
    drawnow;
end
```

Unfortunately, this code will once again produce slow and unsatisfactory results. (If these examples are too slow on your computer, stop the execution with CTRL-C and try changing the t increment to something like 0.1.) We need to get to the root of the problem, which is the fact that deleting and, more importantly, creating graphics requires a lot of overhead. This time we will create only a single line object and update its *XData*, *YData*, and *ZData* properties. The initial creation can be performed with either the **plot3** or **line** command. In addition, we will need to change the *EraseMode* property from its default setting of "normal" to either "xor" or "background". Let's look at the results and then go into the explanation.

```
figure
line_handle = plot3(x(1 ),y(1 ),t(1 ),'co');
set(line_handle,'erasemode','xor');
axis(axislimits);
for indexnumber = 2:length(x)
    set(line_handle ,'xdata',x(indexnumber ),...
                     'ydata',y(indexnumber ),...
                     'zdata',t(indexnumber ));
    drawnow;
end
```

On a 1GHz PentiumIII machine, the **for...end** loop was over eight times faster than the previous approach (over thirty times faster than the first approach) and the axes object did not flicker anymore. Changing the way MATLAB draws and erases the graphics object from "normal" to "xor" has provided a major improvement in the quality of the animation. Let's look at some of the attributes of the various erase modes, listed from the fastest at the top, to the slowest at the bottom.

Erase Mode	Attributes
none	o Object is not erased when it is moved or destroyed (deleted). o 3-D rendering calculations suppressed.
background	o Object drawn and erased by xoring with color of screen beneath. o Damages color of object(s) beneath. o Color guaranteed at all times. o 3-D rendering calculations suppressed.
xor	o Object drawn and erased by xoring with color of screen beneath. o Does not damage color of object(s) beneath. o Color guaranteed only when placed directly over figure background color. o 3-D rendering calculations suppressed.
normal	o Most accurate representation of object.· o Colors and 3-D rendering calculations performed.

Even though the attributes of the erase modes other than "normal" suppress the 3-D rendering calculations for that object, it should be noted that the calculation and therefore the correct rendering order will be performed if the screen is redrawn by a command such as **refresh**.

9.2.3 The Need for Speed

Of the erase modes, "normal" is by far the slowest. To achieve better results in terms of speed, several properties within the object's parent and grandparent (i.e., the axes and figure object) should also be changed. Setting the axes object's *DrawMode* property to "fast" instead of to "normal" and the figure's *BackingStore* property to "off" instead of to "on" will help speed up an animation. As was discussed in Chapter 7, the *BackingStore* should be set to "on" when the simulation is not running and there are surface or image objects in the figure that take some time to render. Remember, when *BackingStore* is "off", the figure will be redrawn every time another Figure Window is selected.

You will also find out that the human eye is fairly slow and you can often speed up a graphics animation by increasing the increments used to translate or rotate an object and still make it look like a smooth simulation. The last speed tip we shall suggest is not to make a practice of changing the number of vertices within the animated object during the simulation. For instance, adding elements to the *XData*, *YData*, or *ZData* will slow an animation considerably.

9.2.4 Animating Lines

Up until now, the line object animated had only one point. Animating a series of coordinates is just as easy. For instance, we can create the appearance of a rope that is being swung around and around like a jump rope with

```
x = 0:(pi/48):pi;
ropeheight = sin(x);
line_handle = plot(x,ropeheight);
axis([0 pi -1.1 1.1]);
grid on;
set(line_handle,'LineWidth',3,'EraseMode','background');
for phi = 0:pi/64:10*pi
    set(line_handle,'ydata',cos(phi)*ropeheight);
    drawnow;
end
```

Next, change the *EraseMode* to "background" (by altering the sixth line of the code above) and run the same animation; notice how sections of the grid lines are being removed by the animated line. After the animation is complete, you can type **refresh** to redraw the screen.

9.2.5 Animated Rotations

A classic example illustrating animation in three dimensions is a spinning wire frame cube. To create the cube, we define the x-, y-, and z-coordinates with

```
x = [0 1 1 0 0 0 1 1 0 0 NaN 1 1 NaN 1 1 NaN 0 0];
y = [0 0 1 1 0 0 0 1 1 0 NaN 0 0 NaN 1 1 NaN 1 1];
z = [0 0 0 0 0 1 1 1 1 1 NaN 1 0 NaN 0 1 NaN 1 0];
```

The NaNs are not necessarily required when drawing lines that have their *EraseMode* property set to "normal". However, if the *EraseMode* is set to "xor", the edges of the cube that are traced over twice can cancel each other out, making it look as if those edges are invisible. Therefore, we use NaNs to lift the "pen" off the paper (see Chapter 8) so that none of the edges of the cube are traced twice.

Now we will plot the cube as shown in Figure 9.3 centered precisely on the origin (0,0,0) with

```
cube_h = plot3(x-0.5,y-0.5,z-0.5);
axis('square');
axis([-1 1 -1 1 -1 1]*2);
view(-37.5,15);
set(cube_h,'erasemode','background');
rotation_increment = 5; % degrees
rotation_axis = [0 0 1];
rotation_origin = [0 0 0];
num_of_incr = 360/rotation_increment;
for loop = 1:num_of_incr
    rotate(cube_h,rotation_axis,...
                  rotation_increment,rotation_origin);
    drawnow;
end
```

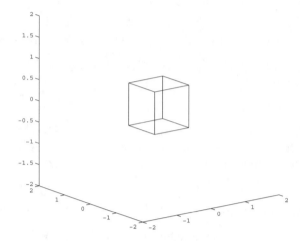

Figure 9.3   Animating a cube.

Some neat interactive rotation animations can be achieved with the **rotate** command using graphical user interfaces and different origins of rotations (e.g., variable rotation_origin) and axes of rotations (e.g., rotation_axis). The **rotate** function performs the calculations for rotating a graphics object about some defined axis of rotation.

This next example demonstrates a truly interactive animation by checking for user entry and rotating the cube accordingly. Copy this function and use the arrow keys to control the rotation of the cube. Press the ESC key to exit the program. Note that you must have the cursor somewhere on the Figure Window before you press the keys on the keyboard.

M
M-File

```
function x=rotcube()
%CUBE is a function to demonstrate run-time animation.
%A cube is drawn with lines.
%Use the arrow keys to control the direction of rotation.
%ESC key to exit.

x = [0 1 1 0 0 0 1 1 0 0 NaN 1 1 NaN 1 1 NaN 0 0];
y = [0 0 1 1 0 0 0 1 1 0 NaN 0 0 NaN 1 1 NaN 1 1];
z = [0 0 0 0 0 1 1 1 1 1 NaN 1 0 NaN 0 1 NaN 1 0];
rot_axis = [0 0 1];
rot_org = [0 0 0];

cube_h = plot3(x-0.5,y-0.5,z-0.5);

axis('square');
axis([-1 1 -1 1 -1 1]*2);
view=[-37,30];

set(cube_h,'erasemode','background');
rotation_increment = 5;
```

```
    rotation_axis = rot_axis;
    rotation_origin = rot_org;

    fig_h=gcf;

    key = 28;

    while key ~= 27 % watch for ESC key
        if waitforbuttonpress == 1;
        key = get(fig_h,'currentcharacter');

        switch key
            case 28     % <- rotate left
                rotation_axis = [0 0 1];
                rotation_increment = -5;
            case 29     % -> rotate right
                rotation_axis = [0 0 1];
                rotation_increment = 5;
            case 30     %  rotate up
                rotation_axis = [0 1 0];
                rotation_increment = 5;
            case 31     %  rotate down
                rotation_axis = [0 1 0];
                rotation_increment = -5;
            case 27 % ESC key
            close(fig_h)
            clear
            return

        end

        rotate(cube_h,rotation_axis,...
            rotation_increment,rotation_origin);
        drawnow;
    end

end

x=key;
```

Try creating several cubes and simultaneously rotating them in different ways with the following code.

```
x = [0 1 1 0 0 0 1 1 0 0 NaN 1 1 NaN 1 1 NaN 0 0];
y = [0 0 1 1 0 0 0 1 1 0 NaN 0 0 NaN 1 1 NaN 1 1];
z = [0 0 0 0 0 1 1 1 1 1 NaN 1 0 NaN 0 1 NaN 1 0];

cube_h = plot3(x-0.5,y-0.5,z-0.5);
axis('square');
axis([-1 1 -1 1 -1 1]*2);
view(-37.5,15);
set(cube_h,'erasemode','background');
rotation_increment = 5; % degrees
rotation_axis = [0 0 1];
rotation_origin = [0 0 0];
num_of_incr = 360/rotation_increment;
```

```
for loop = 1:num_of_incr
    rotate(cube_h,rotation_axis,...
                    rotation_increment,rotation_origin);
    drawnow;
end

cube2_h = line(x+1,y+1,z+1,'erasemode','background');
for loop = 1:num_of_incr
    rotate(cube_h,rotation_axis,...
                    rotation_increment,rotation_origin);
    rotate(cube2_h,rotation_axis+[1 1 0],...
                    rotation_increment,rotation_origin+1);
    drawnow;
end
```

The more objects you manipulate, the slower the animation will become. You will need to experiment to find out what the capabilities of your machine are with respect to animations.

You should always try to look for the simplest way to form the graphic objects. For instance, cubes could also be formed with surface objects as illustrated with the following. Again, some variables are from previous examples.

```
x = [0 0 1 1 0]; y = [0 1 1 0 0]; z = zeros(size(x));
rotation_axis = [0 0 1];
rotation_origin = [0 0 0];
rotation_increment = 5; % degrees
num_of_incr = 360/rotation_increment;
s1_h = surf([x;x]-.5,[y;y]-.5,[z+0.5;z-0.5]);
set(s1_h,'erasemode','background',...
        'facecolor','none',...
        'edgecolor','g');
s2_h = surface([x;x]+1.5,[y;y]+1.5,[z+.5;z-0.5]+1.5,...
        'erasemode','background',...
        'facecolor','none',...
        'edgecolor','r');
s3_h = surface([x;x]+1.5,[y;y],[z+.5;z-0.5],...
        'erasemode','background',...
        'facecolor','none',...
        'edgecolor','b');
axis([-3 3 -3 3 -3 3]);axis('square');
for loop = 1:num_of_incr
    rotate(s1_h,rotation_axis,...
                rotation_increment,rotation_origin);
    rotate(s2_h,rotation_axis+[1 1 0],...
                rotation_increment,rotation_origin+1);
    rotate(s3_h,rotation_axis,...
                rotation_increment,rotation_origin);
    drawnow;
end
```

The *FaceColor* property of the surface objects has been set to "none" to avoid flickering. If you want to graphically animate a solid cube, the top and bottom of each box could be added and the *FaceColor* property could be specified. The following code animates three solid cubes:

```
% Generate vertices for the surface of a single cube.
xx = [0 0 1 1 0 NaN 0 1 NaN 1 0;...
      0 0 1 1 0 NaN 0 1 NaN 1 0];
yy = [0 1 1 0 0 NaN 1 1 NaN 1 1;...
      0 1 1 0 0 NaN 0 0 NaN 0 0];
zz = [1 1 1 1 1 NaN 1 1 NaN 0 0;...
      0 0 0 0 0 NaN 1 1 NaN 0 0];
% Set up rotation variables.
rotation_increment = 5; % degrees
rotation_axis = [0 0 1];
rotation_origin = [0 0 0];
num_of_incr = 360/rotation_increment;
% Generate 3 translated versions of the cube.
s1_h = surf([xx]-.5,[yy]-.5,[zz]-.5);
set(s1_h, 'erasemode','background','facecolor','g');
s2_h = surface([xx]+1.5,[yy]+1.5,[zz]+1,...
        'erasemode','background','facecolor','r');
s3_h = surface([xx]+1.5,[yy],[zz]-0.5,...
        'erasemode','background','facecolor','b');
% Set up the proper proportions.
axis([-3 3 -3 3 -3 3]);axis('square');
% Define the rotation specifications for each cube.
for loop = 1:num_of_incr
   rotate(s1_h,rotation_axis,...
               rotation_increment,rotation_origin);
   rotate(s2_h,rotation_axis+[1 1 0],...
               rotation_increment,rotation_origin+1);
   rotate(s3_h,rotation_axis,...
               rotation_increment,rotation_origin);
   drawnow;
end
```

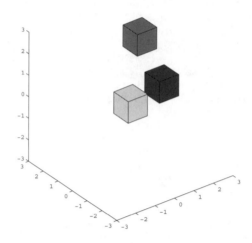

Figure 9.4   Animated surfaces.

Three solid cubes (green, red, and blue in color) as shown in Figure 9.4 will be rendered and then animated about some axis of rotation. As you will see, the surface objects do not render smoothly. There may be some flickering, and if you look closely, you can see that the faces of the cubes that are farthest from the viewer will briefly be visible. Also notice that if you turn **grid on** the grid will be erased where the cubes occlude it.

9.2.6   Forcing a Graphic to Leave a Trail

In some instances you may not want the graphic to be erased from its last position when it is moved to a new location. This can be useful, for instance, if you are tracing a path. This type of graphics animation does not need to update a lot of object vertices and therefore can be very fast. You may find that you need to slow it down with either more steps in the **for...end** loop or by putting a **pause** statement before the **drawnow** as the following code illustrates.

```
x = 0:500;
y = sin(.05*x+cos(x*.1));
figure('backingstore','off');
axes('drawmode','fast','box','on');
axis([min(x) max(x) min(y) max(y)]);
line_handle = line(x(1:2),y(1:2));
set(line_handle,'linewidth',2,'erasemode','none');
for index_counter = 2:length(x);
    set(line_handle,...
            'xdata',[x(index_counter+[0 -1])],...
            'ydata',[y(index_counter+[0 -1])]);
    pause(.5);
    drawnow;
```

```
end

% To keep the trace on the plot for printing or
% if the user resizes the figure window, you may
% want to add the following two lines.
line(x,y,'linewidth',2);
refresh
```

This is similar to the technique used in **comet** and **comet3**. Text, patch, and surface objects can also have their *EraseMode* property set to "none" to create interesting pictures. Remember that after the figure is refreshed, the old locations of the object will be erased. If you want to get a hardcopy of a picture that has been generated with objects that have been moved around without erasing the last object's location, use

```
[imagematrix,map] = getframe(figure_handle);
```

This form of **getframe** essentially takes a snapshot of the object contents (in this case, the figure) and returns an image data matrix and the color map needed to correctly display the colors in the object. The image data can be displayed with an image object. For instance, to display this captured image in a new figure, use

```
newfig = figure;
figpos = get(newfig,'position');
set(newfig,'position',...
            [figpos(1:2) fliplr(size(imagematrix))]);
axes('units','normalized','position',[0 0 1 1]);
image(imagematrix);
colormap(map);
set(gca,'visible','off');
```

Power!

9.3   Choosing the Right Technique

There are several questions that you should ask yourself when deciding whether to use the frame-by-frame movie play back or the on the fly graphic's coordinate manipulation technique to create your graphical animation. In some instances the choice is easy. For example, if you answer yes to either of the following questions, you should consider creating a data matrix with getframe and playing it back with movie:

•   Are large or complex surface objects going to be animated?

•   Is there a small set of still shots that could be used and pieced together?

However, if you answer yes to any of the next few questions, you will want to use the on the fly coordinate manipulation technique:

•   Will there be user interaction during the animation sequence? For instance, will the user have the ability to use the mouse or keyboard to alter the simulation during its progression (see Chapter 6 for information about recognizing user controlled events)?

•   Are the graphics objects to be animated relatively simple?

•   Do you want to mix in sound with your animation?

Other questions that you will need to think about are:

- What constraints in terms of speed and memory are imposed by the machine on which you are running MATLAB?

- Can you store the number of frames needed to generate a movie?

- Can you generate and animate the number of graphics objects that you believe will be required in the animation?

- Experimenting with the different animation techniques and getting a feel for your machine's capabilities with regard to speed and MATLAB animations is the best way to figure out which technique you should use and in what circumstances.

10 ELEMENTS OF GUI DESIGN

IN THIS CHAPTER...
10.1 WHAT IS A MATLAB GRAPHICAL USER INTERFACE? 385
10.2 THE THREE PHASES OF INTERFACE DESIGN 386
10.3 UI CONTROL ELEMENTS...391
10.4 UIMENU ELEMENTS 421
10.5 LOW-LEVEL MATLAB GUI PROGRAMMING TECHNIQUES.....................433
10.6 HIGH-LEVEL GUI DEVELOPMENT – GUIDE 450
10.7 COMMON PROGRAMMING DESIRES WITH UI OBJECTS 461
10.8 THE MATLAB EVENT QUEUE...474
10.9 CREATING CUSTOM USER INTERFACE COMPONENTS 484

10.1 What is a MATLAB Graphical User Interface?

The Graphical User Interface, or GUI, refers to the now universal idea of icons, buttons, etc., that are visually presented to a user as a "front-end" of a software application. Most of us would consider a software application that accepted only keyboard-entered commands as quite archaic, and even down right primitive. We much prefer to point our mouse pointer to a graphical representation of some aspect of the application, click on it (invoking some event), and continue working with the application through interactive cues. We are also accustomed to windows, pull-down menus, slider controls, and check boxes. How slow and boring the software world was before the GUI! There can be many reasons for creating a GUI. For instance, you might wish to automate a function that you use many times, or perhaps you want to share it with others who don't need, want, or care about knowing MATLAB. Perhaps you would like to create an interactive demonstration.

Not to be behind the times, the MathWorks has provided MATLAB programmers with a set of structured event driven components in the form of user interface controls (uicontrols) and menus (uimenus) that can easily be assembled and used to create GUIs. The fundamental power of GUIs is that they provide a means through which individuals can communicate with the computer without programming commands. The components have become quite standardized and developed into a user friendly and intuitive set of tools. These tools can be used to increase the productivity of a user or to provide a window to the sophistication and power of a MATLAB application for people with little or no MATLAB programming experience.

The set of user interface components supplied with MATLAB allows you to design GUIs that match those used in sophisticated software packages. The components are graphics objects just like those we discussed in Chapter 7, with handles and properties, and come in two classes: user interface controls (uicontrols) and user interface menus (uimenus). Considering the great deal of flexibility and usefulness that these objects provide, the programming required to design a fully functional GUI is amazingly simple. The uicontrols and uimenus can be combined with other graphics objects to create informative, intuitive, and aesthetically pleasing GUIs. This chapter is designed to make you aware of all the user interface capabilities and to show you how to program fully functional GUIs that meet your needs. In addition to showing how to create and program uicontrol and uimenu objects, this chapter will also attempt to broaden your programming horizons by showing how you can use the other graphics objects previously discussed to design your own interface components.

In this chapter we will present access to uicontrols and uimenus through two fundamentally different approaches. The first approach is a low-level, bottom-up approach where we use our skills with handle graphics to write M-files that implement any GUI design we wish. The second approach will briefly examine the use of MATLAB's Graphical User Interface Development Environment, or GUIDE for short. (We think of GUIDE as a top-down GUI development approach.) GUIDE is high-level, yet powerful and extremely easy to use; an excellent tool for quickly developing GUIs that takes care of much of the "bookkeeping" usually associated with GUI development. Although not as rapid for quick GUI development as the GUIDE approach, working at a lower level you have complete control over your GUI. We will devote a great deal of this chapter to the low-level approach since it is the approach that gives you the greatest control and also will teach you much about the inner workings of MATLAB GUIs. Even if you end up preferring to use GUIDE for your GUI development, the knowledge of the low-level approach is still very much applicable and will provide you with valuable insight. Let's face it; either way you will be a MATLAB GUI developing fiend when you finish this chapter!

Before we dive right into building GUIs, we believe that a brief discussion on general GUI design is in order. The following discussion is not intended to substitute for a university text on software interface design, but it should give you a basic understanding of what is important in GUI design and how to efficiently proceed with your GUI implementation. After discussing general GUI design, we will present the details of the uicontrol and uimenu objects. Then we will present GUIDE and take you through an example, then look at the low-level programming approach. Finally, we will wrap things up with a discussion of common programming needs and desires.

10.2 The Three Phases of Interface Design

One can make the argument that there are three phases of good GUI design. These are mostly common sense, but it is good for us to present them here in a formal manner. See reference 4 at the beginning of the Appendix for an intelligent yet easy to read treatment of interface design that succinctly covers what's important in a good GUI. In this section we quickly present the three phases of good GUI design and offer some "rules-of-thumb" that are good to apply with your MATLAB, or any other GUIs.

10.2.1 Analysis

Before you start your GUI design, you need to consider who will be using it and how. For instance, if you were creating a computer interface for toddlers, you probably would not use written words, but large, brightly colored clickable pictures would probably work nicely. However, the same approach would probably not be as well received if you were tasked with creating an interface for your company's director of marketing (or maybe it would!). The point is that you need to keep the user in mind. Many MATLAB programmers find themselves as the primary user of their GUIs. This is because they have found that automating tasks and having a convenient GUI is up-front time well spent. You might find yourself as part of a development team and your task is to create a rich yet intuitive to use GUI for functions and data provided by other members. In such a case, the analysis portion of good GUI design could be very important indeed. The analysis process can become very involved, depending on the goals, and could require extensive usability specifications, developing user case scenarios, identifying the expertise of the user, computer system limitations, and plans for future upgrades based on user feedback.

10.2.2 Design

Once you understand your users and the information that is to be interfaced with, you can begin the process of laying out your GUI. In the design phase you still aren't writing the GUI, although you might feel like you want to; instead, you are considering what components, tasks, and sequences are required to make your GUI effective. Unbelievably, pencil and paper is still a great way to explore your GUI design. Again, for major projects, this can become an involved task, but in the course of the GUI development, it is time well spent. We will talk about this again in the next section on Paper Prototyping.

10.2.2.1   User Considerations

Remember, whatever GUI you create has two major components: one is the GUI itself, the other is the user. It is important that you know who your users will be; you would not design a GUI to be used by kindergarteners the same as for a group of scientists at a research laboratory. Human factors specialists consider people from visual, cognitive, and physical perspectives. Of course we are limiting our scope to what we might do with MATLAB, but as you have seen, MATLAB gives you significant graphical capabilities—and as you are about to learn, its GUI capabilities are just as rich.

10.2.2.2   The Reason for the GUI

You should always keep in mind the reason (or reasons) for building a GUI (especially in MATLAB). These reasons stem from the fundamental goal of the GUI of being a useful and reliable tool for accomplishing a larger task. The nature of the tasks you are likely to use GUIs in MATLAB for generally involve automating laborious computations, or searching for or learning about information content in data.

If the GUI is to be used primarily as a tool that helps you accomplish a larger task, then you will want to pay particular attention to methods that:

1. Reduce the demands on the user.

2. Match the user's workflow.

3. Take advantage of accepted interface standards.

When your goal is to expedite a laborious task, keeping things simple should be a rule. Keep the number of windows, decision points, etc., to a minimum. Don't expect a user (or even yourself if you are the user) to learn new ways to do the same old things; put basic pull-downs in the menu bar, use universal accelerators, e.g., CTRL-C for copy, and use accepted language for common descriptors, e.g., "Save" and "Save As...".

If the GUI is to be used for searching for information, such as gleaning data for specific statistical content, looking at data from different perspectives or with different plot types, then it is important that you build in the ability for users to quickly change between different presentations of the data, change resolutions, and dialog with data processing methods. GUIs of this nature should:

1. Provide flexibility.

2. Quickly go back and forth.

3. Not overwhelm the user.

The GUI should be flexible in that the user can select from a list of data searching perspectives and statistical methods. The user should be able to start broadly, and then narrow the search. The user should be able to quickly apply different methods or plot techniques, and "undo" if the selection turned out to be undesirable. Finally, don't overwhelm the user with too many choices. Arrange choices in a logical fashion and limit how much the user must remember. Provide helps and tips where necessary.

10.2.2.3   Cognitive Considerations

Cognition refers to people's ability to think and learn. There are a few rules of thumb you should keep in mind when developing your GUIs that will make using your GUI both intuitive and a generally pleasant experience.

1. Don't require the user to remember many things at once: In general, people can remember about seven new things for about twenty seconds. With MATLAB you can help the user remember by using the *Uicontextmenu* property to include "right-click". Perhaps the easiest to use memory aid you can include in your MATLAB GUIs is the *ToolTipString* (see Section 10.3.2.12).

2. Organize functions and operations into logical groupings: You can use frames in MATLAB (see Section 10.3.1.3) to indicate groupings of user interface control objects (buttons, text fields, etc.) and separators (see Section 10.4.1.9) in pull-downs and other user interface menus.

3. Present information in the proper context: If things don't appear to be where they should be, or if they don't happen when expected, a

user can become frustrated with a GUI. Remember to give your GUI a descriptive title using the figure property *Name*. In addition, label controls and axes as appropriate. If you have to wonder what something is, it probably needs a label.

4. Strive for consistency in your GUIs: Most people know to look to the upper left in a menu bar to find tools that let you save or open a file, print, etc. This is just the standard that we have all become accustomed to; don't be arrogant and force your users to look in places that are not standard.

10.2.2.4 Physical Considerations

Don't lose sight of the fact that you (or your users) must interact physically with your GUI. That means they will have to use their eyes, hands, and possibly their ears. (Yes, you can use sound in MATLAB but we do not explore that in this text.) Whatever your GUI accomplishes, the user must use the keyboard, mouse, and monitor to effectively interact with the computer. Some rules of thumb here are:

Keep accelerator key combinations simple, e.g., CTRL+SHIFT+Character requires three fingers so should probably be avoided (unless you want to make the action very deliberate). Don't mix mouse and keyboard commands without careful consideration. It is best to keep the interface predominantly one or the other. If the text entries are always the same, then consider using a list box (see Section 10.3.1.5); if they are always different consider using editable text (see Section 10.3.1.2). The visual display should not be too busy or have too many colors as this can obscure the presentation of data and interface controls.

10.2.3 Paper Prototyping

Perhaps the most effective GUI development process you can do before actually creating your GUI is to create a paper prototype. Simply put, take a sheet of paper, and sketch just how you want the GUI to appear to the user. Of course, this is done after you have determined what the goals of the GUI are to be. The paper prototype is a design mockup that lets you explore the layout of your user interface objects, buttons, dialogs, etc., and data presentation components, e.g., plots. You will be trying to optimize the position and organization of your GUI to best accomplish your goal. If your task is large, or if you are part of an organized software (or analysis) team effort, your paper prototype can also be used to communicate your understanding of the GUIs goals with the rest of the team.

10.2.3.1 Appearance

Soon in this chapter, we will be developing a GUI. In this GUI, we want the user to be able to specify an arbitrary function and arbitrary range over which the function will be plotted. We also want the user to be able to easily change some of the plot features. Since this GUI is simple, we can probably assume a single window with a single axes object and some uicontrols to let the user quickly change things. Figure 10.1 shows our paper prototype.

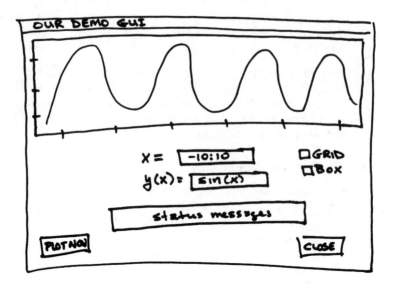

Figure 10.1 Paper prototype of a GUI we will build.

This paper prototype is simple, since we will use this GUI to demonstrate many things. Regardless of the complexity, the paper prototyping approach is always a good way to start.

10.2.4 Construction

Ah, here is the part for which you are waiting! Now that you know how you want to use your GUI, what information is presented through it, what features you will need, how you will arrange your objects, etc., you can start building something that works. The bulk of this chapter deals with uicontrol and uimenu objects and their properties and how to use them in constructing MATLAB GUIs.

Depending on the complexity of the GUI task you are undertaking, you might find the need to prototype the GUI. (This can be particularly easy with MATLAB's GUIDE.) Your prototype can help you identify flaws in your design before you have invested too much time in implementation. By prototyping, we mean creating the user interface portion without detailing the functions that respond to the user actions (callbacks). First, we will explore the uicontrol and uimenu objects and their properties, then use MATLAB's Graphical User Interface Development Environment (GUIDE) to get a GUI running quickly, and finish this chapter with a look at some specific GUI applications that demonstrate GUI capabilities.

10.3 UI Control Elements

Most of the MATLAB user interface control, or uicontrol, elements are created with the purpose of performing an action or setting up the options for a future action. The action is executed or the option is set when the user selects the uicontrol with the mouse pointer. As you will see, there are different methods of selecting the various uicontrol objects. However, the act of selecting usually consists of moving the mouse pointer directly over the object and clicking the mouse button.

This section has been designed to introduce the set of uicontrol object styles, the type of actions each style is normally used for, and the properties that are associated with every uicontrol object. This will be accomplished by means of descriptions, tables, and examples. It is essential that you have a good understanding or at least are familiar with the various properties, so that the advanced programming techniques discussed in later sections are clear and easy to follow.

10.3.1 The Styles

The ten styles of MATLAB uicontrol objects along with a brief description are listed below.

UI Control	*Style* value	Description
Check Box	'checkbox'	indicates the state of an option or attribute
Editable Text	'edit'	user editable text box
Frame	'frame'	used to visually group controls
Pop-up Menu	'popup'	provides a list of mutually exclusive options
List Box	'listbox'	shows a scrollable list of selections
Push Button	'pushbutton'	invokes an event immediately
Radio Button	'radio'	indicates an option that can be selected
Toggle Button	'toggle'	only two states, "on" or "off"
Slider	'slider'	used to represent a range of values
Static Text	'text'	displays a string of text in a box

Each style will be discussed along with example illustrations of its appearance in different states.

10.3.1.1 Check Boxes

The check box uicontrol (*Style* property set to "checkbox") is a useful means of representing two states of an option that you may want to provide. The two states will be referred to as "on" or "off" for simplicity, but can just as easily indicate true/false, yes/no, or some other bipolar combination. In its off state, the check box will consist of an empty (Macintosh or MS-Windows) or unfilled (X-Windows) square with some type of label located on the right-hand side of the box.   The label should be descriptive enough that the user understands the implications of setting the box to its on or off state. In the on state, the check box's square will contain a "✓".

The state of a check box can be changed by clicking the mouse over any portion of the uicontrol. You are not restricted to clicking in the actual square as illustrated below. If the user had clicked on the text or the shaded gray region, the check box's state would have toggled as well. By following the arrows in Figure 10.2, we see the manner in which the appearance of the check box changes from one state to the next. The intermediate states appear only during the time that the user is pressing the mouse button when the mouse is over the uicontrol object.

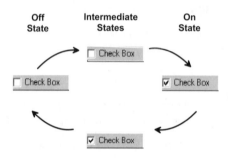

Figure 10.2 Checkbox states.

If you need more than the two bipolar choices a check box offers, look at pop-up menus or radio buttons (both are discussed later in this chapter) as an alternative uicontrol object. Multiple check boxes are convenient in situations where the user may have several options that can be simultaneously selected. In addition, it is recommended that when you can group a set of check boxes in terms of some type of similarity (e.g., function or importance), you should visually group them with their physical location and, where appropriate, with a frame object (see Frames later in this chapter).

10.3.1.2   Editable Text

The editable text style (*Style* property set to "edit") is used in situations that require the user to enter strings of characters or numbers. The strings, in turn, are used by the application for which the interface has been built.   For instance, rather than prompting the user at the command line for a string, you can create an editable text uicontrol. Later, you will see that this uicontrol can be appropriately sized to contain one or more text lines.

The editable text item can be initialized with a string or string matrix that the user can delete, edit, or leave alone. Clicking anywhere within this object will change the mouse from a pointer to a text insertion indicator. Once the text insertion indicator is available, characters can be inserted by typing the desired keys or deleted by using the delete or backspace key. Portions or all of the text can be highlighted by click and dragging within the uicontrol item, to allow for quick string replacement or deletion. Highlighted text will be replaced with the next keyboard character that is pressed.

In Figure 10.3, the editable text uicontrol has been initialized with "Editable Text". The portion of the string "able" is highlighted and then replaced with "ed".

Figure 10.3 Editable text

Editable text elements are often used in conjunction with a static text uicontrol (see Static Text later in this chapter) so that the user is aware of what he or she is providing the application. It is a good idea, whenever possible, to initialize the editable text uicontrol with the default value of the string so that the user does not always need to type in the most likely string.

10.3.1.3   Frames

The frame object (*Style* property set to "frame") serves no purpose in terms of action-related responses to a user's mouse click. However, it is usually used to serve as an important visual aid. Other uicontrol items may be visually grouped with a frame so that the appearance of the GUI guides the user's actions. It is an extremely effective method of organizing the GUI in a logical and intuitive fashion.

The frame makes the GUI more aesthetically pleasing by providing a solid background that helps blend a set of uicontrols into one complete and cohesive interface. If the colors remain in their default values or are appropriately chosen, the edges of other uicontrol objects like static text, check boxes, and radio buttons will no longer be distinctly visible. Figure 10.4 shows how several miscellaneous uicontrol objects can be combined with two frame objects into one interface.

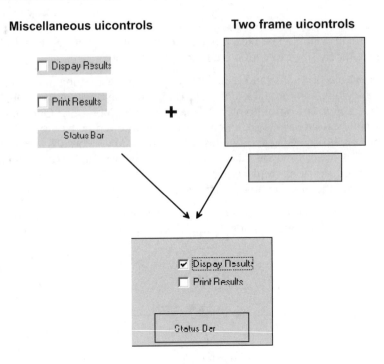

Figure 10.4 Use frames to create logical groupings.

10.3.1.4   Pop-Up Menus

A pop-up menu (*Style* property set to "popup") is usually used in situations where multiple choices need to be available to the user. The current selection is displayed in an unopened pop-up menu. However, when the user clicks and holds down the mouse button anywhere within the object, a list of choices will appear. Another choice can be made by dragging the mouse over to any of the choices and releasing the mouse button. The example found in the following illustration shows that first the "Pop-up Menu" choice is displayed. The user then clicks and drags the mouse pointer down to the choice represented by "Option 2" and releases the mouse. In the final state, we see that "Option 2" is the current selection.

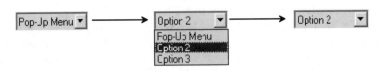

Figure 10.5 Use pop-up menus to pick one of many choices.

A pop-up menu is readily recognized by the down pointing triangle symbol appearing on the right-hand side of the object.

10.3.1.5   List Boxes

List Boxes (*Style* property set to "listbox") are very similar to pop-up menus. Essentially they are used to provide users with a set of options from which they can choose one. The main difference with a list box is that you can make the set of options visible to the user at all times (depending on the size of the box you make and the number of items in the box). Then, depending on the size of the box, the user may need to scroll through the list to find the option he or she desires. Once the item is found, the user must click on it to select it. With the current version of MATLAB you are not able to select more than one item.

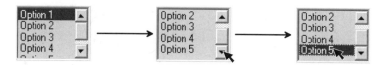

Figure 10.6 List boxes let you display as many choices as you wish.

If any of the items that the user can select from are wider than the box, a scroll bar will be placed on the bottom edge of the box. Finally, if the number of items are all visible in the space provided by the box, the scroll bar on the right hand side becomes disabled.

10.3.1.6   Push Buttons

The push button (*Style* property set to "pushbutton") is perhaps the most prevalent uicontrol style. It is used primarily to indicate that a desired action should immediately take place. Since push buttons represent actions, they are usually labeled with a verb, e.g., start, run, install, etc., that describes the action that will take place if the user clicks on the button.

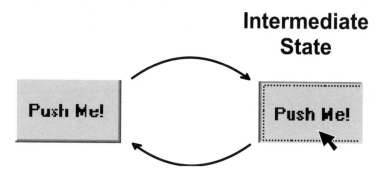

Figure 10.7 Push buttons are for immediate actions.

Push buttons have a 3-dimensional look that makes it appear as if they are being pressed when the user clicks on the object. In addition, they are very similar in appearance on all computing platforms.

10.3.1.7   Toggle Buttons

The toggle button (*Style* property set to "toggle") looks just like a push button, except there is no intermediate state. Rather, the button will remain in its selected or not selected state after the user clicks on it. Functionally, it is very similar to a check box user interface, since there are two states associated with it.

Figure 10.8 Toggle buttons are binary.

The toggle button is considered to be selected when it looks as if it is pressed in, and unselected when it looks like it is raised out of the screen.

10.3.1.8   Radio Buttons

The radio button uicontrol style (*Style* property set to "radio") is similar to the check box in that there are two states associated with each button. The difference lies in the manner in which they are normally used. Usually two or more radio buttons are "linked" together as a group. They are linked in the sense that only one of the buttons will be in its selected (i.e., on) state.

The individual radio button consists of a circular- (Macintosh and MS-Windows) or diamond- (X-Windows) shaped symbol with an accompanying label. The label should be descriptive enough that the user understands the implications of setting the radio button to its on or off state. In its off state, the radio button will be empty. In the on state, the circle will contain a dot (Macintosh or MS-Windows) or the diamond will be filled in (X-Windows).

The following figure indicates the appearance of the radio button as it transitions from its off state to its on state and back.

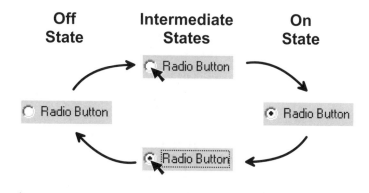

Off State **Intermediate States** **On State**

Figure 10.9 Radio buttons are either "off" or "on".

If your GUI has more than one set of linked radio buttons, you should separate them with enough space or with multiple frames so that the group is visually distinct from another group of radio buttons.

10.3.1.9 Sliders

Sliders (*Style* property set to "slider") are useful in representing to users that they have a fixed range of values from which to choose. In its most common form, the slider is comprised of a trough, an indicator bar, and a set of arrows. The trough represents the range of values, while the location of the indicator bar within the trough represents the current value specified by the slider. The arrows are supplied to assist in moving the bar in one direction or another.

The user moves an indicator bar to specify a desired value from within the allowable range. This bar can be moved in one of several ways. The first is accomplished by click and dragging the indicator bar to a new location within the trough. The second is accomplished by clicking within the trough on the side of the indicator bar that corresponds with the direction in which the bar should move. This will shift the bar by approximately 10% of the total range specified by the trough. The final method is to click on the arrow that points in the direction in which it is desired that the bar move. This will shift the bar by approximately 1% of the total range specified by the trough. In the example below, the user clicks and holds the right-hand arrow until the indicator bar has shifted to the desired setting.

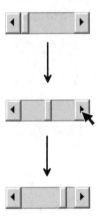

Figure 10.10   **Use sliders to select a fixed range.**

Depending on its size, the slider may consist of only a trough and indicator bar when used with an X-Windows version of MATLAB. If either the length-to-width or width-to-length proportion is less than four to one, the X-Windows slider will consist of only a trough and indicator bar. Macintosh and MS-Windows sliders will always consist of all three slider components.

The slider has no way of explicitly indicating the numeric value that the slider represents; therefore, it is recommended that an editable text or static text style uicontrol accompany the slider. The text uicontrol will allow the user to see the numeric value to which the slider is set. Furthermore, editable text would allow the user to manually type in an exact value. It is also recommended that the limits of the range be shown with one of these text uicontrol elements. Later on in this chapter, you will learn how to create a GUI that contains this type of slider/text uicontrol combination. The circumstances and ultimate purpose of the GUI will most likely dictate the requirements regarding the appearance and amount of information that needs to be presented to the user.

10.3.1.10 Static Text

The static text style (*Style* property set to "text") of uicontrol is available for creating labels, status messages, or other information pertinent to the user. The text graphics objects (i.e., those objects created with the text command) cannot be placed on top of frames. Therefore, if you are using frame objects and want to create labels, you will need to use the static text style.

Static text does not perform any action if the user clicks on any part of the object. In addition, the user cannot edit the information that is displayed.

10.3.2 UI Control Properties

Just as with all other MATLAB graphics objects, uicontrol objects have a set of properties that can be manipulated to suit your needs and help you obtain the look you want for your GUI. The following table lists all of the properties associated with a uicontrol object. Each row contains the property's name, the read-only status, the property values (the default value is contained in "{}"), and the format of the value. Note that these objects also contain the universal properties discussed in Chapter 3. Some of the universal properties, such as *ButtonDownFcn*, are shown below and discussed in a following section since they have a special or somewhat different meaning with uicontrol objects.

Property	Read Only	ValueType/Options	Format
BackgroundColor	No	[Red Green Blue] or color string	RGB row
ButtonDownFcn	No	string	row
CData	No		matrix
CallBack	No	string	row
Enable	No	[on | {off} | inactive]	row
Extent	Yes	[0,0,width,height]	row
FontAngle	No	[{normal} | italic | oblique]	row
FontName	No	string	row
FontSize	No	number	1 element
FontUnits	No	{points} | normalized | inches | centimeters | pixels	row
FontWeight	No	[light | {normal} | demi | bold]	row
ForegroundColor	No	[Red Green Blue] or color string	RGB row
HorizontalAlignment	No	[left | {center} | right]	row
Interruptible	No	{on} | off	row
ListBoxTop	No	number	1 element
Max	No	number	1 element
Min	No	number	1 element
Position	No	[left bottom width height]	4-element row
String	No	string	string matrix
Style	No	[{pushbutton} | radiobutton | togglebutton | checkbox | edit | text | slider | frame | popupmenu | list box]	row
SliderStep	No	number	1 element
TooltipString	No	string	row
Units	No	[inches | centimeters | normalized | points | {pixels}]	row
UIContextMenu	No	handle	1 element
Value	No	number	1 element
Tag	No	string	row
UserData	No	string(s) or number(s)	matrix
Visible	No	[{on} | off]	row

10.3.2.1   Uicontrol BackgroundColor

The *BackgroundColor* property defines the color of the region defined by the uicontrol object's *Position* property. You may define the value with either an RGB intensity triplet vector or a legal color specification string (e.g., 'red', 'r', 'white', etc.). By default the background color will be a light gray whose RGB intensity triplet is [0.8314   0.8157   0.7843].

10.3.2.2   Uicontrol ButtonDownFcn

The *ButtonDownFcn* (button down function) property is a string of one or more legal MATLAB expressions that specify the action that should take place if the user clicks the mouse button down on top of a narrow strip that runs along the object's perimeter. Please make a distinction between this property and the *CallBack* property. The action stored in the *ButtonDownFcn* is not evaluated when the mouse button pointer location is within the region of the uicontrol defined by its *Position* property. This string is evaluated as if the command

```
eval(buttondownfcnstring)
```

had been typed in at the command line, where *buttondownfcnstring* is the string stored in the *ButtonDownFcn* property. Therefore, if it requires any stored variables, the variables must be available at the base MATLAB workspace (not the function workspace). Finally, if you are not sure whether or not a *ButtonDownFcn* string is considered legal, see Section 10.5.1 ("Strings of MATLAB Statements and Expressions").

10.3.2.3   Uicontrol CData

Just as we saw *CData* with image, surface, and patch objects, uicontrol objects have a *CData* property as well. The value of the *CData* property is an M-by-N-by-3 matrix of RGB values that specify an image that can be on both push buttons and toggle buttons.

10.3.2.4   Uicontrol CallBack

The *CallBack* property specifies the action that is performed when the user clicks within the uicontrol boundary as defined by its *Position* property. Just as with the *ButtonDownFcn*, the *CallBack* property stores a string that is evaluated in the base MATLAB workspace. As long as the string can be evaluated error free with the command

```
eval(callbackstring)
```

from the command line (i.e., all variables it requires exist in some fashion), there will be no error messages invoked when the uicontrol button is activated by the user. The *CallBacks* of frame and static text uicontrols will never be evaluated. They were not designed with the purpose of performing an action if the user clicks on them. Editable text objects can be activated in the following instances:

- the string is altered and the user moves the pointer outside the editable text region, or

- the user presses the return key in a single line editable text object, or

- the user presses the control-return (X-Windows or MS-Windows) or command-return (Macintosh) keyboard combination

All other uicontrol object styles will be activated when the user clicks down and releases the mouse button anywhere within the object's perimeter as defined by the *Position* property.

Examples of the *CallBack* will be provided later when we create and program the GUI.

10.3.2.5   Uicontrol Enable

The *Enable* property can be set to "on", "off", or "inactive". If it is set to "off" or "inactive", the user will not be able to activate the uicontrol and, correspondingly, no callback action will occur as a result of a mouse click on the object. In its default value of "on", the uicontrol will perform the action defined by its *CallBack* when the user clicks on the object.

As illustrated in the figure below, the text that is displayed in a uicontrol object will become *dim* when the *Enable* property is set to "off". When the uicontrol is in its inactive or "on" setting it will look the same (shown on the left below); however, in the inactive state, the user cannot execute the callback by clicking on the uicontrol.

Figure 10.11   **Enable property can deactivate a control.**

If a uicontrol object's *Enable* property is "on", clicking with the left mouse button causes MATLAB to perform the following actions in the order shown:

1. Set the figure's *SelectionType* property.

2. Execute the control's *CallBack* routine.

3. MATLAB will not set, i.e., update, the figure's *CurrentPoint* property and will not execute either the control's *ButtonDownFcn* or the figure's *WindowButtonDownFcn* callback.

If a uicontrol object's *Enable* property is set to either "inactive" or "off", then left-clicking on it causes the following to take place in the order shown:

1. Sets the figure's *SelectionType* property.

2. Sets the figure's *CurrentPoint* property.

3. Executes the figure's *WindowButtonDownFcn* callback.

4. Executes the control's ButtonDownFcn callback.

5. Executes the selected context menu item's Callback routine.

6. Does not execute the control's Callback routine.

The previous also occur as shown anytime you right-click on a uicontrol object, regardless of the setting of *Enable*. If you right-click on the object, the objects context menu (which will be discussed later in this section) will be shown if one has been associated with it.

A particular use of setting this property to "inactive" or "off" is to enable you to implement object dragging or resizing using the *ButtonDownFcn* callback routine.

10.3.2.6   Uicontrol Extent

The *Extent* property is a read-only four-element vector that specifies the size and position of the character string used to label the uicontrol. It is of the form [0,0,width,height] where the first two elements are always zero and *width* and *height* are the dimensions of the rectangle. These are in units specified by the *Units* property.

Since the *Extent* property is defined in the same units as the uicontrol itself, it is particularly useful in determining the proper sizing for the uicontrol with regard to its label. You can do this by first defining the *String* property and setting the font using the relevant font properties, then get the value of the *Extent* property. All you need to do then is set the *Position* property to be slightly larger than the *width* and *height* values of *Extent*.

If you have more than one line of strings, the *Extent* rectangle encompasses all the lines of text. For single line strings, the *Extent* is returned as a single line, even if the string wraps when displayed on the uicontrol object.

10.3.2.7   Uicontrol ForegroundColor

The *ForegroundColor* property specifies the color of the label and symbols (e.g., the square in a check box) of an uicontrol object. You may define the value with either a RGB intensity triplet vector or a legal color specification string (e.g., 'red', 'r', 'white', etc.). By default, the foreground color will be black, i.e., [0 0 0].

10.3.2.8   Uicontrol Font Angle, Name, Size, Units, and Weight

These properties allow you to change the font characteristics of the text label associated with each uicontrol object. These are basically the same set of properties that affect the appearance of text objects. The *FontAngle* is by default set to "normal", but you can also set it to "italic" or "oblique".

Figure 10.12   FontAngle

The *FontName* and *FontSize* properties can respectively be the name of your favorite font that your system supports and a value corresponding to how big or small you want your fonts. The *FontUnits* property is used to specify the units used by the *FontSize* property. The default is "points" where 1 point is 1/72 of an inch. When set to "normalized" the unit is set to a fraction of the height of the uicontrol so that if you resize the uicontrol the font size will change accordingly.The *FontWeight* property can be set to "normal", "light", "bold", or "demi" to give you the following look to a label:

Figure 10.13   FontWeight

10.3.2.9   Uicontrol HorizontalAlignment

The *HorizontalAlignment* dictates how the text label is displayed on the uicontrol. The next figure illustrates left, center, and right alignment of the label in a static text uicontrol.

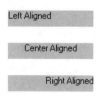

Figure 10.14   HorizontalAlignment

As a note, the push button labels are always center aligned, while the check boxes, radio buttons, editable text, and pop-up menu labels will come up, by default, left aligned. (You can change the default by editing the root objects *defaultUicontrolHorizontalAlignment* property.) There is no property available that allows you to align the label in the vertical direction.

10.3.2.10 Uicontrol Min, Max, and Value

The significance of these three properties is different for each of the uicontrol object styles. However, in all uicontrols the values of the *Min* and *Max* properties must be scalars and are by default set to zero and one, respectively. Even though the *Value* property can take on a row or column vector, it will always revert to a scalar after the uicontrol has been activated.

Power!

For check boxes and radio buttons, the number that is stored in the *Min* property will be used to set the value of the *Value* property when the check box or radio button transitions to its "off" state. The number that is stored in the *Max* property will be used to set the value of the *Value* property when the check box or radio button transitions to its "on" state.

The difference in the values stored in the *Max* and *Min* properties determines whether or not an editable text uicontrol can contain a single or multiple lines of text. If the difference is greater than one, the editable text uicontrol can have multiple lines of text; otherwise, it will have only a single line of text. Unfortunately, it does not specify how many lines can be entered by the user. As long as

$$Max - Min > 1$$

holds true, the user can enter as many lines as he or she desires.

For push buttons, the value stored in the *Max* property will be transferred to the *Value* property for the period during which the *CallBack* is being evaluated. After the *CallBack* is completed, the *Value* property will once again be set to the value stored in the *Min* property.

The range of values that governs the trough of a slider uicontrol object is defined by the *Min* and *Max* properties. The *Value* property will contain the numeric value that corresponds to the position of the indicator bar. If the slider is drawn horizontally, the *Min* property value will correspond to the value of the slider when the indicator bar is as far left as it can go. In addition, the *Max* property value will correspond to the value of the slider when the indicator bar is as far right as it can go. In the event that the slider is drawn vertically, the *Min* property value is associated with the bottom of the trough and the *Max* property value is associated with the top of the trough.

The *Min* and *Max* properties have no meaning for pop-up, static text, and frame uicontrols. The *Value* property does indicate which pop-up menu choice is being displayed (an integer indicating the first, second, etc. item); however, it has no meaning for static text and frame objects.

10.3.2.11 Uicontrol SliderStep

The *SliderStep* property affects only slider uicontrol objects. It is a 2-element row where the first element (by default 0.01) specifies how far the bar should move as a fraction of the entire length of the trough when the user clicks on one of the arrows on either end of the slider. The second element (by default 0.1) specifies how far the bar should move as a fraction of the entire length of the trough when the user clicks in the trough.

Additionally, *SliderStep* sets up the granularity of the values to which the slider bar can be moved and the values that are stored in the *Value* property of the slider uicontrol object. For example, this feature can be used to relieve you of having to round or manipulate the value returned. If you created a slider object with

```
slider_handle = uicontrol('style','slider',...
            'sliderstep',[.2 .25],'max',10,'min',0,...
            'position',[10 10 200 10]);
```

and then you moved the slider bar around and did a

```
value = get(slider_handle,'value');
```

you are guaranteed that the value will be one of the values in the matrix [0 2 2.5 4 5 6 7.5 8 10].

10.3.2.12 Uicontrol TooltipString

The *TooltipString* property stores a string that is displayed whenever a user allows the mouse cursor to loiter over the uicontrol. This property is very useful in providing brief explanations or reminders to help the user in using a GUI.

10.3.2.13 Uicontrol Position

The location of the uicontrol object in the figure object is specified with the *Position* property. The format of the position vector is the usual [left bottom width height] vector that we have seen with other graphics objects. Since the parent object of the uicontrol is the figure, the position vector is defined with respect to the lower left-hand corner of the figure. The units of this vector are specified by the *Units* property.

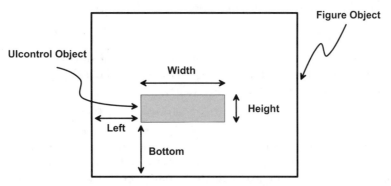

Figure 10.15   **The location of uicontrol objects given by Position.**

10.3.2.14 Uicontrol String

The displayed text labels and choices that appear on the uicontrol are specified by means of the *String* property. This property can be set to a string matrix or vector. If there is more than one row in the matrix, only the first row will be used to label push buttons, check boxes, radio buttons, and static text. Editable text (*Style* = "edit") and pop-up menus make use of the additional rows. If the editable text object has been set up to allow multiple lines (see Section 10.5.8, "UIcontrol *Min*, *Max*, and *Value*"), the additional rows will correspond to the displayed lines of text; otherwise, only the first row of the string matrix will be used. For pop-up menus, each row corresponds to a choice in the list of items that appears when the user clicks the mouse pointer on top of the object. Frames and sliders do not make use of the String property.

Speed!

As a helpful speed hint, you do not necessarily need to create string matrices with the usual matrix format or with **str2mat**. When setting the *String* property of a uicontrol object, MATLAB recognizes the character "|" as the end of a row. For instance, if you wanted a string matrix that had the words "Apple," "Banana," and "Pear" on different lines, you could create a cell array by typing

```
string_value = {'Apple';'Banana';'Pear   '};
```

or

```
string_value = str2mat('Apple','Banana','Pear');
```

However, the simplest and quickest way is to type

```
string_value = ['Apple|Banana|Pear'];
```

which does not require you to count characters or to use another MATLAB function. Please note that this works only when you want to create a string matrix for uicontrol objects. The first two forms actually store a string matrix (i.e., in this case a 3-by-6 matrix of characters) in the variable *string_value*; however, the method that uses the "|" stores a row vector (i.e., a 1-by-17 vector of characters) in the variable *string_value*. But once you

```
set(uicontrol_handle,'string',string_value)
```

FAQ

where *uicontrol_handle* is the graphics handle of a uicontrol object, the *String* property will contain the same string matrix regardless of which one of the methods was used to create *string_value*.

10.3.2.15 Style

The *Style* property specifies whether the user interface control object will be a check box, editable text, frame, pop-up menu, list box, push button, toggle button, radio button, slider, or static text component. The default *Style* value creates a push button. The table in Section 10.4.1 shows the value that you need to use for the *Style* property to create each type of control object.

10.3.2.16 ListBoxTop

List boxes have their own special property called *ListBoxTop*. For a given set of items that are specified in the *String* property of the list box uicontrol object, you can specify which item is at the top of the visible portion of the list. This is applicable only if there are more items than fit in the space provided by the Position property. For example, if you typed

```
u = uicontrol('style','listbox',...

'string','Option1|Option2|Option3|Option4|Option5',...
     'position',[10 10 75 50]);
```

you would get the following list box.

Figure 10.16   **A typical list box.**

Without clicking on the arrows, you can specify that "Option3" be at the top of the visible list of items by setting the *ListBoxTop* property with

```
set(u,'listboxtop',3)
```

to get the list shown below.

Figure 10.17   **Same list box with ListBoxTop set to 3.**

The actual order of the list box has not changed, but rather the list box has been positioned such that item 3 is at the top.

10.3.2.17   Uicontrol Units

The location of the uicontrol object is specified by the *Position* property in units specified by the *Units* property. By default, the units will be in pixels; however, you may also choose them to be in inches, centimeters, points, or normalized.

Pixels, inches, centimeters, and pixels are referred to as absolute units, while normalized units are considered a relative unit of measurement. If a uicontrol object uses absolute units, its size will be independent of the size of the figure object parent. In fact, if the figure is too small, the uicontrol objects may be located outside the boundaries of the figure and will not be seen by the user. However, if the units are "normalized", the control objects will scale proportionately with respect to the figure object parent (the lower left-hand corner is considered to be (0,0) and the upper right-hand corner is (1,1)).

10.3.2.18   Uicontrol Interruptible

The *Interruptible* property will be discussed in detail later in this chapter (see Section 10.8.3, "Interruptible vs. Uninterruptible"). Before this property can be fully understood, it is essential for you to be familiar with the various types of MATLAB events and how they are processed. For now, suffice it to say that this property controls whether or not the execution of a *CallBack* can be interrupted by another event, such as clicking the mouse button on a uicontrol object.

The *Interruptible* property can be set to "on" or "off". By default the value is "on" which means that the *CallBack* can be interrupted to execute the action associated with another event. Only after the interrupting action has been completed, can the interrupted *CallBack* be completed. On the other hand, a value of "off" means that a uicontrol's *CallBack* execution cannot be interrupted, so must be completed before another event (such as the *CallBack* of another uicontrol) object can be executed.

10.3.2.19   Uicontrol Tag

This property is extremely useful when programming GUIs in a fashion that requires MATLAB to search for the handle of a specific uicontrol object. The *Tag* property can contain a string row vector of your choice. It is usually assigned a descriptive name that uniquely identifies a particular uicontrol object from all the other uicontrol objects. This property does nothing in terms of the appearance or the action associated with the uicontrol object. Using the *Tag* property in conjunction with **findobj** has already been mentioned in Section 7.8; further examples that illustrate this property's usefulness will be presented in Section 10.5.2.4 of this chapter.

10.3.2.20   Uicontrol UserData

The *UserData* property is essentially used as a storage facility that accompanies the object until it is deleted. It is unaffected by the **clear** command and therefore is a safe place to store matrix data that you want associated with a uicontrol or that you want to be able to access regardless of the state of MATLAB's base workspace. By default, this property contains the empty matrix, [ ]. You can store any valid MATLAB data type in this property. You can use structures or cell-arrays to store a mix of data types.

10.3.2.21 Uicontrol Visible

By default, the *Visible* property is set to "on" so that the object can be seen by the user. However, in certain circumstances, you may wish to make the uicontrol object invisible. The uicontrol object can be made invisible by setting this property to "off". This becomes useful when you want to have layers of uicontrol objects in the same figure.

Power!

The *Visible* property can also be used to limit the number of uicontrol objects that are displayed at once. For instance, you may want to program a GUI so that the state of a particular uicontrol object dictates whether or not other uicontrol objects are available to the user. This can be quite important to the readability of the GUI. It helps reduce the chance that the user will be overwhelmed with too many controls and options to an application. Over time, as the user becomes familiar with the application, he or she can explore additional features by bringing them to view.

10.3.2.22 Other UI Control Properties

The *Type* property specifies the kind of MATLAB graphics object and is always set to "uicontrol" for an uicontrol object. This is a read-only property.

In terms of the object's family tree, the parent of a uicontrol object will always be the handle to the figure in which the object is drawn. This handle will be stored in the *Parent* property. The *Children* property will contain the empty matrix because uicontrols have no children.

10.3.3 Creating Uicontrol Objects

UIcontrol objects are created with the **uicontrol** command. In this section of the chapter we merely present a few examples of creating uicontrol objects since here we are purely interested in the appearance of the GUI, not the functionality. Defining or coding the *CallBack* property will be discussed in Section 10.5 ("Low-Level GUI Programming Techniques").

There are a few basic forms of the uicontrol function that you can choose from. The simplest is

```
handle = uicontrol('Property1Name',Property1Value,...
                   'Property2Name',Property2Value,...
                        .
                        .
                        .
                   'PropertyXName,PropertyXValue)
```

where you specify the attributes of the uicontrol object with as many *PropertyName/PropertyValue* pairs required to fully describe the object. In this sample form there are X pairs. The *PropertyName* can be any one of the properties listed in the table found in Section 10.3.2 and *PropertyValue* is any legitimate value that can be assigned to the property.   This form of the uicontrol command will create the uicontrol object in the current figure.   If there is no current figure object, a new figure will be created.

You may also use the form

```
handle = uicontrol(figure_handle,...
                'Property1Name',Property1Value,...
                'Property2Name',Property2Value,...
                        .
                        .
                        .
                'PropertyXName,PropertyXValue)
```

which forces the uicontrol object to be created in the figure with graphics handle *figure_handle*. In both forms, it is at your discretion to decide whether or not the uicontrol object's graphics handle needs to be stored.   In the examples above, the control objects graphics handle is stored in the variable *handle*.

10.3.3.1   Uicontrol Object Layering

The order in which you create your control objects is very important, especially if your GUI will contain frame objects. The first uicontrol that is created will lie at the "bottom," while every additional uicontrol object will be closer to the viewer than the previous.   All uicontrols will be drawn in front of all other graphics object types. The next example illustrates both of these points.

```
figure('position',[100 200 250 160]);
uicontrol('style','frame',...
        'position',[10 50 160 80]);
uicontrol('style','pushbutton',...
        'string','Close Figure',...
        'position',[30 70 80 20],...
        'callback','close');
uicontrol('style','frame',...
        'position',[80 10 70 130]);
axes
```

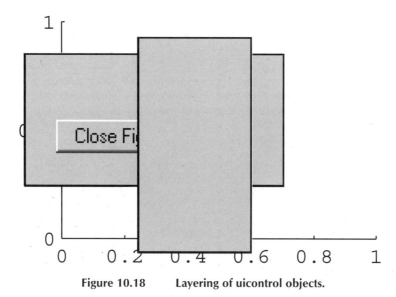

Figure 10.18   **Layering of uicontrol objects.**

In Figure 10.18 above, we see that the axes object is drawn below all other objects even though it was created last. The other point seen is that the uicontrol objects, on the other hand, are layered in terms of the order in which they were created. A useful feature of MATLAB is that the layering of uicontrol objects can be controlled by manipulating the order of their handles in the figure object's *Children* property. The handle of a uicontrol object with a lower index in the figure object's *Children* property will be drawn closer to the user (i.e., will be drawn above the other uicontrol objects).

10.3.3.2   Framing Objects

In the previous example, the default units (pixels) were used to define the positions of the objects. Pixels or any other absolute unit of measurement is often useful when defining the distance between two objects when creating a GUI. Since different monitors have different pixel spacings, if you use pixels as the unit of measurement, you can ensure that there is enough space so that the edges of one object remain distinguishable from any neighboring object's edges.

For example, let's say you want to create a set of radio buttons that are encompassed by a frame and that the figure should be only fairly compact. You first need to decide on the size and positions of the buttons and the static text label. This is done somewhat experimentally at first, until you gain a feel for how much space you need for the longest label. In this case, we find out that 100 pixels in width and 20 in height is sufficient. We could then decide that each button will be separated vertically with five pixels and that we want ten pixels between an edge of a button and the figure boundary. Start adding up the pixels and you should be able to determine that a figure object that is 120 by 100 pixels will do the job. The figure object is created with these dimensions and its *Resize* and *NumberTitle* properties are turned off. The *Resize* property is important because we do not want the user to change the size of the figure in this example. After the figure is created, create the frame, add a descriptive static text uicontrol object, and finally, create the three radio buttons. The code and the final result (Figure 10.19) are presented below.

```
h_fig = figure('position',[200 200 120 100],...
               'resize','off',...
               'numbertitle','off')
% Create frame object that covers entire figure region.
h_frame = uicontrol(h_fig,'style','frame',...
                          'position',[0 0 120 100])
% Create overall label.
h_stext = uicontrol(h_fig,'style','text',...
                         'string','Waveform Type',...
                         'position',[10 75 100 20]);
% Create set of three radio buttons.
h_radio(1) = uicontrol(h_fig,'style','radio',...
                            'string','Square Wave',...
                            'position',[10 55 100 20],...
                            'value',1);
h_radio(2) = uicontrol(h_fig,'style','radio',...
                            'string','Saw Tooth Wave',...
                            'position',[10 30 100 20]);
h_radio(3) = uicontrol(h_fig,'style','radio',...
                            'string','Sinusoidal Wave',...
                            'position',[10 5 100 20]);
```

Figure 10.19   Framed static text and radio button controls.

Please note that the code provided above performs no function and the radio buttons are not mutually exclusive, since we have not added any *CallBacks* to the uicontrols. To learn how to make mutually exclusive radio buttons, see Section 10.7.1.

10.3.3.3   A Stretchable GUI

If you are going to allow the user to resize the figure, it is recommended that you use normalized units (*Units* property set to "normalized") for all uicontrol objects within the figure. Using pixels or one of the other absolute units of measurement inside a resizable figure can lead to a situation where the user has resized the figure to such a degree that some of the uicontrol objects are no longer visible. With normalized units, the uicontrols will scale themselves with respect to the figure boundary. In the worst case, the user might shrink the figure to a point at which it is difficult to read labels; however, at least the user will be aware that the uicontrol objects exist and can always increase the size of the figure until the controls are once again readable.

Power!

In the next example we will create a resizable figure that has a pop-up menu, some static text, and some editable text (see Figure 10.20). From this example you can see how objects can be positioned with normalized units and how an editable text string can be created with multiple lines.

```
% Create the figure.
figure('position',[150 100 200 150],...
    'MenuBar','none',...
    'Color','white');
% Create the uicontrol objects with normalized units.
h_frame = uicontrol('style','frame',...
            'units','normalized',...
            'position',[0 0 1 1]);
h_stext_font = uicontrol('style','text',...
            'units','normalized',...
            'position',[.05 .1 .25 .15],...
            'string','Font:');
h_popup_font = uicontrol('style','popup',...
            'units','normalized',...
            'position',[.3 .1 .65 .15],...
            'string','Helvetica|Times|Courier|Symbol');

h_stext_color =  uicontrol('style','text',...
            'units','normalized',...
            'position',[.05 .3 .25 .15],...
            'string','Color:');
h_edit_color =  uicontrol('style','edit',...
            'units','normalized',...
            'position',[.3 .3 .65 .15],...
            'string','white');
% Create a multiple line editable text object
% by setting the Max property to a value greater
% than 1 plus the Min property (Min default = 0).
h_edit_multi = uicontrol('style','edit',...
        'units','normalized',...
        'position',[.05 .5 .9 .45],...
        'string',['Line Number 1|Line # 2|and line number
3'],...
        'max',2)
```

After you have created the previous GUI, resize the Figure Window so that you understand what happens to uicontrols with normalized units.

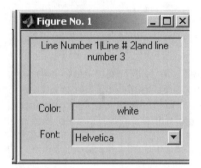

Figure 10.20   A stretchable GUI.

10.3.3.4   Predefined GUIs and Dialog Boxes

There are some functional GUIs that come with MATLAB that you may find useful. There are four "canned" dialog box generating functions that use some of the uicontrol objects presented in this chapter, namely: **errordlg** (error dialog), **helpdlg** (help dialog), **msgbox** (message box), **warndlg** (warning dialog), **inputdlg** (input dialog), and **questdlg** (question dialog).

The first four commands are essentially the same in the sense that you can display a message in a small Figure Window that also contains a push button labeled "OK". The figure will disappear when the user presses the "OK" push button. All three functions take at least two arguments. The first is the message that you want to have displayed in the dialog box, and the second is the name of dialog box figure (there is a default name supplied if you do not provide one). You may also pass the **errordlg** function a third string argument, 'on'. This will make sure that there is only one figure with the name provided in the second argument string (i.e., if another dialog figure has this name, its message window will be updated). The **warndlg** function also takes a third argument, 'replace', which will replace an existing warning dialog box that has the same window name with the new information. If a help dialog has the same name as the one you are creating with the **helpdlg** function, MATLAB will replace its message with the new string. All of these dialog functions will wrap your string as needed to fit the dialog box size; however, a cell array is preferred so that you define the string for each line with a new cell element. Some examples of these commands are given below

```
h_wfig = warndlg('Warning Message String',...
'Warning Dialog');
h_efig = errordlg('Not a valid input',...
'Input Error','on');
h_helpdlg = helpdlg('Try again!');
```

The difference between the commands is the procedure by which they deal with existing dialog boxes that have the same name (i.e., the figure *Name* property). Both the **helpdlg** and **warndlg** commands will bring an existing named dialog box to the front of the screen without updating its contents.

This is different from the way the **errordlg** command works with the third argument as described above.

The message dialog box, in addition to the message string and the window name string, can take more arguments. You can provide a string as the last argument to the **msgbox** function that contains 'modal', 'non-modal', or 'replace'. These specify the behavior of the message dialog box and whether or not it should replace any existing message box. The default is non-modal which means that the user can click on other windows while the message box is active. Modal means that the user must acknowledge the message box by clicking on the "OK" button before he can select another window with the mouse.

The **msgbox** function can also take on a set of arguments to define an icon that will be displayed in the box. By default, no icon is displayed; however, you can display an error icon, help icon, warning icon, or your own custom image icon.

Typing

```
msgbox('My Error Message','Error Window Name','error');
```

will produce:

Figure 10.21   An error dialog window.

Typing

```
msgbox('My Help Message','Help Window Name','help');
```

will produce:

Figure 10.22   A help dialog window.

The code

```
msgbox('My Warning Message',...
'Warning Window Name','warn');
```

will produce:

Figure 10.23   A warning dialog window.

You can also use the form

```
msgbox('My Message',' Window Name',...
        'custom',iconData,iconCmap);
```

where iconData is a matrix containing image data and iconCmap is the image's color map.

You can use the **questdlg** command to create a question message with either two or three push-button answer options. The format for a two-button question dialog box is

```
string_returned = questdlg(QuestionString,...
                        ButtonString_1,...
```

```
                                     ButtonString_2);
```

where the three arguments are all strings. After the user clicks on one of the two push buttons, the string of that push button will be returned and stored in the *string_returned* variable. A three-push-button question dialog box is created in the following manner:

```
string_returned = questdlg(QuestionString,...
                           WindowNameString,...
                           ButtonString_1,...
                           ButtonString_2,...
                           ButtonString_3,...
                           DefaultString);
```

where DefaultString must be ButtonString_1, ButtonString_2, or ButtonString_3.

The question dialog box is modal, that is, the user must press one of the answers before control is returned to the source, such as the command line, function, or script that originally called the **questdlg** function. For example, you may want to question the user after a figure was generated to find out whether or not a print-out of the plot is desired. This can be done with the following code:

```
question_ans = questdlg('Do you want a hard copy?',...
                 'OUTPUT','Yes','No','No')
if strcmp(question_ans,'Yes')
   print
end
```

Figure 10.24 shows the resulting question dialog box.

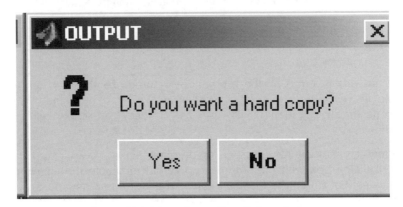

Figure 10.24   A question dialog window.

The **inputdlg** function is a very good way to quickly ask a user for information without having to do text-based questions and answers or generate a GUI.

Creating a set of questions as a cell array (where each question is in one cell of the array) provides the number of lines you want the user to be able to answer the questions on, and the default answers. You can quickly create a convenient method for prompting a user for information.

For example,

```
answers = inputdlg({'My first question',...
                    'My 2nd question',...
                    'My 3rd question',},...
                    'Window.Name',[1 2 1],...
                    {'defAns1','defAns2','defAns3'});
```

will produce the following GUI:

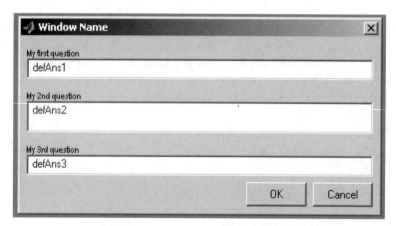

Figure 10.25   An input dialog window.

The answers typed in by the user will be returned by the function (to the variable *answers* in this example) as a cell array, where each cell index, i, is the answer to the ith question. If the user hits "Cancel", an empty cell array is returned.

Besides creating simple message or question dialog boxes, you can also create a dialog box with the command **uigetfile** to obtain the filename and directory path of a file. The complete syntax used for this command is

```
[filename, pathname] = uigetfile('FilterString',...
                                 'Dialog Box Title',...
                                 left, bottom);
```

'FilterString' specifies the extension that the file must have in order to be listed in the dialog box and 'Dialog Box Title' is the string that will appear as a title to the interface. The variables left and bottom are used to specify the location of the figure in terms of the distance in pixels from the lower left corner of the screen. The left and bottom arguments do not work on all platforms. In addition, you are not required to specify all four arguments. However, the order in which they are specified cannot be changed. If you want to provide the user with a list of all the M-files in the directory, you can use

```
[filename,pathname] = uigetfile('*.m',...
                                'UIGETFILE TITLE',...
                                100,100);
```

which will create the interface shown in Figure 10.26. In this figure, if the demopopup M-file was chosen (either by double clicking on the file or by highlighting the file and pressing "Open"), the *filename* variable would equal 'demopopup.m' and the *pathname* would be a string identifying the path (directory or folder) in which the demopopup.m file is located. If the user had instead selected "Cancel," a zero would be returned in the *filename* and *pathname* variables.

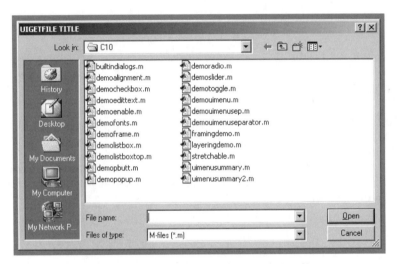

Figure 10.26   An example use of uigetfile.

To request the name of a new file from the user you can use the command **uiputfile**. The syntax format of **uiputfile** is very similar to that of **uigetfile**, the only difference being that instead of specifying the extension of the files to be listed, you specify the default name of the new file. For example,

```
[filename,pathname] = uiputfile('Default.m',...
                                'UIPUTFILE TITLE');
```

will produce the dialog box shown in Figure 10.27.

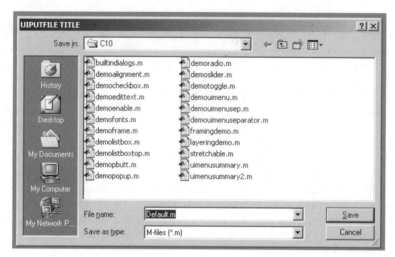

Figure 10.27   An example uiputfile GUI.

The function **uisetfont** allows a user to select the font, size, and style of text objects. To change the font attributes of text and/or axes objects, pass their graphics handles to the function with

```
uisetfont(object_handles)
```

Be aware that the variable *object_handles* may contain only the handles to uicontrol, axes, and text objects. Figure 10.28 shows a typical font dialog box. To find out more about this function (like how to use it with a font structure) type **help uisetfont**.

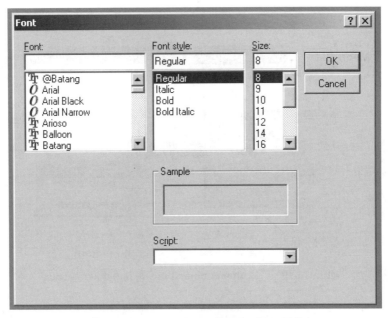

Figure 10.28   A sample uisetfont GUI.

Finally, there is one last built-in GUI that can be used to change the color of any graphics object. This is the function **uisetcolor**. This function is used by passing the handles of the objects whose color you would like to change. To learn more about this function, type **help uisetcolor**.

10.4 Uimenu Elements

User interface menus can exist within figures, other menus, or context menus. This takes the form of pull-down menus such as the familiar menu bar at the top of a Figure Window on X- and MS-Windows systems, and at the top of the screen on a Macintosh system. The menu bar will have one or more menu titles from which the user can choose. If a user clicks and holds down the mouse button when the pointer is located on top of a title, a list of menu items will appear. The user can then drag the pointer over any of the menu items. Those menu items that have arrowheads on the right-hand side are called submenus. If a user selects a submenu, another list of items will appear. You are not limited to the number of submenu levels (see Figure 10.29). When the user releases the mouse button after highlighting a menu item, that item's *CallBack* will be executed just as a uicontrol object's *CallBack* is executed after it has been activated. Menu titles, menu items, and submenus are all uimenu (user interface menu) objects that are created with the uimenu function.

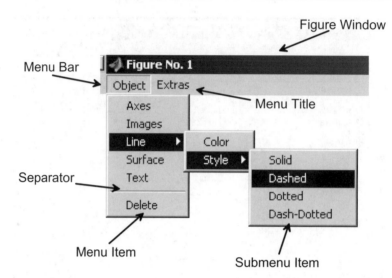

Figure 10.29   **Uimenu objects create pull-down menus.**

In addition to the differences in the location of the menu bar among the different platform versions of MATLAB, there are several other differences. Since there are no system default menu items in X-Windows figures, the Figure Window will contain only a menu bar if a menu object has been created. MS-Windows and Macintosh versions of MATLAB have default menu titles and, therefore, will always have a menu bar. By default, the MS-Windows figure will contain the "File," "Edit," "Windows," and "Help" menu titles. If you want to turn off these system default pull-down menus, set the *MenuBar* property of the Figure Window to "none". If you don't turn them off, new menu bar menus will be placed after the defaults.

10.4.1 Uimenu Properties

Just as with all other MATLAB graphics objects, uimenu objects have a set of properties that can be manipulated to suit your needs and help obtain the look you want for your GUI. The following table lists all of the properties associated with a uimenu object. As before with our property tables, each row contains the properties name, the read-only status, the property values (the default value is contained in "{}"), and the format of the value. Note that these objects also contain the universal properties discussed in Chapter 7. The universal properties that you are likely to find useful have been included at the end of the table.

Property	Read Only	ValueType/Options	Format
Accelerator	No	string	row
CallBack	No	string	row
Checked	No	[on | {off}]	row
Children	Yes	object_handles	column
Enable	No	[on | {off} | inactive]	row
ForegroundColor	No	[Red Green Blue] or color string	RGB row
Label	No	string	row
Position	No	[left bottom width height]	4-element row
Separator	No	[on {off}]	row
Interruptible	No	{on} | off	row
Tag	No	string	row
UserData	No	string(s) or number(s)	matrix
Visible	No	[{on} | off]	row

10.4.1.1   Uimenu Accelerator

The *Accelerator* property defines the keyboard strokes that the user can use to activate the uimenu object. This provides the user with an alternative to the point, click, and select method of activating the object. When users become familiar with the GUI, they tend to look for shortcuts to reduce the time and effort it takes to accomplish an action, and accelerators provide this. It is good practice to always provide an accelerator for a uimenu object.

The manner in which an accelerator key is defined depends on the platform on which you are running MATLAB. For MS-Windows, the accelerator defaults to the first character in the *Label* property if the *Accelerator* property is left blank. For labels on the menu bar, the accelerator is activated by pressing the ALT+*character* keyboard combination. If you have labels on the menu bar that begin with the same letter, subsequent presses of the *character* while still holding down ALT will stepthrough the labels. For submenus, where the *Accelerator* property has been specified, the keyboard combination is CTRL+*character* where *character* is the first character of the string set in *Accelerator*. The user will know that a menu item has an accelerator by the fact that the text "<Ctrl>-character" (where character is the letter that must be pressed along with the control key) appears to the right of the object's label as shown in Figure 10.30.

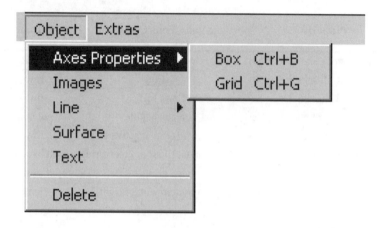

Figure 10.30   Press the Ctrl key to activate an accelerator.

10.4.1.2   Uimenu CallBack

The *CallBack* property specifies the action that is performed when the uimenu object is activated.   There is a difference between menu (and submenu) titles and menu (submenu) items in terms of when the uimenu object is activated and the *CallBack* is processed. Uimenus that contain menu and submenu titles will be activated when the user clicks down on these objects.   A menu item's *CallBack* is processed when the user releases the mouse button over the uimenu object.   Just as with uicontrol objects, the *CallBack* property stores a string that is evaluated in the base MATLAB workspace.   As long as the string can be evaluated error free with the command **eval**(*callbackstring*) from the command line (i.e., all variables it requires exist in some fashion), there will be no error messages invoked when the uimenu is activated by the user.

Examples of uimenu CallBack coding will be provided later when we create and program the GUI.

10.4.1.3   Uimenu Checked

The *Checked* property specifies whether or not the uimenu object will have a check mark (Macintosh or MS-Windows) or open box (X-Windows) symbol placed to the left of the displayed label. The symbol will appear only when the *Checked* property is set to "on" as shown in Figure 10.31. By default, this property is set to "off". The check mark is typically used to indicate whether or not a specific attribute regarding the application is turned on or off.

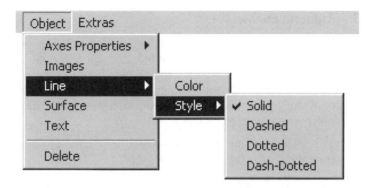

Figure 10.31   The Checked property set to "on" marks a label.

10.4.1.4   Uimenu Children

The menu items that appear below a menu title or to the right of a submenu title are the children of that menu title or submenu title object. Figure 10.32 depicts this relationship. The circled uimenu objects are the children of the object pointed to by the arrow leading from the circle.

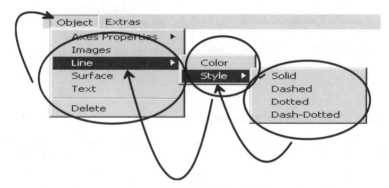

Figure 10.32   Parent / Child relationship between uimenus.

The *Children* property is a read-only property that lists the graphics handles of an object's children in a column vector. The order in which the handles are listed is from most recent to earliest created uimenu. The only type of object that a uimenu object can have as a child is another uimenu object. Menu or submenu items will have no children and the *Children* property will be the empty matrix.

10.4.1.5   Uimenu Enable

The *Enable* property can be set to either "on" or "off". If it is set to "off", the user will not be able to activate the uimenu and, therefore, no action will occur as a result of a mouse release over the object. In addition, if a submenu title is not enabled, the user will not be able to see that object's children. In its default value of "on", the uimenu will perform the action defined by its *CallBack* property when selected by the user.

As illustrated below, the text that is displayed in a uimenu object will be "dimmed" when the *Enable* property is set to "off".

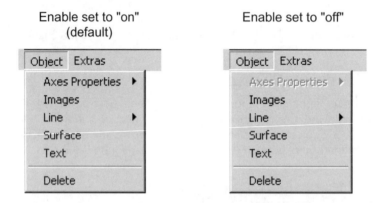

Figure 10.33   **Uimenu Enable property in "on" and "off" states.**

10.4.1.6   Uimenu ForegroundColor

The *ForegroundColor* property specifies the color of the label and symbols (e.g., the arrowheads, check marks, or boxes) of a uimenu object. You may define the value with either a RGB intensity triplet vector or a legal color specification string (e.g., 'red', 'r', 'white', etc.). By default, the foreground color will be black or [0 0 0].

10.4.1.7   Uimenu Label

The descriptive text that appears on the uimenu is stored in the *Label* property. This vector must be a string row vector.

As stated earlier (see Section 10.4.1.1, "UImenu Accelerator"), the uimenu mnemonic for MS-Windows versions of MATLAB is specified in the *Label* property. Any character that exists in the label and is not already used as an mnemonic to another menu object can be used as the mnemonic by inserting the "&" character in front of the desired character. The user can then simply press the Alt-*character* keyboard combination to activate and execute the *CallBack* associated with that uimenu object. For example, if the label is 'Grid', you can define the letter "G" as the accelerator by setting the Label property to '&Grid'. If instead you want to use the letter "d", set the Label property to 'Gri&d'.

If you use a uimenu to bring up another figure which also contains a GUI, it is good practice to add three dots (...) to the end of the uimenu label. This has become the conventional way of indicating to the user that there is more than meets the eye with this menu object selection. For example, the label 'Save As...' is usually used to indicate that more information (such as a new file name) will be requested from the user.

10.4.1.8 Uimenu Position

By default, menu items appear in the order in which they are created beneath their respective parents. You can alter this order by setting the *Position* property to the integer value that coincides with the desired order. Menu titles in the menu bar are ordered from left to right and menu items are ordered from top to bottom by increasing *Position* value. Figure 10.34 indicates the value stored in the *Position* property of the uimenu objects shown.

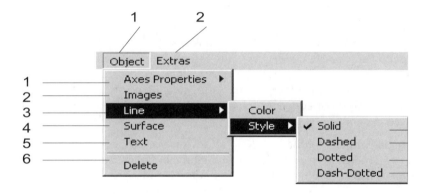

Figure 10.34   **Position numbers of the various uimenu objects.**

10.4.1.9 Uimenu Separator

The *Separator* property specifies whether or not a uimenu object will have a horizontal line drawn along its top edge. The *Separator* property's default setting is "off" which means that there will not be a line between the menu item and the menu item directly above it. This property does not apply to uimenu objects located in the menu bar nor to the first choice in a list of menu items. Figure 10.35 shows the affect that the separator has on the appearance of the menu list (the menu with label "Print" and "Quit" have their *Separator* properties set to "on" in the right-hand side of the illustration).

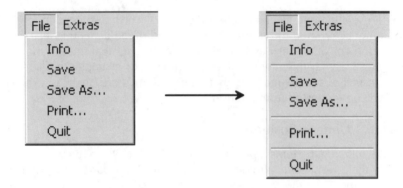

Figure 10.35   Using Separator to emphasize logical groupings.

10.4.1.10   Uimenu Interruptible

The *Interruptible* property will be discussed in great detail later in this chapter (see Section 10.8.3 "Interruptible vs. Uninterruptible"). Before this property can be fully understood, it is essential for you to be familiar with the various types of MATLAB events and how they are processed. Basically, this property controls whether or not the execution of a *CallBack* can be interrupted by another event such as when the user clicks the mouse button to activate a uimenu object.

The *Interruptible* property can be set to "on" or "off". By default the value is "on" and means that a CallBack's execution can be interrupted to execute the action associated with another event. On the other hand, a setting of "off" means that the *CallBack* must be completed before another event (such as the *CallBack* of another uimenu or uicontrol object) can be executed; only after the interrupting action has been completed can the interrupted *CallBack* be completed.

10.4.1.11   Uimenu Tag

This property is extremely useful when programming GUIs in a fashion that requires MATLAB to search for the handle of a specific uimenu object. The *Tag* property can contain a string row vector of your choice. It is usually assigned a descriptive name that uniquely identifies a particular uimenu object from all the other graphics objects. This property does nothing in terms of the appearance or action associated with the pull-down menu. We will see this property's usefulness in Section 10.5.2.4 of this chapter.

10.4.1.12   Uimenu UserData

The *UserData* property is essentially used as a storage facility that accompanies the object until it is deleted. It is unaffected by the command clear and, therefore, is a safe place to store matrix data that you want associated with a uimenu or that you want to be able to access regardless of the state of MATLAB's base workspace. You can use cell arrays or structures to hold arbitrary mixes of data in this property.

10.4.1.13 Uimenu Visible

Normally, the *Visible* property is set to "on" so that the object can be seen by the user. However, in certain circumstances, you may wish to make the uimenu object invisible. The uimenu object can be made invisible by setting this property to "off". This becomes useful when you want to limit the amount of information that the user is subjected to or when there are items from a list that do not apply to the particular situation under which the pull-down menu was selected.

If the *Visible* property is set to "off" for a menu title or submenu title, then that title uimenu object along with all of its children will not be visible.

10.4.1.14 Other Uimenu Properties

The *Type* property specifies the kind of MATLAB graphics object and is always set to uimenu for a uimenu object. This is a read-only property.

In terms of the object's family tree, the parent of a uimenu object found in the menu bar (i.e., a menu title object) will always be the handle to the figure that the object is drawn in (X- and MS-Windows) or associated with (Macintosh). This handle will be stored in the *Parent* property. The parent of the uimenu objects found in the menu will be the handle to the uimenu object that had to be selected to make the object visible.

The *ButtonDownFcn*, *Clipping*, *Selected*, and *DestroyFcn* properties are available to the uimenu object. The *Clipping* and *Selected* properties are both by default set to "off" and even in their "on" state do absolutely nothing to the appearance and performance of the uimenu object. The *ButtonDownFcn* property is by default set to the empty string matrix (['']) and has no effect on the performance or appearance of the uimenu object.

10.4.2 Creating Uimenus

Uimenu objects are created with the **uimenu** command. In this section of the chapter we will look at only a few examples of uimenu object creation. Here we are purely interested in the appearance of the pull-down menus, not the functionality. Defining or coding the *CallBack* property will be left for a later section.

10.4.2.1 Top Level Uimenu

There are a couple of basic forms of the **uimenu** function from which you can choose. The difference is only in the manner in which the parent of the uimenu object is specified. The simplest way to create a uimenu object in the menu bar (sometimes called a menu title or a top level menu) is with

```
handle = uimenu('Property1Name',Property1Value,...
                'Property2Name',Property2Value,...
                    .
                    .
                'PropertyXName,PropertyXValue)
```

where you specify the attributes of the uimenu object with as many *PropertyName/PropertyValue* pairs as required to fully describe the object. In this sample form there are X pairs. The *PropertyName* can be any one of the properties listed under Section 10.3.2 and *PropertyValue* is any legitimate value that can be assigned to the property. Even though there are a lot of different properties to choose from, menu titles are often created by specifying the *Label* and in some instances, the *Tag*, *CallBack*, and/or *UserData* properties. For example, a menu title labeled "Help" could be created with

```
h_uimenu_title1 = uimenu('label','Help');
```

This form of the **uimenu** command will create the uimenu object in the current figure. If there is no current figure object available, a new figure will be created.

You might also use the form

```
h_uimenu_title1 = uimenu(figure_handle,...
                    'Property1Name',Property1Value,...
                    'Property2Name',Property2Value,...
                           .
                           .
                           .
                    'PropertyXName,PropertyXValue)
```

which forces the top level uimenu object to be created in the figure with graphics handle *figure_handle*. In both forms, it is at your discretion to decide whether or not the uimenu object's graphics handle needs to be stored (in both forms shown, the control objects graphics handle would be stored in the variable *h_uimenu_title1*).

10.4.2.2   Menu Items and Submenu Titles

Menu items and submenu titles are created with a form of **uimenu** that is almost identical to the second form used to create top level menu titles. The only difference is that a uimenu object's handle must be supplied as the first argument to the uimenu command as shown with

```
h_uimenu_item = uimenu(uimenu_handle,...
                  'Property1Name',Property1Value,...
                  'Property2Name',Property2Value,...
                         .
                         .
                         .
                  'PropertyXName,PropertyXValue)
```

This new menu item will become visible (assuming its *Visible* property has not been set to "off") whenever the user selects the uimenu with the graphics handle *uimenu_handle*. The menu object whose handle is stored in *h_uimenu_item* will become a submenu title if another uimenu object uses *h_uimenu_item* as its parent.

10.4.2.3 Summary

To summarize menu object creation, all you really need to remember is that when you create a menu object, you specify the parent with the first argument to the **uimenu** command. A uimenu object's parent can only be either a figure (in the case of a top level menu) or another uimenu object (in the case of a menu item). A menu item becomes a submenu title whenever another uimenu object uses its graphics handle. In addition, you should also remember that unless you want to redefine the *Position* properties, the order in which you create the menu objects is important. The menu titles in the menu bar will be created from left to right, while menu items will be created from top to bottom. The next example illustrates some of the code needed to create the structure of a small portion of a pull-down menu interface as shown in Figure 10.36.

```
%Create a figure window and title it.
h_fig = figure('MenuBar','none','Color','white',...
    'Name','Uimenu Demo','NumberTitle','off');

% Create top level menus.
h_menu_props = uimenu(h_fig,'label','Properties');

% Create menu items.
h_menu_axes = uimenu(h_menu_props,'label','Axes');
h_menu_line = uimenu(h_menu_props,'label','Line');
h_menu_patch = uimenu(h_menu_props,'label','Patch');
h_menu_surface = uimenu(h_menu_props,'label','Surface');
h_menu_text = uimenu(h_menu_props,'label','Text');

% Create some submenu items to the line object.
h_menu_line_col = uimenu(h_menu_line,'label','Colors');
h_menu_line_sty = uimenu(h_menu_line,'label','Styles');
h_menu_line_thk = uimenu(h_menu_line,'label','Width');

% Create submenu items to Styles.
h_menu_line_solid =
uimenu(h_menu_line_sty,'label','Solid');
h_menu_line_solid =
uimenu(h_menu_line_sty,'label','Dashed');
h_menu_line_solid = uimenu(h_menu_line_sty,...
        'label','Stars','separator','on');
h_menu_line_solid = uimenu(h_menu_line_sty,...
        'label','Crosses');
      .


      .


      .
```

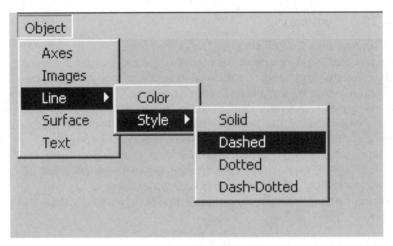

Figure 10.36   A simple uimenu.

This example was provided purely to show how the relationship (parent/child) between one menu item and the next is created. There is no functionality to the interface at this point because the *CallBack* properties have not been specified.

Designing and making intelligent use of all the uimenu object's attributes can lead to the creation of an intuitive, easy to use, robust pull-down menu interface. In particular, combinations of the *Separator*, *Enable*, and *Checked* properties can be used to guide the user to the choice he or she is most likely looking for. Consider a situation in which you are designing an interface that allows the user to select graphics objects and then manipulate their attributes. You might have a menu, such as the one in the previous figure, called "Properties" and beneath it, the names of the various objects as submenu titles to those object's attributes. If the user selected a line object you could make the other object submenu titles dim (i.e., set *Enable* "off").

Perhaps in this same example, you have a menu called "Edit" that has Undo, Cut, Copy, Paste, and Clear as menu items. You could segregate Undo from the rest with a separator. Furthermore, if a graphics object was not selected, you might make the Cut, Copy, and Clear items dim as shown in Figure 10.37.

Figure 10.37   Uimenu labels dimmed by setting Enable to off.

10.5 Low-Level MATLAB GUI Programming Techniques

At this point you have seen the inner workings of and how to create GUI's controls and pull-down menus. However, even though your interface may look nice, it will not perform any actions. Therefore, the next stage of GUI development involves coding the *CallBack* properties of the individual uicontrol and uimenu objects. Whenever you activate a GUI object, that object invokes MATLAB code that we call the callback. The callback can be in the form of an **eval** compatible string, or can refer to a function name that is stored in the object's *Tag* property. If the callback is a function, then it will have the following syntax:

```
function varargout =
objectTag_Callback(h,eventdata,handles,varargin)
```

Where *objectTag* is the name stored in the object's *Tag* property, h is the handle of the object that called the callback, eventdata is a reserved argument not currently used, handles is a structure of all the objects in the GUI, and varargin is the variable-length list of arguments you want to pass to the callback function.

For uicontrol, uimenu, and uicontextmenu objects, you can use the *Callback* property to define the function that is invoked when these objects are activated. This all sounds quite complicated at first blush, but is really quite straightforward once you see it in action. Let's proceed with discussions and examples.

10.5.1 Strings of MATLAB Statements and Expressions

Before we look closely at the *CallBack* property, which we are primarily interested in when programming uicontrol and uimenu objects, there are other object properties such as the *ButtonDownFcn*, *WindowButtonDownFcn*, *WindowButtonMotionFcn*, *WindowButtonUpFcn*, and *KeyPressFcn* that can be used to add features and capabilities to your interface. All of these properties can be set to a string vector. This string can, in turn, contain legal MATLAB expressions and statements that are interpreted and evaluated with MATLAB's **eval** function when some type of user induced event occurs (e.g., the user selects an uicontrol or uimenu, the user moves the mouse pointer across a figure, the user clicks the button down on a line object, etc.).

The easiest way to create an **eval** compatible string is to first type the code you want evaluated as if you were creating a script. (Remember to put a semicolon (;) at the end of every statement!) Follow along with the discussion and code here as we take you through creating strings of code that will work with **eval**; we will put together a *CallBack* string that will create a figure, plot a simple line, and put a title over the plot. Use the Editor/Debugger to follow along with your own entry of the following code.

```
figure;
plot(1:10);
title('A very complicated plot');
```

Next, add a quote at the beginning of every line and whenever there is a quote already in the code, add an extra one. Now you should have something like

```
'figure;'
'plot(1:10);'
'title(''A very complicated plot'');'
```

Then, change this into a single string vector by adding an open square bracket at the beginning and a closed square bracket at the end, while tacking on an ellipse (...) to the end of all but the last line.

```
['figure;'...
'plot(1:10);'...
'title(''A very complicated plot'');']
```

Finally, reformat the lines and specify the name of the variable that this string will be stored with

```
callback_string = ['figure;'...
            'plot(1:10);'...
            'title(''A very complicated plot'');'];
```

Now, we realize that this is a very simple example and that you could probably put this into a one-line string with no problems; however, if you follow this technique with even the most complicated code, you will be able to generate an evaluatable string! As you become more familiar with this process, you will be able to create strings more efficiently and quickly.

Cell arrays provide a useful ability to store your strings for generating your callbacks. The previous example could be accomplished in the following manner:

```
callback_string{1}='figure; plot(1:10);...
                    title(''A very complicated plot'')'
```

As you can see, using a cell array has a couple of advantages both having to do with creating lists of strings to be evaluated. One advantage is that you can choose a string by indexing the cell array and the other is that you can create lists without concern for matrix padding.

10.5.2 Programming Approaches in MATLAB

An important concept to understand is that the strings in the *CallBack*, *ButtonDownFcn*, *WindowButtonDownFcn*, *WindowButtonMotionFcn*, *WindowButtonUpFcn*, and *KeyPressFcn* properties are evaluated in the MATLAB's command or "base workspace." The base workspace is the workspace that is used when you execute M-file scripts or when you type in commands at the command line (assuming command line control is not a result of a keyboard command from within a function or a debugging state). A very simplified conceptual way of looking at workspaces is shown in Figure 10.38. When you execute a function, called "A" for example, from the base MATLAB workspace, a temporary workspace is created for function A. This temporary workspace contains all the variables and information that are "local" to that function. Function A can, in turn, sequentially (as defined by the code in function A's M-file) call function B or any number of other functions, and so on. Information can be passed between functions by means of input and output arguments specified in the function's calling syntax. After a function has been executed, its workspace is removed (i.e., all the local variables are cleared) so that the next time it is called, a fresh temporary workspace will once again be created.

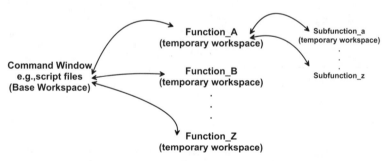

Figure 10.38   **The MATLAB Workspace**

The importance of this with respect to GUIs is that if you create graphics objects during the execution of a function, all of the locally stored information, such as graphics handles, that might have been available in the function's workspace, will be lost when the function has finished executing unless the information is either globally available (i.e., the variables are global variables), passed back and stored in the base workspace, or stored in a graphics object (i.e., a graphics object's *UserData* property). Since we have so many ways of storing and retrieving information, there are several approaches to programming a GUI.

The three most common techniques will be discussed in the following sections. With MATLAB's flexibility, you are certainly likely to think of some others as well. However, we are quite certain that you will feel comfortable with one of the following techniques and will be designing robust, easy to use GUIs in no time.

In order to make comparisons between the three techniques, we will design the GUI shown in Figure 10.39 with the different methods. In this GUI we want to give the user the ability to specify the x data values and a function y(x) that will be plotted in the axes object above the GUI. The user should also be able to turn the grid and box attributes of the axes on and off. To inform the user as to what his actions have done and if there are any errors, a status message window is included at the bottom of the GUI.

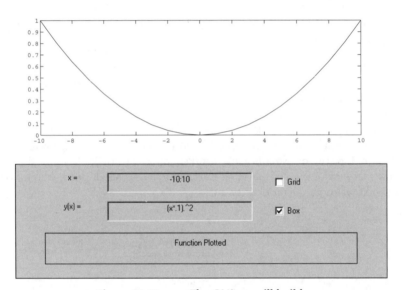

Figure 10.39   The GUI we will build.

10.5.2.1   Creating All Graphics Elements in the Base Workspace

The first approach we will present is perhaps the easiest way low-level way to build a GUI. In this approach we create all of the graphics objects in the base MATLAB workspace. With this method you can store all of the graphics handles needed when the objects are created, and the code in the *CallBack* properties of the objects that will perform an action can refer to these handles when necessary. Please be forewarned that there are several disadvantages with this structure, and they will be presented after the code that creates the GUI in Figure 10.39 is presented and discussed.

```
% M-File: fun_plt1.m
% All UIcontrol items are in normalized units so
% that the user can resize the screen as desired.
```

```
% Create the figure object and store its handle.
h_fig = figure('MenuBar','none');

% Create the axes object in the upper half of the
% figure.
axes('position',[.07 .5 .86 .4],'box','on')

% Create the two frames. The first lies below all
% uiobjects
% while the second is used to make a border for the
% status/message window.
h_frame_1 = uicontrol(h_fig,...
          'Position',[ 0 0 1 0.4 ],...
          'Style','frame',...
          'Units','normalized');
h_frame_2 = uicontrol(h_fig,...
          'Position',[0.08 0.05 0.84 0.11 ],...
          'Style','frame',...
          'Units','normalized');

% Create the callback for check box labeled "Box".
% This callback will determine the value of the
% checkbox object, whose handle is stored in h_box,
% and then set the current axes property accordingly.
% Finally, it displays a message by setting the
% string of the static text uicontrol whose handle
% is stored in h_status (created later).
box_clbk_str = ['boxstatus = get(h_box,''value'');'...
               'if boxstatus == 0;'...
               '  set(gca,''box'',''off'');'...
               'else;'...
               '  set(gca,''box'',''on'');'...
               'end;'...
               'boxstatus = get(gca,''box'');'...
               'set(h_status,''string'',' ...
               '[''The box property is ''
boxstatus]);'];
% Create the check box, setting its value to 1
% since we initialized the axes figure this way.
h_box = uicontrol(h_fig,...
          'CallBack',box_clbk_str,...
          'Position',[ 0.7 0.2 0.16 0.07 ],...
          'String','Box',...
          'Style','checkbox',...
          'Units','normalized',...
          'Value',[ 1 ]);

% Create the callback for the check box labeled "Grid"
% This callback will determine the value of the
% checkbox object, whose handle is stored in h_grid,
% and then use the grid function accordingly.
% Finally it displays a message by setting the
% string of the static text uicontrol whose handle
% is stored in h_status (created later).
grid_clbk_str = ['gridstatus = get(h_grid,''value'');'...
                'if gridstatus == 0;'...
                '  grid off;'...
```

```matlab
                        'else;'...
                        '   grid on;'...
                        'end;'...
                        'gridstatus = get(gca,''xgrid'');'...
                        'set(h_status,''string'',' ...
                        '[''The grid is '' gridstatus]);'];

% Create the grid check box.
h_grid = uicontrol(h_fig,...
                'CallBack',grid_clbk_str,...
                'Position',[ 0.7 0.3 0.16 0.07 ],...
                'String','Grid',...
                'Style','checkbox',...
                'Units','normalized');

% Create the callback that will plot the function any
% time the x data values or y function has been altered
% by the user.  Some error checking is performed just
% in case the user types in values or a function that
% cannot be plotted.
plot_clbk_str = [...
    'err_ind = 0;'...
    'eval([''x = '' get(h_xdata,''string'') '';''],'...
    '        ''err_ind=1;'');'...
    'if err_ind == 0;'...
    ' eval([''y = '' get(h_ydata,''string'') '';''],'...
    '        ''err_ind=2;'');'...
    'end;'...
    'if err_ind == 0;'...
    '   plot(x,y);'...
    '   boxstatus = get(h_box,''value'');'...
    '   if boxstatus == 0;'...
    '     set(gca,''box'',''off'');'...
    '   else;'...
    '     set(gca,''box'',''on'');'...
    '   end;'...
    '   gridstatus = get(h_grid,''value'');'...
    '   if gridstatus == 0;'...
    '     grid off;'...
    '   else;'...
    '     grid on;'...
    '   end;'...
    '   set(h_status,''string'',''Function Plotted'');'...
    'elseif err_ind == 1;'...
    '   set(h_status,''string'',...
        ''Error defining x'');'...
    'elseif err_ind == 2;'...
    '   set(h_status,''string'',...
        ''Error defining y(x)'');'...
    'end'];
% Create the edit boxes for the x and y data.  Both of
% these edit boxes will use the previous callback. In
% addition, initialize them with valid inputs.
h_ydata = uicontrol(h_fig,...
                'CallBack',plot_clbk_str,...
                'Position',[ 0.25 0.2 0.39 0.07 ],...
                'String','(x*.1).^2',...
                'Style','edit',...
```

```matlab
                    'Units','normalized');
    h_xdata = uicontrol(h_fig,...
                    'CallBack',plot_clbk_str,...
                    'Position',[ 0.25 0.3 0.39 0.07 ],...
                    'String','-10:10',...
                    'Style','edit',...
                    'Units','normalized');

    % Create a static text object that will be used
    % to display messages to the user.
    h_status = uicontrol(h_fig,...
                    'CallBack','guiplot1(''h_uic_12'');',...
                    'Position',[ 0.1 0.07 0.8 0.07 ],...
                    'String','Status Window',...
                    'Style','text',...
                    'Units','normalized');

    % Create the "x = " and "y(x)=" static text objects.
    % We do not need to store their handles since these
    % objects are neither manipulated nor queried by other
    % object callbacks.
    uicontrol(h_fig,...
                    'Position',[ 0.08 0.3 0.15 0.07 ],...
                    'String','x =',...
                    'Style','text',...
                    'Units','normalized');
    uicontrol(h_fig,...
                    'Position',[ 0.08 0.2 0.15 0.07 ],...
                    'String','y(x) =',...
                    'Style','text',...
                    'Units','normalized');

    % Initialize the plot with the initial x and y data
    % by evaluating the callback string that would be
    % evaluated if the x or y data changes.

    eval(plot_clbk_str);
```

This method has several drawbacks; the first and most important is that if the user clears the workspace with the command **clear** and then uses the GUI, the user will encounter error messages and find that the GUI no longer operates as expected. This is because some of the *CallBack* properties refer to variables that will not exist after the **clear** command is issued.

Another drawback is that it can be annoying and time consuming to generate the strings required in the callbacks, particularly if there are strings within strings that need to be manipulated. These strings also become difficult to read and follow, even for the individual who originally created the callback strings.

In addition, the strings make it difficult to modify and add additional functionality and features to the GUI. The larger the string becomes, the more likely it becomes that a single quotation mark, ', is missing or that there are too many. You might spend a lot of time trying to find the extraneous mark or the location that requires an additional one.

Finally, the execution speed of the callback may suffer as the string grows in length and complexity. This is because every time the uicontrol or uimenu callback is evaluated, each statement or expression is re-interpreted. If we could somehow put these MATLAB statements and expressions inside a function, the callback would execute considerably faster. The reason for this is that functions only need to be interpreted and compiled the first time they are encountered by the MATLAB compiler.

So, why don't we just put the code callback statements into a function? For example, let's put the callback for the "Grid" checkbox into a function called setgrid.m as follows

FAQ

```
function setgrid(h_grid,h_status)
gridstatus = get(h_grid,'value');
if gridstatus == 0
  grid off;
else
  grid on
end
gridstatus = get(gca,'xgrid');
set(h_status,'string',['The grid is ' gridstatus]);
```

and change the "Grid" check box's *CallBack* property to

```
'setgrid(h_grid,h_status)'
```

If you have not issued a **clear** command and have not closed the GUI's figure, you can do this at the command prompt by typing

set(h_grid,'setgrid(h_grid,h_status)')

Warning!

This will work since the graphics handles of the "Grid" check box and the "Status/Message" static text objects are passed to the function. However, the problem is that if you need to design callbacks that query or manipulate a lot of graphics objects during callback execution, you will have to pass a lot of arguments (i.e., the variables that store the graphics handles) to the functions. The next problem is that you may have many uicontrol or uimenu objects that all have callbacks! You will need to consider whether or not you really want to have a lot of M-files cluttering up your directory or folder.

The next three sections are aimed at resolving these problems. There are several key strategies and goals that you should keep in mind when you design your GUI.

- For simplicity and compactness, create a stand-alone function that is designed to create the GUI and specify the callback actions.

- For readability, design the function so that input arguments can be strings that describe the action that will occur. This can make it easy for you or other programmers to edit the function and add to or modify the GUI's appearance and features.

- For speed, design the GUI so that the user does not become frustrated or inconvenienced with the response time associated with performing an action after the user has activated a callback.

With the ease of M-file cross-platform portability, you must realize that other systems or machines might not be as fast as yours and users may be impatient with the response time, and click multiple times on a GUI or click on several GUIs. Try, when possible, to keep this behavior from leading to undesirable results (i.e., keep the number of interruptible callbacks to a minimum; see Section 6.6.3 to learn more about interruptible GUIs).

- For robustness, make the application's GUI resistant to undesirable user actions. It is very likely that the user will unintentionally perform actions that you did not specifically plan on occurring. While it is impossible or at least very difficult to make a GUI that is completely foolproof, there are steps that you can take to help minimize the chance of a GUI crashing or not performing as expected. For example, design the GUI so that it is able to withstand an accidental **clear** command. If there should not be more than one instance of the GUI, design the creation portion so that it checks to see if the GUI already exists in a Figure Window. In addition, use status bars wherever applicable. How much effort you put into making the GUI robust to these types of user actions depends primarily on your knowledge of the end users and the training they will receive.

With these thoughts in mind, let's look at the next GUI programming approach.

10.5.2.2   Storing Handles as Global Variables

In this approach we will put all of the previous example's code into one function. In addition, we will try to follow some of the strategies that were identified in the last section by making the graphics object handles that this function requires global variables. Finally, we will partition the code based on the possible input arguments with which the function may be called. This will make a big difference with respect to the ease at which the code can be read.

When the user types either

```
fun_plt2
```

or

```
fun_plt2('initialize')
```

the GUI figure will be created. In the initialization section of the code, a check is made to see if the GUI already exists. If the GUI exists, the figure containing the GUI is brought to the front. If it does not exist, the GUI is created. The callbacks of the uicontrol objects call the **fun_plt2** function with a 'Set Grid', 'Set Box', or 'Plot Function' string argument that specifies which section of the **if...elseif...end** should be executed. This could also be achieved, and is simpler to make use of the **switch...case** construct. The callbacks are easy to read and modify. You may also notice that the 'Plot Function' section that is executed after the user alters either the x data or y(x) expression is much cleaner and simpler than in the code found in the previous section.

```matlab
function fun_plt2(command_str)
% FUN_PLT2
%
% This function demonstrates how global variables
% can be used to create a GUI in a function.

if nargin == 0
  command_str = 'initialize';
end

% DEFINE VARIABLES THAT WILL STORE THE HANDLES AS GLOBAL
global h_box h_grid h_ydata h_xdata h_status

% INITIALIZE THE GUI SECTION.
if strcmp(command_str,'initialize')
    % Make sure that the GUI has not been already
    % initialized in another existing figure.
    h_figs = get(0,'children');
    fig_exists = 0;
    for fig = h_figs'
      fig_exists = strcmp(get(fig,'name'),...
                        'Function Plotter');
      if fig_exists
          figure(fig);   % Bring figure to front of screen.
          return;   % No need to reinitialize, exit
function.
      end
    end

  h_fig = figure('name','Function Plotter');

  axes('position',[.07 .5 .86 .4])

  % Create the two frames.
  uicontrol(h_fig,...
          'Position',[ 0 0 1 0.4 ],...
          'Style','frame',...
          'Units','normalized');
  uicontrol(h_fig,...
          'Position',[0.08 0.05 0.84 0.11 ],...
          'Style','frame',...
          'Units','normalized');

  % Create the "Box" check box.
  h_box = uicontrol(h_fig,...
          'CallBack','fun_plt2(''Set Box'');',...
          'Position',[ 0.7 0.2 0.16 0.07 ],...
          'String','Box',...
          'Style','checkbox',...
          'Units','normalized',...
          'Value',[ 1 ]);
  % Create the check box labeled "Grid".
  h_grid = uicontrol(h_fig,...
          'CallBack','fun_plt2(''Set Grid'');',...
          'Position',[ 0.7 0.3 0.16 0.07 ],...
          'String','Grid',...
          'Style','checkbox',...
          'Units','normalized');
```

```
   % Create the edit boxes for the x data.
   h_ydata = uicontrol(h_fig,...
           'CallBack','fun_plt2(''Plot Function'');',...
           'Position',[ 0.25 0.2 0.39 0.07 ],...
           'String','(x*.1).^2',...
           'Style','edit',...
           'Units','normalized');
   % Create the edit boxes for the y data.
   h_xdata = uicontrol(h_fig,...
           'CallBack','fun_plt2(''Plot Function'');',...
           'Position',[ 0.25 0.3 0.39 0.07 ],...
           'String','-10:10',...
           'Style','edit',...
           'Units','normalized');

   % Create a static text object that will be used
   % to display messages to the user.
   h_status = uicontrol(h_fig,...
           'Position',[ 0.1 0.07 0.8 0.07 ],...
           'String','Status Window',...
           'Style','text',...
           'Units','normalized');
   % Create the "x = " and "y(x)=" static text objects.
   uicontrol(h_fig,...
           'Position',[ 0.08 0.3 0.15 0.07 ],...
           'String','x =',...
           'Style','text',...
           'Units','normalized');
   uicontrol(h_fig,...
           'Position',[ 0.08 0.2 0.15 0.07 ],...
           'String','y(x) =',...
           'Style','text',...
           'Units','normalized');

   % INITIALIZE the plot with the initial x and y data.
   fun_plt2('Plot Function');

% CALLBACK FOR THE "Box" CHECK BOX.
elseif strcmp(command_str,'Set Box')
   boxstatus = get(h_box,'value');
   if boxstatus == 0;
     set(gca,'box','off');
   else
     set(gca,'box','on');
   end
   set(h_status,'string',...
       ['The box property is ' get(gca,'box')]);

% CALLBACK FOR THE "Grid" CHECK BOX.
elseif strcmp(command_str,'Set Grid')
   gridstatus = get(h_grid,'value');
   if gridstatus == 0
     grid off
   else;
     grid on
   end
   set(h_status,'string',...
```

```
                ['The grid is ' get(gca,'xgrid')]);

    % CALLBACK FOR THE X and Y(X) EDIT BOXES.
    elseif strcmp(command_str,'Plot Function')
        err_ind = 0;
        eval(['x = ' get(h_xdata,'string') ';'],'err_ind=1;');
        if err_ind == 0;
         eval(['y = ' get(h_ydata,'string') ';'],'err_ind=2;');
        end

        if err_ind == 0
          plot(x,y);
          fun_plt2('Set Box');
          fun_plt2('Set Grid');
          set(h_status,'string','Function Plotted');
        elseif err_ind == 1
          set(h_status,'string','Error defining x');
        elseif err_ind == 2
          set(h_status,'string','Error defining y(x)');
        end

    end % END command_str comparison checks.
```

We have used the **elseif** programming construct for the callback sections in this example so that you can easily read the code regardless of your programming background. However, the **switch...case** construct is a more organized technique and once you are familiar with it is even more readable. Furthermore, it runs more efficiently. For example, we could have written the callback code for "Box" check box using **switch...case**,

```
    switch command_str
    % CALLBACK FOR THE "Box" CHECK BOX.
    case 'Set Box'
        boxstatus = get(h_box,'value');
       if boxstatus == 0;
         set(gca,'box','off');
       else
         set(gca,'box','on');
       end
       set(h_status,'string',...
           ['The box property is ' get(gca,'box')]);
       .
       .
       .
       .
    case 'Set Grid'
       .
       .
```

Whether you program this GUI using **elseif** or **switch...case** constructs, there are some fundamental problems with an approach that relies on global variables. The first is that errors can still occur if the user issues either the

```
    clear all
```
or
```
    clear global
```

commands. The second problem is that the code which makes sure that only one instance of the GUI exists is absolutely necessary for functions that use global variables for storing graphics handles. If this code was not in place and the user created two instances of the GUI by typing **fun_plt2('initialize')** a second time, then only the most recently created GUI's setting will control the performance of the application. This would occur because the values stored in the global variables that contain the graphics object's handles would be updated with the most recently created object handles. The older object handles would no longer be stored.

10.5.2.3 Storing Handles in the UserData Properties

So we have seen problems with both approaches presented so far (although the GUI does work). The approach presented in this section overcomes the two problems (i.e., user clears the global variables, or you want to have multiple instances of a GUI) that are associated with using global variables as a means of storing graphics handles. However, everything has a price. We will see that this technique adds a little more processing overhead. The overhead is associated with the process of storing and retrieving the graphics handles. The relative price depends on how readable you want the M-file function to be. With the power of modern computers and the improved efficiency of the latest MATLAB version, you may very well prefer readability to eeking out the last drop of performance.

A solution to the problems of the previous example is presented in the next listing of code and illustrates how the graphics handles can be stored in the figure object's *UserData* property. Since much of the code is identical to that shown in the previous section, we have indicated where this occurs with a vertical ellipse (...) and suggest that you either look at the previous example's code if you want to see these sections again or look at the file (fun_plt3.m) on the book's web page.

```
function fun_plt3(command_str)
% FUN_PLT3
%
% This function demonstrates how graphics handles
% can be stored in the figure's UserData property.

if nargin == 0
  command_str = 'initialize';
end

if ~strcmp(command_str,'initialize')
% RETRIEVE HANDLES AND REDEFINE THE HANDLE VARIABLES
   handles = get(gcf,'userdata');
   h_box = handles(1);
   h_grid = handles(2);
   h_ydata = handles(3);
   h_xdata = handles(4);
   h_status= handles(5);
end

% INITIALIZE THE GUI SECTION.
if strcmp(command_str,'initialize')
        .
```

```
                          .
                            .
                              .
    {CODE THAT CREATES THE GRAPHICS }
    { OBJECTS HAS BEEN SNIPPED OUT  }
    {    SEE M-FILE OR EXAMPLE IN   }
    {         IN SECTION 10.6.2.2   }
                              .
                            .
                          .

        % STORE THE HANDLES IN THE FIGURE'S USERDATA.
        handles = [h_box h_grid h_ydata ...
                   h_xdata h_status];
        set(h_fig,'userdata',handles);

        % Initialize the plot with the initial x and y data.
        fun_plt3('Plot Function');
                          .
                            .
                              .
    {CODE THAT DEFINES THE CALLBACKS }
    {     FOR THE OBJECTS HAS BEEN   }
    {     SNIPPED OUT, SEE M-FILE OR }
    {     EXAMPLE IN SECTION 10.6.2.2 }
                              .
                            .
                          .
        end % END command_str comparison checks.
```

This format provides you with the choice of allowing single or multiple instances of your GUI to be created. For the GUI that we have created with this structure, you could comment out or delete the section of the code that checks to see if another instance already exists. If multiple instances were allowed, each GUI instance would operate fine and without errors.

There are some additional comments that need to be made about this coding approach. In the GUI we created for illustrative purposes, there are not a lot of uicontrol handles that need to be stored; with larger, more complex GUIs, there may be many uicontrol, uimenu, and other graphics handles that need to be stored. Redefining all of the handle variables (i.e., *h_box = handles(1), h_grid = handles(2),* etc., which was done in the example above) every time the function is called as a result of a uicontrol callback may be inefficient. The inefficiency grows when there are large numbers of graphics objects, since it is unlikely that all of the coded callbacks require all of the stored handles. One solution to this problem is to redefine the handle variables that are used within the section of the code needed to execute the callback. For instance, let's say we removed the "RETRIEVE HANDLES AND REDEFINE THE HANDLE VARIABLES" section of the code. Then we would need to code the "elseif strcmp(command_str,'Set Box')" section with something like

```
    % CALLBACK FOR THE "Box" CHECK BOX.
    elseif strcmp(command_str,'Set Box')
      % REDEFINE NEEDED HANDLE VARIABLES.
      handles = get(gcf,'userdata');
```

```
h_box = handles(1);
h_status = handles(5);
boxstatus = get(h_box,'value');
if boxstatus == 0;
  set(gca,'box','off');
else
  set(gca,'box','on');
end
set(h_status,'string',...
    ['The box property is ' get(gca,'box')]);
```

Another solution might consist of retrieving the graphics handles and storing them in the handles variable as before, while forgetting about redefining the individual object handle variables (i.e., *h_box*, *h_grid*, etc.). Then the code would need to make direct use of the *handles* variable. For example, the "elseif strcmp(command_str,'Set Box')" section of the code would then look like

```
% CALLBACK FOR THE "Box" CHECK BOX.
elseif strcmp(command_str,'Set Box')
  boxstatus = get(handles(1),'value');
  if boxstatus == 0;
    set(gca,'box','off');
  else
    set(gca,'box','on');
  end
  set(handles(5),'string',...
      ['The box property is ' get(gca,'box')]);
```

With a lot of references to the *handles* variable, the code can become slightly unreadable. In addition, any changes made to the appearance of the GUI, such as removing one or more of the graphics objects, might require you to go back and renumber a lot of indices. This can make the job of GUI modification a painfully tedious experience.

Unless you are noticing unbearable response times, it is recommended that you use the format in which you get the handles from the figure's *UserData* and then redefine all of the handle variables at once in the beginning of the function. The process of altering your GUI becomes a fairly simple process, and you still have MATLAB code that is easy to follow.

The only other potential problem which you need to be aware of can occur if you are designing GUIs that have multiple Figure Windows. If you need to provide the user with the ability to click on a uicontrol (or any other graphics object) in one Figure Window and have it call a function that was used to create a GUI in another window, you will need to throw in additional coding hooks. Remember that the method presented in this section retrieved the graphics handles from the current figure (i.e., *handles = get(gcf,'userdata')*). The process of clicking in a Figure Window makes that figure the current figure. Therefore, when the function that created the other GUI is called, the wrong handles will be retrieved. It then becomes almost inevitable that you receive error messages, unless you have been careful enough to design for this with additional code or a form similar to that shown in the next section's code. You get around this problem by providing the command **gcbf** (get handle to current callback figure). This function gets the handle of the figure that contains the object whose callback is currently being executed.

10.5.2.4   Utilizing Tags and the FINDOBJ Command

In the sections of this book that discussed the creation of the various graphics objects, it was mentioned that every graphics object has a property called *Tag*. This property can store a string vector, which means that you can assign a unique name to every single graphics object, if you so desire. This property also provides an alternative to the technique of storing the variables in the *UserData* property of the figure.

To retrieve a handle to a particular object, we will use the **findobj** command (see Chapter 7). The same arguments and discussions about the advantages and disadvantages of retrieving the object handles up front in a function instead of only when they are needed apply with this method, just as they did with the *UserData* technique. The only thing that should be added to the discussion is that the **findobj** is a search routine and there may be a substantial amount of processing associated with it. The amount of processing depends on the number of graphics objects and the syntactical form of **findobj** that is used. However, it should also be pointed out that **findobj** is a built-in function that has been optimized to provide a rapid response.

Once again we will revisit our famous function plotting GUI. Sections of the code that are unchanged will be snipped out of the provided code, but can be found in earlier examples or in the file fun_plt4.m found on the book's web site.

```
function fun_plt4(command_str)
% FUN_PLT4
%
% This function demonstrates how graphics handles
% can be retrieved with the findobj command.

if nargin == 0
  command_str = 'initialize';
end

if ~strcmp(command_str,'initialize')
% RETRIEVE HANDLES AND REDEFINE THE HANDLE VARIABLES.
  % Assume that the current figure contains the
  % fun_plt4 GUI.
  h_fig = gcf;
  if ~strcmp(get(h_fig,'tag'),'fun_plt4_figure')
    % If the current figure does not have the right
    % tag, find the one that does.
    h_figs = get(0,'children');
    h_fig = findobj(h_figs,'flat',...
                    'tag','fun_plt4_figure');
    if length(h_fig) == 0
      % If the fun_plt4 GUI does not exist
      % initialize it. Then run the command string
      % that was originally requested.
      fun_plt4('initialize');
      fun_plt4(command_str);
      return;
    end
```

```
      end

         % At this point we know that h_fig is the handle
         % to the figure containing the GUI of interest to
         % this function.  Therefore we can use this figure
         % handle to cut down on the number of objects
         % that need to be searched for tag names as follows:
         h_box = findobj(h_fig(1),'tag','h_box');
         h_grid = findobj(h_fig(1),'tag','h_grid');
         h_ydata = findobj(h_fig(1),'tag','h_ydata');
         h_xdata = findobj(h_fig(1),'tag','h_xdata');
         h_status= findobj(h_fig(1),'tag','h_status');

         % We could have just as easily replaced the previous
         % five lines with the technique that retrieves the
         % handles from the figure userdata, assuming the
         % handles are stored there.
      end

   % INITIALIZE THE GUI SECTION.
   if strcmp(command_str,'initialize')
         % Make sure that the GUI has not been already
         % initialized in another existing figure.
         % NOTE THAT THIS GUI INSTANCE CHECK IS NOT REQUIRED,
         % UNLESS YOU WANT TO INSURE THAT ONLY
         % ONE INSTANCE OF THE GUI IS CREATED!
         h_figs = get(0,'children');
         h_fig = findobj(h_figs,'flat',...
                     'tag','fun_plt4_figure');

      if length(h_fig) > 0
            figure(h_fig(1));
            return
      end

      h_fig = figure('name','Function Plotter',...
                     'tag','fun_plt4_figure');
                     .
                     .
                     .

   {CODE THAT CREATES THE OBJECTS }
   { HAS BEEN SNIPPED OUT. TAGS    }
   { ARE ADDED TO THE REQUIRED     }
   { GUI OBJECTS, SEE M-FILE OR     }
   { EXAMPLE IN SECTION 10.6.2.2   }
                     .
                     .
                     .

      % Initialize the plot with the initial x and y data.
      fun_plt4('Plot Function');
                     .
                     .
                     .

   {CODE THAT DEFINES THE CALLBACKS }
   {    FOR THE OBJECTS HAS BEEN    }
   {   SNIPPED OUT, SEE M-FILE OR    }
   {   EXAMPLE IN SECTION 10.6.2.2  }
                     .
```

```
                    .
                    .
  end % END command_str comparison checks.
```

The findobj technique of obtaining the graphics handles makes it easy to design robust GUI functions. The code presented in this section showed how you can program the GUI creation function so that the figure containing the GUI and the handles of the graphics object does not necessarily need to be the current figure. This allows you to have a complex GUI that is comprised of several Figure Windows and a situation in which an activated graphics object in one window calls the function that was used to create a different window. Another advantage of the technique that was illustrated in this section is that the *UserData* property of the figure does not need to be used to store graphics handles.

Finally, to wrap up this section on GUI programming approaches, you should consider combining features from all of the previous techniques. Each one has its merits, and if you can exploit the advantages of the individual techniques, you may find that your GUI programming style will satisfy all your requirements in addition to meeting the typical goals of

- simplicity
- compactness
- readability
- speed
- robustness

as previously mentioned.

10.6 High-Level GUI Development – GUIDE

Now that we have given you a sound foundation in graphics objects and their properties and looked at low-level GUI development, we now shall look at MATLAB's Graphical User Interface Development Environment (GUIDE). The low-level development approach consists of more "hand-programming" to create our GUIs, whereas GUIDE is more point and click. Which approach is best is a decision left to you as the programmer. Each method has its advantages.

Since this section deals with GUIDE, we shall restrict our discussion to it, at least where it makes sense to do so. GUIDE takes much of the tedium out of GUI development so results can be achieved quickly. However, you do lose a little visibility of the low-level creation of the graphics objects. Also, instead of an M-FIle (or collection of M-Files) there is also a FIG-File. We will discuss the roles of both with respect to using GUIDE in this chapter. Once you have explored this section on GUIDE we will again return to the low-level approach so that we can best explain some specific GUI desires.

GUIDE is itself a suite of tools that let you lay out your GUI by clicking and dropping objects, then dragging and resizing them in the manner in which you want them to appear. GUIDE includes a Property Inspector, that allows you to see and edit most of the properties belonging to an object. You can then invoke the GUI from GUIDE to test it. We will discuss these in the following subsections, and reproduce our example GUI from the previous section (for the last time) with GUIDE.

10.6.1 The Layout Editor

You start GUIDE by simply typing **guide** at the command prompt. You can also start it by selecting GUIDE from the Launch Pad. Either way, when you start GUIDE, you will see a window like that shown in Figure 10.40.

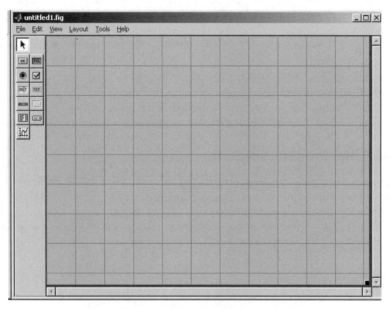

Figure 10.40   The GUIDE Layout Editor.

The Layout Editor is where you will begin your GUI. It is comprised of a component palette on the left, various toolbar selections across the top, and a large gridded area where you will lay out your GUI objects. You can go directly from your paper prototype to the Layout Editor, selecting the different uicontrol and uimenu objects from the component palette and placing them in the layout area. Simply select the object you want and drag it with the mouse to the location of your choice in the layout area. You can then use the mouse to move and resize the object. You can also use the toolbar to access various tools in the GUIDE toolset, such as alignment, menu editor, and object browser tools.

We know you are anxious to try GUIDE, so let's get started building the GUI we introduced in the previous section. You should be aware that at this point you will actually be creating a FIG-File. This is where all your user interface components, i.e., objects, will reside. Notice that at no time will you be specifying the *Position* property for your objects as you did with the low-level approach. At this point, if you haven't started GUIDE, do so now by typing

```
guide
```

at the command prompt, or by selecting GUIDE from the launch pad. You might find it convenient to look at   Figure 10.39 and use it as your paper layout.

Once you have GUIDE running, all you need to do is to click on the object you want to create and drag it into the layout area. (If you aren't sure what an object is by its icon in the component palette, if you let the mouse pointer dwell a moment on the icon, an information box will tell you.) Do that now for each of the objects in the GUI. You should have an axis object, a static text objects for "x=", "y(x)=", and the status message area, and two editable text areas, one for the value of x and the other for the function y(x). You also need two check box objects, one for "Grid" and one for "Box". Don't worry about the labels on any of the objects just yet. Also, don't save or activate the figure at this point. When you do save or activate the figure, the second thing GUIDE will do is create an M-File to functionalize all your callbacks. There's a few things you need to know before we are ready to let GUIDE do its magic.

10.6.2 The Property Inspector

If you select **View** → **Property Inspector** from the GUIDE toolbar, or if you double-click (on Windows computers) on one of the objects in your layout, the Property Inspector will be invoked. (You can also right-click on an object, which will bring up a selection menu from which you can select the Property Inspector.) It will look like that shown in Figure 10.41.

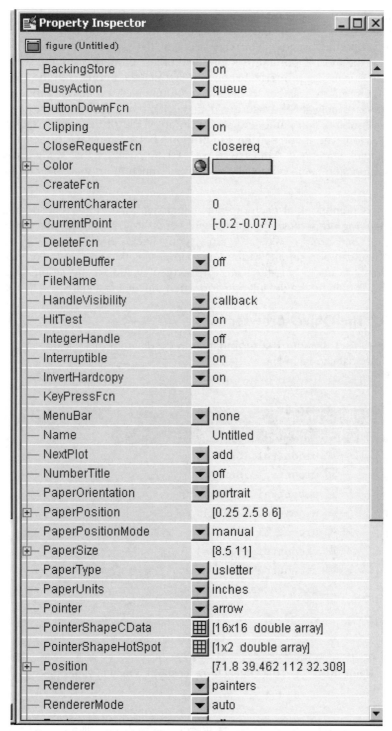

Figure 10.41   **The Property Inspector lets you view and edit properties for an object.**

The Property Inspector will show the properties for the object, and will change as you click on different objects. You should be quite familiar with these properties by now. You can also edit the property values as needed. Perhaps the most important user interface object property from the standpoint of GUIDE is its *Tag* property. When you first placed an object in the layout, its *Callback* property was set to the string "automatic". What will happen when you save or activate the figure is that GUIDE will convert this string into a callback subfunction name and save it in an M-File of the same name as the Figure. However, if you set the *Tag* property before you save or activate to a meaningful name, e.g., "status_window", GUIDE will use that string instead as the subfunction name potentially resulting in much more readable code in the M-File.

At this point, look at each of your objects with the Property Inspector, and set their *Tag* properties to a meaningful name. If you want to use the names we used they are, "x_value", "y_fun", "grid_check", and "box_check". At this point you can also set any of the *Value* properties that you wish. For instance, you might want to set the *Value* property of y_fun to the string '(x*.1).^2' which will then be the default function when the GUI first starts.

10.6.3 The Object Browser

The Object Browser is a tool that shows a diagram of your objects in your GUI, like that shown in Figure 10.42.

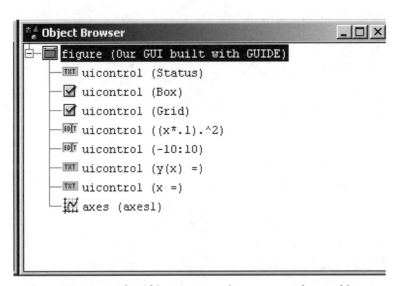

Figure 10.42   The Object Browser shows a map of your objects.

Clicking on any of the objects in the Object Browser will open the Property Editor to that object's properties.

10.6.4 The Menu Editor

The Menu Editor is available by selecting **Tools → Menu Editor** from the toolbar. From here you can specify what you want to appear on your GUI's menu bar, and also specify any uicontextmenu values.

10.6.5 Saving the GUI

Once you have all your objects positioned and sized in the layout area, specified names for the *Tag* property of them, and set any *Label* or other properties you want, you are ready to let GUIDE create the M-File where we will next spend some time. You have two options at this point; one, you can simply select **File → Save As...** from the menu bar and give your GUI a name, or you can select **Tools → Activate Figure** in which case GUIDE will prompt you for a name. When you do this, GUIDE will save your GUI to a file with the name you gave it with the extension .fig, and also create an M-File of the same name. In this example we named our GUI fun_pltg, so GUIDE created fun_pltg.fig and fun_pltg.m in our current directory. We will now discuss what GUIDE creates in these two files.

10.6.5.1 The GUIDE Created FIG-File

When GUIDE created the FIG-File from your layout, it created a binary file that contains a serialized figure object, i.e., a complete description of the figure object and all of its children. MATLAB uses this FIG-File to reconstruct the figure and all of its children when you open the file. All of the objects' property values are set to the values they had when the figure was created. When you run your GUIDE created GUI, the FIG-File is used in conjunction with the M-File of the same name which has the callback functions. The M-File created by GUIDE will use the function **openfig** to display the GUI.

Figure 10.43 shows our sample GUI in the GUIDE Layout Editor. Next, we will examine the M-File.

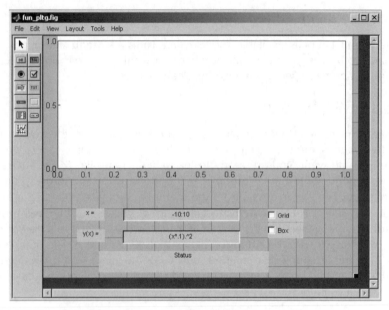

Figure 10.43   Our GUI in the Layout Editor.

10.6.5.2   The GUIDE Created M-File

The default method GUIDE uses to create your GUI automatically creates the FIG-File and M-File. You do have the option to change this to create only the FIG-File, to allow only one instance of the GUI to run, etc. In the default mode, GUIDE will create an M-File with *callback function prototypes*. Aside from other *overhead* items needed to run the GUI, these callback functions will be the heart of your GUI. Since you have already saved your GUI, look at the M-File now in the Editor. Hopefully you used the *Tag* property to give meaningful names to your objects so the callback functions should be easy to spot. They should have the form

```
function varargout = objectTag_Callback(h, eventdata,
handles, varargin)
```

where *objectTag* is the string you specified in the object's *Tag* property. The other variables are used in processing the callback and are given in the following table.

Argument	Purpose
h	the handle of the object whose callback is executing
eventdata	this argument is not presently used
handles	the structure containing the handles of all objects in the GUI whose names are specified by the objects' Tag property
varargin	a list of arguments that you can pass

Following is the M-File generated by GUIDE for our simple GUI. All that you see here, until the callback subfunctions which start with a comment that has a

```
%-->
```

at the start, was generated by GUIDE. Yours should look very similar.

```
function varargout = fun_pltg(varargin)
% FUN_PLTG Application M-file for fun_pltg.fig
%    FIG = FUN_PLTG launch fun_pltg GUI.
%    FUN_PLTG('callback_name', ...) invoke the named callback.

% Last Modified by GUIDE v2.0 07-Aug-2002 11:14:24

if nargin == 0  % LAUNCH GUI

    fig = openfig(mfilename,'reuse');

    % Use system color scheme for figure:
    set(fig,'Color',get(0,'defaultUicontrolBackgroundColor'));

    % Generate a structure of handles to pass to callbacks, and
store it.
    handles = guihandles(fig);
    guidata(fig, handles);

    if nargout > 0
            varargout{1} = fig;
    end

elseif ischar(varargin{1}) % INVOKE NAMED SUBFUNCTION OR CALLBACK

    try
            if (nargout)
                    [varargout{1:nargout}] = feval(varargin{:}); %
FEVAL switchyard
            else
                    feval(varargin{:}); % FEVAL switchyard
            end
    catch
            disp(lasterr);
    end

end


%| ABOUT CALLBACKS:
%| GUIDE automatically appends subfunction prototypes to this
file, and
%| sets objects' callback properties to call them through the
FEVAL
%| switchyard above. This comment describes that mechanism.
%|
%| Each callback subfunction declaration has the following form:
%| <SUBFUNCTION_NAME>(H, EVENTDATA, HANDLES, VARARGIN)
%|
%| The subfunction name is composed using the object's Tag and
the
%| callback type separated by '_', e.g. 'slider2_Callback',
%| 'figure1_CloseRequestFcn', 'axis1_ButtondownFcn'.
%|
%| H is the callback object's handle (obtained using GCBO).
%|
%| EVENTDATA is empty, but reserved for future use.
%|
%| HANDLES is a structure containing handles of components in GUI
using
```

```
%| tags as fieldnames, e.g. handles.figure1, handles.slider2.
This
%| structure is created at GUI startup using GUIHANDLES and
stored in
%| the figure's application data using GUIDATA. A copy of the
structure
%| is passed to each callback.  You can store additional
information in
%| this structure at GUI startup, and you can change the
structure
%| during callbacks.  Call guidata(h, handles) after changing
your
%| copy to replace the stored original so that subsequent
callbacks see
%| the updates. Type "help guihandles" and "help guidata" for
more
%| information.
%|
%| VARARGIN contains any extra arguments you have passed to the
%| callback. Specify the extra arguments by editing the callback
%| property in the inspector. By default, GUIDE sets the property
to:
%| <MFILENAME>('<SUBFUNCTION_NAME>', gcbo, [], guidata(gcbo))
%| Add any extra arguments after the last argument, before the
final
%| closing parenthesis.


%--> The following were stubbed by GUIDE, then filled
%out to implement our GUI.
% --------------------------------------------------------------
-----
function varargout = x_value_Callback(x_value, eventdata,
handles, varargin)
fun_pltg('y_fun_Callback',handles.y_fun, eventdata, handles,
varargin);




% --------------------------------------------------------------
-----
function varargout = y_fun_Callback(y_fun, eventdata, handles,
varargin)

err_ind = 0;
   eval(['x = ' get(handles.x_value,'string') ';'],'err_ind=1;');
   if err_ind == 0;
    eval(['y = ' get(y_fun,'string') ';'],'err_ind=2;');
   end
   if err_ind == 0
    plot(x,y);
    set(handles.box_check,'Value',1);
     fun_pltg('box_check_Callback',handles.box_check, eventdata,
handles, varargin);
      set(handles.grid_check,'Value',1);
     fun_pltg('grid_check_Callback',handles.grid_check,
eventdata, handles, varargin);

    set(handles.status,'string','Function Plotted');
   elseif err_ind == 1
    set(handles.status,'string','Error defining x');
   elseif err_ind == 2
    set(handles.status,'string','Error defining y(x)');
   end
```

```
% ---------------------------------------------------------------
-----
function varargout = grid_check_Callback(grid_check, eventdata,
handles, varargin)

gridstatus = get(grid_check,'value');
   if gridstatus == 0
     grid off
   else;
     grid on
   end
   set(handles.status,'string',...
       ['The grid is ' get(gca,'xgrid')]);


% ---------------------------------------------------------------
-----
function varargout = box_check_Callback(box_check, eventdata,
handles, varargin)

 boxstatus = get(box_check,'value');
   if boxstatus == 0;
     set(gca,'box','off');
   else
     set(gca,'box','on');
   end
   set(handles.status,'string',...
       ['The box property is ' get(gca,'box')]);
```

10.6.6 Executing a GUI

Once you have saved your GUI, typing its name at the command prompt will invoke it by starting the M-File. The line

```
fig = openfig(mfilename,'reuse');
```

in the GUIDE generated M-File opens the FIG-File by using **openfig** with the command **mfilename**. The command **mfilename** will return the name of the most recently run M-File, or when called from within an M-File as it is here, it returns the name of that M-File, and so determines its own name. The option "reuse" opens the FIG-File named, but only if a copy is not currently open; if it is already opened it will ensure that the existing copy is still completely on screen and visible, raising it above all other windows.

The next code segment

```
handles = guihandles(fig);
guidata(fig, handles);
```

retrieves all the handles in the figure. This is very important to understand because you will need to able to access these handles in order to make your uicontrols do what you want. In our example, we named the check box for turning the axis box on or off "box_check". The handle for that uicontrol object is accessed by

```
handles.box_check
```

Later in the code, in our y_fun_Callback, we access this uicontrol and affect it with

```
set(handles.box_check,'Value',1);
fun_pltg('box_check_Callback',handles.box_check,
eventdata, handles, varargin);
```

Notice that we are calling our GUI within this subfunction after we have set box_check's *Value* to one; this will cause the box_check_Callback subfunction to turn on the box around the axis.

Take some time playing with this example and notice how the callback process is working. Later we will discuss the event queue and learn more about how MATLAB determines what to do when in a GUI.

10.6.7 Editing a Previously Created GUI

When using GUIDE, it is good practice to select tag and filenames before activating or saving your GUI. Undoubtedly, there will be times when you will want to change or add to what you have done, in which case you will need to edit your previously created GUI. Let's say that you want to add two buttons to the GUI you just made, one to force plotting the function, and another to exit the GUI. You can do this readily by simply by opening the existing GUI in GUIDE and then dropping the desired buttons in the layout area, as shown in Figure 10.44.

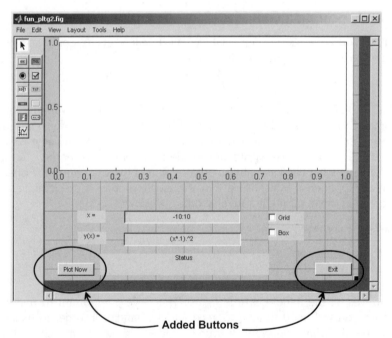

Figure 10.44   Editing an existing GUI in GUIDE.

Remember that GUIDE will automatically assign a string to any uicontrol *Tag* properties if you do not specify it, and it will use that string to construct the name of the callback functions it generates.

Additionally, when the GUI is run, a field will be added to the *handles* structure with the new name, but GUIDE can not generate a new subfunction if you change the value of a *Tag* property after you have saved or activated a GUI.

The best approach to avoiding problems related to *Tag* property settings, is to set *Tag* whenever you add new uicontrols to your GUI.

If you put buttons like that shown in Figure 10.44 in your GUI via the layout editor, then save, GUIDE will add callback subfunctions something like,

```
% -----------------------------------------------------
function varargout = Exit_Button_Callback(h, eventdata,
handles, varargin)

% -----------------------------------------------------

function varargout = Plot_Now_Button_Callback(h,
eventdata, handles, varargin)
```

Here is the code we added to finish the Exit and Plot Now buttons:

```
% -----------------------------------------------------
function varargout = Exit_Button_Callback(h, eventdata,
handles, varargin)

exit_button=questdlg('Exit Now?','Exit
Program','Yes','No','No');
switch exit_button
case 'Yes'
    delete(handles.figure1)
case 'No'
    return
end

% -----------------------------------------------------
function varargout = Plot_Now_Button_Callback(h,
eventdata, handles, varargin)

fun_pltg('y_fun_Callback',handles.y_fun, eventdata,
handles, varargin);
```

10.7 Common Programming Desires with UI Objects

We strongly feel that the best way to learn something new is to experience it yourself. Therefore, six examples are included in this section. These examples illustrate the implementation of several typical desired GUI features using the common coding techniques previously discussed. In some cases, only the portions of the coded function that contain the important and key features that are relevant towards achieving the GUI's goal are presented.

When sections of a function have been removed, you will see a vertically drawn ellipse (...). For a complete listing, you can visit this book's web site. For the most part, we have used the "Store the graphics handles in UserData" technique so that users of older versions of MATLAB can run the routines. If you are running one of the more modern versions of MATLAB in which graphics objects contain the *Tag* property and the **findobj** command is available, you may want to rewrite some of the code to practice the other techniques.

10.7.1 Creating Exclusive Radio Buttons

Radio button uicontrol objects are usually used in a fashion that allows a user to select from only one of the choices that a group of radio buttons offers at a time. Typically, a radio button group is visually separated with a frame object that encompasses the group. You might be wondering,

"How can I make a set of radio buttons mutually exclusive?"

This is the usual question that comes up after the radio buttons have been positioned in the GUI. In order to make a radio button group behave in a mutually exclusive fashion, you need to provide the MATLAB code that accomplishes this functionality. Each radio button's *CallBack* property has to be programmed as shown below (shown from parts of the gui_wave.m M-file).

```
function gui_wave(command_str,Argument2)
% GUI_WAVE
% examples/chap6/gui_wave.m
%
% Example of mutually exclusive radio button coding.
        .
        .
        .
if ~strcmp(command_str,'initialize')
   handles = get(gcf,'userdata');
   h_radio = handles(1:3);
end

if strcmp(command_str,'initialize')
        .
        .
        .
   % Create set of three Radio buttons.
   h_radio(1) = uicontrol(h_fig,'style','radio',...
               'callback',guiwave(''Waveform
Change'',1),...
               'string','Square Wave',...
               'position',[10 55 100 20],...
               'value',1);
   h_radio(2) = uicontrol(h_fig,'style','radio',...
               'callback',guiwave(''Waveform
Change'',2),...
               'string','Saw Tooth Wave',...
               'position',[10 30 100 20]);
   h_radio(3) = uicontrol(h_fig,'style','radio',...
```

```
                    'callback',guiwave(''Waveform
Change''',3),...
                    'string','Sinusoidal Wave',...
                    'position',[10 5 100 20]);
   handles = [h_radio];
   set(h_fig,'userdata',handles);
                         .
                         .
                         .

elseif strcmp(command_str,'Waveform Change')
   num_buttons = length(h_radio);
   button = Argument2;
   if get(h_radio(button),'value') == 1
      set(h_radio([1:(button-
1),(button+1):num_buttons]),...
          'value',0);
   else
      set(h_radio(button),'value',1);
   end

end % END command_str comparison check.
```

Running **gui_wave** will produce the window shown in Figure 10.45.

Figure 10.45   **Exclusive radio buttons created with gui_wave.**

The CallBack property of each radio button calls the **gui_wave** function with two arguments. The first argument forces MATLAB to run the "Waveform Change" section. The second argument identifies which button has been activated (clicked on by the user). If the activated button's *Value* property is equal to one, the code will set the *Value* property of the other radio buttons to zero (the off state). The reason behind this is that the *Value* property is automatically toggled before the *CallBack* is executed when a user clicks on a radio button. In other words, recognize that in order for the activated button's *Value* property to equal one, its *Value* had to be zero before being clicked on, since the process of activation toggles the *Value* property in radio buttons. In addition, this also means that one of the other radio button's *Value* property equaled one and hence all of the other radio buttons' values are set to zero. Furthermore, if instead the activated button's *Value* property is zero (for this to happen, the user had to select the button when it was already in its on state), all we need to do is set the *Value* back to one so that the button remains in its on state.

The exact same structure could be used to create mutually exclusive checked uimenu objects. You could create the functionality for uimenus using virtually the same code. All you need to do is change the uicontrol objects (h_radio) to uimenu (h_menu) objects. Then instead of setting the *Value* property, set the *Checked* property of the uimenus in the same fashion.

10.7.2 Linking Sliders and Editable Text Objects

"How do I get the value of a slider bar to show up in a text item after the user clicks on the slider bar?"

In this section we will look at a simple example that links together an editable text and slider uicontrol object. When the user slides the slider bar indicator, the editable text will be updated with the new value of the slider. The user can also type a value in the editable text uicontrol to specify a new value of the slider bar. The indicator bar will move to the value in the editable text after the user presses the return key or clicks the mouse button in another window or somewhere else within the same Figure Window. Figure 10.46 shows the GUI we will create.

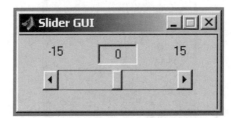

Figure 10.46   **Linking text and slider controls.**

```
function gui_sldr(command_str)
% GUI_SLDR
% examples/chap10/gui_sldr.m
%
% Example of creating slider GUIs.

if nargin < 1
   command_str = 'initialize';
end

if ~strcmp(command_str,'initialize')
   handles = get(gcf,'userdata');
   h_sldr = handles(1);
   h_val = handles(2);
end
if strcmp(command_str,'initialize')

   h_fig = figure('position',[100 200 200 75],...
```

```
                'resize','off',...
                'numbertitle','off',...
                'name','Slider GUI',...
            'MenuBar','none');

    h_frame = uicontrol(h_fig,...
            'style','frame',...
            'position',[0 0 200 75]);

    h_sldr = uicontrol(h_fig,...
            'callback','gui_sldr(''Slider Moved'');',...
            'style','slider',...
            'min',-15,'max',15,...
            'position',[25 20 150 20]);

    h_min = uicontrol(h_fig,...
            'style','text',...
            'string',num2str(get(h_sldr,'min')),...
            'position',[25 45 25 20]);

    h_max = uicontrol(h_fig,...
            'style','text',...
            'string',num2str(get(h_sldr,'max')),...
            'position',[150 45 25 20]);

    h_val = uicontrol(h_fig,...
            'callback','gui_sldr(''Change Value'');',...
            'style','edit',...
            'string',num2str(get(h_sldr,'value')),...
            'position',[80 45 40 20]);

    handles = [h_sldr h_val];
    set(h_fig,'userdata',handles);
elseif strcmp(command_str,'Change Value')
    user_value = str2num(get(h_val,'string'));
    if ~length(user_value)
     user_value = (get(h_sldr,'max')+get(h_sldr,'min'))/2;
    end
    user_value = min([user_value get(h_sldr,'max')]);
    user_value = max([user_value get(h_sldr,'min')]);
    set(h_sldr,'value',user_value);
    set(h_val,'string',num2str(get(h_sldr,'value')));

elseif strcmp(command_str,'Slider Moved')
    set(h_val,'string',num2str(get(h_sldr,'value')));

end
```

We recommend that you practice your GUI programming skills by attempting to alter this code so that the slider's minimum and maximum values can also be changed with editable text uicontrol objects. To get started, change the *h_min* and *h_max* uicontrols to editable text objects, and store their handles in the figure's *UserData* with the other handles. Then create two more callbacks that make the controls operate correctly.

10.7.3 Editable Text and Pop-Up Menu

In some instances you may want to allow the user to add items to a pop-up menu if the option is not already available in the pop-up menu list. This can be done with an editable text and pop-up menu uicontrol object. In this example, we will allow the user to change the *FontSize* property of a text object by selecting a size from a pop-up-menu. If the size the user wants does not exist, he or she can type it into the editable text object, at which point it will be added to the list of pop-up menu choices. Depending on the size of the symbol used in the pop-up menu on your platform, you may need to alter the position property so that only the symbol appears when the user does not select the pop-up menu.

Figure 10.47   **Entering a new number will add it to the pop-up.**

```
function gui_size(command_str)
% GUI_SIZE
% examples/chap10/gui_size.m
%
% Example of user on-the-fly defined pop-up-menu.
if nargin < 1
   command_str = 'initialize';
end

if ~strcmp(command_str,'initialize')
   handles = get(gcf,'userdata');
   h_text = handles(1);
   h_editsize = handles(2);
   h_popsize = handles(3);
end

if strcmp(command_str,'initialize')
    h_fig = figure('position',[200 200 200 100],...
                'resize','off',...
                'name','String Sizer',...
                'numbertitle','off',...
```

```
                    'MenuBar','none');

   h_ax = axes('position',[0 .5 1 .5],...
    'visible','off',...
    'xlim',[0 1],'ylim',[0 1]);

   h_text = text(.5,.5,0,'String',...
    'FontSize',10,...
    'HorizontalAlignment','center',...
    'VerticalAlignment','middle');

   h_editsize = uicontrol(h_fig,...
    'callback','gui_size(''Sized by Edit'');',...
    'style','edit',...
    'position',[70 15 30 20],...
    'string','10');

   h_popsize = uicontrol(h_fig,...
    'callback','gui_size(''Sized by Popup'');',...
    'style','pop',...
    'position',[110 15 30 20],...
    'string',' 5|10|15|20',...
    'value',2);

   handles = [h_text h_editsize h_popsize];
   set(h_fig,'userdata',handles);

elseif strcmp(command_str,'Sized by Popup')
   option_sizes = get(h_popsize,'string');
   choice = get(h_popsize,'value');
   set(h_editsize,'string',option_sizes(choice,:));
   set(h_text,'fontsize',str2num(option_sizes(choice,:)));

elseif strcmp(command_str,'Sized by Edit')
   option_sizes = str2num(get(h_popsize,'string'));
   size_choice = floor(str2num(get(h_editsize,'string')));

   % MAKE SURE THE USER'S INPUT IS A LEGAL FONT SIZE.
   if size_choice > 0
    if any(option_sizes == size_choice)
    % IF THE USER'S CHOICE EXISTS IN THE LIST, USE IT.
      choice = find(option_sizes == size_choice);
      set(h_popsize,'value',choice);

set(h_editsize,'string',num2str(option_sizes(choice)));
      set(h_text,'fontsize',option_sizes(choice));
    else
    % OTHERWISE CREATE A NEW OPTION IN THE MENU LIST,
    % PUTTING IT IN THE RIGHT SORTED POSITION.
      option_sizes = [option_sizes; floor(size_choice)];
      [new_opt_sizes,ind] = sort(option_sizes);
      choice = find(ind == length(new_opt_sizes));

      new_pop_str = sprintf('%3d',new_opt_sizes);
      new_pop_str = reshape(new_pop_str,...
                      3,length(new_opt_sizes))';

      set(h_popsize,'string',new_pop_str);
```

```
      set(h_popsize,'value',choice);
      set(h_editsize,...
            'string',num2str(new_opt_sizes(choice)));
      set(h_text,'fontsize',new_opt_sizes(choice));
    end
  else
    choice = get(h_popsize,'value');
    set(h_editsize,'string',num2str(option_sizes(choice)));
  end

end % END command_str comparison checks.
```

10.7.4 Windowed Frame and Interruptions

Previously, we learned that axes objects and axes children cannot be placed on top of uicontrol objects. For instance, if you create a frame that covers the entire Figure Window and then create a plot, you will not be able to see the plot. Therefore, it would be nice to create a window in a frame object through which the plot could be seen. Unfortunately, there is no way to cut a hole in a frame object. The solution is to create four static text objects that are positioned to cover the regions around the desired location of the window. The only purpose such a "simulated windowed frame" serves is purely for aesthetic reasons, which can be important in certain situations. In the next example we will create the GUI shown in Figure 10.48.

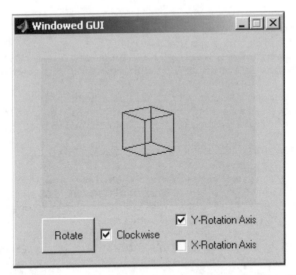

Figure 10.48   Creating a user interruptible animation GUI.

In addition to creating the windowed frame, this example will also show how you can interrupt an object's *CallBack*. This question comes up a great deal, and quite often there is an alternative means of getting the task accomplished without resorting to interrupting a *CallBack*. However, in some cases allowing interrupts is the easiest solution. In this example, the user will be able to rotate the cube by clicking the "Rotate" push button. The user can also change the direction in which the cube rotates and the axis around which it rotates either before or while the cube is rotating. The cube will rotate for 720° and then come to a rest until the push button is once again selected.

```matlab
function gui_wind(command_str)
% GUI_WIND
% examples/chap10/gui_wind.m
%
% Example of creating windowed GUIs.

if nargin < 1
   command_str = 'initialize';
end

if ~strcmp(command_str,'initialize')
   handles = get(gcf,'userdata');
   h_cube = handles(1);
   h_dir = handles(2);
   h_xrot = handles(3);
   h_yrot = handles(4);
end
if strcmp(command_str,'initialize')

   h_fig = figure('position',[100 200 300 250],...
           'resize','off',...
           'numbertitle','off',...
           'name','Windowed GUI',...
        'MenuBar','none');

   h_s(1) = uicontrol('style','text',...
                'position',[0 0 1 .25],...
                'units','normalized');
   h_s(2) = uicontrol('style','text',...
                'position',[0 0 .1 1],...
                'units','normalized');
   h_s(3) = uicontrol('style','text',...
                'position',[0 .9 1 .1],...
                'units','normalized');
   h_s(4) = uicontrol('style','text',...
                'position',[.9 0 .1 1],...
                'units','normalized');
   h_push = uicontrol(h_fig,...
                'style','pushbutton',...
                'position',[.1 .05 .2 .15],...
                'units','normalized',...
                'string','Rotate',...
                'interruptible','on',...
                'callback','gui_wind(''Rotate'');');

   h_dir = uicontrol(h_fig,...
                'style','checkbox',...
                'position',[.32 .07 .25 .11],...
                'units','normalized',...
                'string','Clockwise',...
                'value',1,...
                'callback','gui_wind(''Change
Rotation'');');
```

```
   h_xrot = uicontrol(h_fig,...
               'style','checkbox',...
               'position',[.6 .02 .35 .11],...
               'units','normalized',...
               'string','X-Rotation Axis',...
               'callback','gui_wind(''Change
Rotation'');');
   h_yrot = uicontrol(h_fig,...
               'style','checkbox',...
               'position',[.6 .13 .35 .11],...
               'units','normalized',...
               'string','Y-Rotation Axis',...
               'callback','gui_wind(''Change
Rotation'');');

   h_ax = axes('position',[.1 .25 .8 .65],...
                    'userdata',0);

   x = [0 1 1 0 0 0 1 1 0 0 NaN 1 1 NaN 1 1 NaN 0 0];
   y = [0 0 1 1 0 0 0 1 1 0 NaN 0 0 NaN 1 1 NaN 1 1];
   z = [0 0 0 0 0 1 1 1 1 1 NaN 1 0 NaN 0 1 NaN 1 0];
   h_cube = line(x-0.5,y-0.5,...
               z-0.5,'erasemode','background');
   axis('square');
   axis([-1 1 -1 1 -1 1]*1.5);
   axis('off')
   view(-37.5,15);

   handles = [h_cube h_dir h_xrot h_yrot];
   set(gcf,'userdata',handles);

elseif strcmp(command_str,'Change Rotation')
   direction = sign(get(h_dir,'val')-.5);
   rotation_axis = [get(h_xrot,'value') ...
                    get(h_yrot,'value')1];
   set(gca,'userdata',[1 direction rotation_axis]);

elseif strcmp(command_str,'Rotate')
   rotation_increment = 5*sign(get(h_dir,'value')-.5);
% degrees
   rotation_axis = [get(h_xrot,'value') ...
                    get(h_yrot,'value')1];
   rotation_origin = [0 0 0];
   num_of_incr = 720;
   angle_swept = 0;
   rotate_counter = 0;
   while abs(angle_swept) < 720
      rotate(h_cube,rotation_axis,...
               rotation_increment,rotation_origin);
      rotate_counter = rotate_counter + 1;
      angle_swept = angle_swept + rotation_increment;
      if rotate_counter > 5
         command_issued = get(gca,'userdata');
         if command_issued(1) > 0
            rot_dir = command_issued(2);
               rotation_increment =
abs(rotation_increment)*rot_dir;
```

```
               rotation_axis = command_issued(3:5);
             set(gca,'userdata',0);
          end
          rotate_counter = 0;
      end
      drawnow;
   end
end
```

The three check boxes all execute **gui_wind**('Change Rotation') and the push button executes **gui_wind**('Rotate'). If the *Interruptible* property of the push button had not been set to "on", the object would rotate in the direction and about the axis determined by the state of the check boxes at the moment the user activated the push button and would remain unaffected by any user actions during the course of its execution. However, since the *Interruptible* property was set to "on", the user can click any of the check boxes which will cause the current axes *UserData* property to be altered with the new inputs. Since the current axes *UserData* property is polled once every six passes through the while loop, the user's actions can be recognized and the appropriate changes in the direction and axis of rotation can be made.

10.7.5 Toggling Menu Labels

MATLAB provides the opportunity to change your pull-down menu labels "on-the-fly." Since the *CallBack* of a menu bar title or submenu title is executed before the menu items are displayed to the user, your code can change the attributes of the menu items before they are displayed. The menu bar that is created with the code below does nothing except illustrate a capability. A set of different menu items will be made available to the user that depend on the manner in which he or she last clicked in the Figure Window. The different types of clicks are a normal single click, a quick double click, a shift-click (press the shift key before clicking), and a control-click (press the control key before clicking).

```
function [name] = gui_togm(command_str)
% GUI_TOGM
% examples/chap10/gui_togm.m
%
% Example of a GUI that toggles menus.

if nargin < 1
   command_str = 'initialize';
end

if ~strcmp(command_str,'initialize')
   handles = get(gcf,'userdata');
   h_menu_opt = handles(1);
   h_menu = handles(2:4);
end

if strcmp(command_str,'initialize')

   h_fig = figure('position',[200 200 200 50],...
                  'resize','off',...
```

```
                    'numbertitle','off',...
                    'menubar','none');
    h_menu_opt = uimenu('label','Options',...
           'callback','gui_togm(''Set Menu Labels'');');
    h_menu(1) = uimenu(h_menu_opt,'label','Properties...');
    h_menu(2) =
uimenu(h_menu_opt,'label','','visible','off')
    h_menu(3) = uimenu(h_menu_opt,'label','',...
                       'visible','off');

    handles = [h_menu_opt,h_menu];
    set(gcf,'userdata',handles);
    gui_togm('Set Menu Labels');

elseif strcmp(command_str,'Set Menu Labels')
    seltyp = get(gcf,'selectiontype');
    if strcmp(seltyp,'normal')
      set(h_menu(1),'label','Properties','visible','on');
      set(h_menu(2),'label','','visible','off');
      set(h_menu(3),'label','','visible','off');
    elseif strcmp(seltyp,'alt')
      set(h_menu(1),'label','Alternate Properties',...
                    'visible','on');
      set(h_menu(2),'label','Delete Alternates',...
                    'visible','on');
      set(h_menu(3),'label','Copy Alternates',...
                    'visible','on');
    elseif strcmp(seltyp,'extend')
      set(h_menu(1),'label','Cut','visible','on');
      set(h_menu(2),'label','Copy','visible','on');
      set(h_menu(3),'label','Paste','visible','on');
    elseif strcmp(seltyp,'open')
      set(h_menu(1),'label','Open 1','visible','on');
      set(h_menu(2),'label','Open 2','visible','off');
      set(h_menu(3),'label','Open 3','visible','on');
    end
  end
```

10.7.6 Customizing a Button with Graphics

You are not limited to buttons in MATLAB that only have text on them. You can take advantage of a button's *CData* property to place your own custom graphic on the button. This can be very useful by making the button more eye-catching or even convey information itself. All you need to do to put a graphic on a button is to provide an image in RGB form as the value of the *CData* property. The most challenging part of this is actually getting the appropriate image you want to put on the button. Most likely, you will generate this image with your own favorite image editing software; however, in this example, we will generate an image from a MATLAB plot and use that to adorn our button.

First let's create an plot to make our image. We can make one quickly with MATLAB's **membrane** function:

```
membrane
```

Next we can grab a snapshot of the plot that will eventually become our button image.

```
F=getframe(gcf)
```

will get the snapshot and store an M-by-N-by-3 (RGB) image in F.cdata. (F.colormap is also created by **getframe** but is empty on systems that support true color.)

At this point we will go ahead and create our button.

```
h_fig=figure('Position',[100 100 200 200],...
            'menubar','none')
h_button = uicontrol('style','pushbutton',...
            'tooltipstring','bitmap on a button',...
            'position',[30 70 140 125])
```

From here we see that our image needs to be no more than 140-by-125. If we type

size(F.cdata)

we see that the image is 420-by-560. If we use f.cdata as it is, our button will only show a portion of the image, so we need to scale the image in some way. The following code will (crudely) scale the image by simply throwing out a number of rows and columns as specified.

```
scale=4;
[m n p]=size(f.cdata);
rows = 1:scale:m;
X=f.cdata(rows,:,:);
cols = 1:scale:n;
bimage=X(:,cols,:);
```

results in an image *bimage* that is 105-by-140, which will do.

Now all that is needed to do is to set the button's *CData* property with the RGB image *bimage*.

```
set(h_button,'Cdata',bimage);
```

Figure 10.49 shows the result.

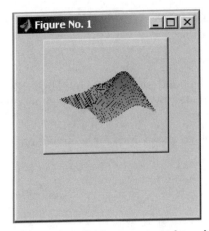

Figure 10.49   A plot image mapped to a button.

Placing images on buttons has a lot of possibilities. By having a callback change a button's *CData*, you can have the image on the button change dynamically. For instance, you might do this where the button shows the next image (like a thumbnail) to be displayed.

10.8 The MATLAB Event Queue

The intent of a GUI is to provide the user with a means of interacting with an application. A user's actions are not predictable, and therefore the interface must be programmed to react appropriately to an undetermined number and order of sequential events. When a user clicks the mouse button or moves the pointer in an interface that has a lot of features and capabilities, many different callbacks can be triggered. In addition to user-induced events, there are also MATLAB commands that trigger events. Having knowledge about the order and the circumstances in which the callbacks are scheduled is very important when deciding on how to program a GUI.

10.8.1 Event Scheduling and Execution

The events that will be discussed in this section are the user-invoked events such as the mouse click, up and down, and mouse pointer motion. Since a single action such as a click could trigger several callbacks (the Window-*ButtonDownFcn*, *WindowButtonUpFcn*, *CallBack*, and *Button-DownFcn* of any graphics object that exists below the region that was clicked on), these events need to be scheduled. In addition to evaluating and processing the actual callback string, MATLAB also needs to update all the properties that store information about the action. All of the actions that MATLAB needs to perform are placed in what will be called an "event queue" so that they can be acted on in a logical and consistent order. Once the callback event queue has been formed, additional user actions that attempt to schedule a set of new callbacks are ignored unless the Interruptible property has been manipulated. For simplicity, right now we will assume that the callbacks are not interruptible.

During execution of an individual callback string, for efficiency, MATLAB stores all events that affect the appearance of any or all graphics on the screen, so that they can be executed at once. These events are stored in what will be called a "graphics event queue." The events in the graphics event queue are evaluated and are only updated under the following circumstances:

- all callbacks in the "callback queue" are finished executing and control is passed back to the command line

- a drawnow command is encountered

- a figure command is executed

- a getframe command is executed

- execution is temporarily halted because a pause

waitforbuttonpress, or **input** command is issued.

In the event that a **drawnow discard** command is evaluated, the graphics event queue will be flushed (cleared) so that none of the graphics commands that were in the queue will be drawn on the screen. This does not mean that if one of the commands in the event queue is to set an existing graphics object's property that changes that object's appearance, the **set** command is not issued. Rather, the appearance will just not be displayed on the screen until a command that forces the figure to be redrawn is issued, such as **refresh** or another plotting command.

10.8.2 Execution Order of Events

The order in which graphics object information is updated and callbacks are evaluated can be seen in the flow charts found in the next several sections. The best way to learn is to write a little script and experiment with the different possibilities by clicking and dragging in different parts of the two figures. The following code is provided to you just for that.

```
h_fig_1 = figure('position',[100 100 100 100],...
    'menubar','none',...
    'windowbuttondownfcn','disp(''Fig1 WBDF'')',...
    'windowbuttonupfcn','disp(''Fig1 WBUF'')',...
    'windowbuttonmotionfcn',...
        'disp(''Fig1 WBMF'')',...
    'buttondownfcn','disp(''Fig1 BDF'')');
h_ui = uicontrol('style','pushbutton',...
    'position',[25 25 50 50],...
    'callback','disp(''UI CallBack'')',...
    'buttondownfcn','disp(''UI BDF'')');
h_fig_2 = figure('position',[200 100 100 100],...
    'menubar','none',...
    'windowbuttondownfcn','disp(''Fig2 WBDF'')',...
    'windowbuttonupfcn','disp(''Fig2 WBUF'')',...
    'windowbuttonmotionfcn',...
        'disp(''Fig2 WBMF'')',...
    'buttondownfcn','disp(''Fig2 BDF'')');
```

10.8.2.1   Mouse Button Pressed Down

When the user presses the mouse button down within the area defined by a figure's boundaries, MATLAB processes the following sequence of actions.

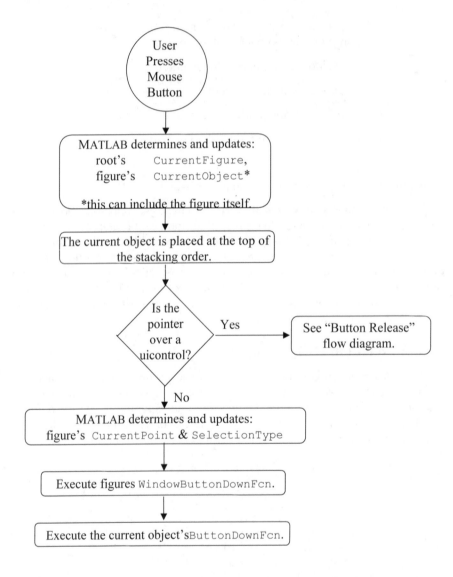

As is noted in the diagram, in the event that the user clicks the mouse down over an uicontrol button, the appearance of the uicontrol may change; however, the *CallBack* is not evaluated until the mouse button is released over that uicontrol. This gives the user the opportunity to back out of an accidental choice by moving the mouse away from the uicontrol and releasing the button over another region of the figure.

10.8.2.2   Mouse Button Released

When the user releases the mouse button within the area defined by a figure's boundaries, MATLAB processes the following sequence of actions:

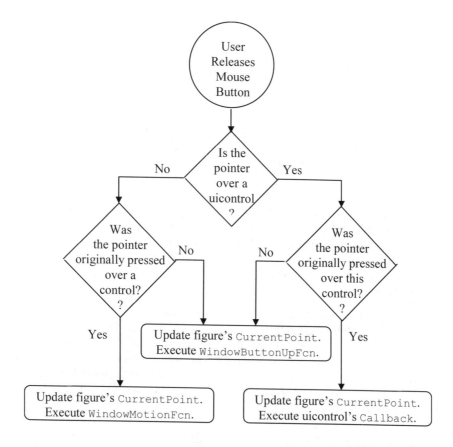

The *WindowButtonMotionFcn* can be executed at the time a mouse button is released under the circumstance indicated by the flow diagram. It occurs only if the *WindowButtonMotionFcn* is defined. If the user clicks down in one figure and then moves over to another figure before releasing the mouse button, the *WindowButtonUpFcn* property of the figure in which the mouse was clicked down will be evaluated. The other figure that the pointer was moved into will not have its *WindowButtonUpFcn* evaluated, but its *WindowButtonMotionFcn* will be evaluated. In other words, in order for the *WindowButtonUpFcn* to be evaluated, the mouse button has to be pressed down in that figure.

10.8.2.3   Mouse Pointer Moved

When the user moves the mouse pointer within the area defined by a figure's boundaries, MATLAB processes the sequence of actions shown in the following figure. This is true only if the *WindowButtonMotionFcn* is defined for that figure. If it is not defined, MATLAB does not waste time updating the figure's *CurrentPoint* and the root's *PointerLocation* and *PointerWindow* properties until they are requested.

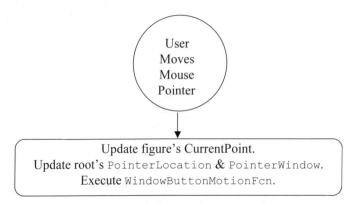

If the mouse button is pressed and held down while the pointer is moved into a figure whose *WindowButtonMotionFcn* properties are also defined, only the *WindowButtonMotionFcn* of the figure that the mouse was pressed down in will be evaluated. If the mouse is not held down and the user moves the mouse into a different figure, normal operation will ensue. In other words, MATLAB will evaluate the *WindowButtonMotionFcn* for the figure in which the pointer is located.

Each slight movement of the mouse is an action event that can schedule a *WindowButtonMotionFcn* callback. Since the number of these events that are processed in a given amount of time depends on your machine's speed and the rate at which the mouse pointer is being moved across the screen, MATLAB evaluates only the most recent *WindowButtonMotionFcn* callback; the rest are discarded, otherwise a machine could become seriously bogged down in evaluating callbacks.

10.8.3 Interruptible vs. Uninterruptible

One of the properties found in every single graphics object is the *Interruptible* property. By default, this property is set to "on", which means that if an object's callback is being evaluated (no matter where it is defined: *CallBack, ButtonDownFcn, WindowButtonDownFcn*, etc.), it can be interrupted by any other object's callback. It is useful to set this property to "off" when you want to ignore all user-invoked actions (mouse clicks or pointer movement) that may occur while a MATLAB program is being executed.

There are many situations in which you want the user to be able to interrupt a callback. For example, if you would like to program the *CallBack* of an uicontrol, let's say button A, to bring up another GUI that the user must respond to before button A's *CallBack* can be completed, you will want to keep A's Interruptible property set to "on".

In the example GUI provided in Section 10.7.4, we saw that while an animation was running, the user could manipulate the uicontrols and see an immediate effect on the animation. This was because the push button that started the simulation (i.e., the one labeled "Rotate") has its *Interruptible* property set to "on".

In order to interrupt an object's callback, there are two requirements

1. The object's *Interruptible* property must be set to "on" (the default).

2. The callback must contain a **drawnow**, **getframe**, **figure**, **input**, **pause**, or **waitforbuttonpress** (actually, any of the **waitfor...**) command.

There are situations where it does not matter what you have set the *Interruptible* property to and the executing callback will be interrupted any way; these are:

1. when the interrupting callback is a DeleteFcn or CreateFcn callback

    or

2. when a figure is executing a CloseRequest or ResizeFcn callback

10.8.4 Common Mouse Action Examples

There are a couple of examples that we can offer to teach and reinforce some of the ideas learned in this section. First, we will demonstrate a capability that allows the user to use the mouse button to move and resize objects, such as text and axes objects, so that the changes do not need to be made at the command line before printing out a hard copy of a figure. The second is being able to create a dynamic box when the user clicks and drags the mouse. Both of these examples were presented in a similar fashion in the earlier editions of this book, and although modern MATLAB provides some functions, e.g., **selectmoveresize** and **dragrect**, and the figure property *WindowStyle*, we still feel that these examples are educational and will help in understanding the event queue and the nature of mouse-related operations.

10.8.4.1 Moving Objects with the Mouse

Although you can easily move text in a plot using the plot editing mode in the Figure Window, here we are going to develop a function that allows you to move any graphics objects with the mouse. When the user clicks the mouse, we determine the current object's *Type* property, and then set the *WindowButtonMotionFcn* and *WindowButtonUpFcn* properties appropriately to allow the user to move the selected object. The user is notified that the object has been selected by a box that appears when the *Selected* property is set to "on". In addition, the type of operation (move/resize) is identified by the pointer type. The task of programming such a routine relies on knowing the points where position and location data is measured with respect to, and structuring a function to respond to, the user's actions as summarized in Figure 10.50.

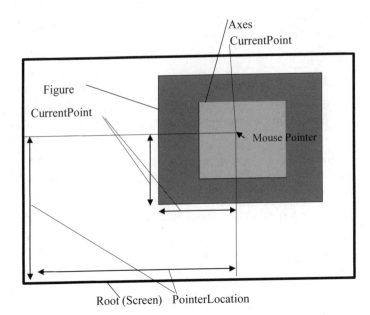

Figure 10.50   **The root, figure, and axes objects keep track of the pointer's location.**

```
function mvrs_obj(command_str,Argument);
% MVRS_OBJ
% chap10/mvrs_obj.m
%
% Used to move and resize axes objects
% and move text objects.
% Start capability by issuing
%  mvrs_obj
% Then click and hold and drag to
% move an object (axes objects will be
% moved from lower-left corner).
% To resize an axes object hold the control
% or alt key before click hold and dragging
% near desired corner of axes object.

global CUR_OBJ CUR_OBJ_TYPE FIX_PT

if nargin == 0
   command_str = 'initialize';
end
if strcmp(command_str,'initialize')
   set(gcf,'windowbuttondownfcn','mvrs_obj(''Set Up'')');
elseif strcmp(command_str,'Set Up')
   CUR_OBJ = get(gcf,'currentobj');
   if CUR_OBJ ~= gcf
     CUR_OBJ_TYPE = get(CUR_OBJ,'type');

     if strcmp(get(gcf,'selectiontype'),'normal')
     % SET UP MOVING OBJECT ROUTINE
         set(gcf,'pointer','fleur');
         if strcmp(CUR_OBJ_TYPE,'text')
```

```
                    set(CUR_OBJ,'erasemode','xor');
            elseif strcmp(CUR_OBJ_TYPE,'axes')
                    set(gcf,'units','pixels');
                    set(0,'units','pixels');
                    set(CUR_OBJ,'units','pixels');
                    cur_obj_loc = get(CUR_OBJ,'position');
                    fig_pos = get(gcf,'position');
                    set(0,'pointerlocation',fig_pos(1:2)+...
                            cur_obj_loc(1:2));
            end

    set(gcf,'windowbuttonupfcn','mvrs_obj(''Done'')');
            set(gcf,'windowbuttonmotionfcn',...
                'mvrs_obj(''Move Object'')');
            set(CUR_OBJ,'selected','on');
        elseif strcmp(get(1,'selectiontype'),'alt')
            % SET UP RESIZE OBJECT
            if strcmp(CUR_OBJ_TYPE,'axes')
                    set(gcf,'units','pixels');
                    set(0,'units','pixels');
                    set(CUR_OBJ,'units','pixels');
                    cur_obj_loc = get(CUR_OBJ,'position');
                    fig_pos = get(gcf,'position');
                    corner_loc = [cur_obj_loc(1:2); ...
                            cur_obj_loc(1:2)+...
                    [0 cur_obj_loc(4)];...
                            cur_obj_loc(1:2)+...
                    [cur_obj_loc(3) 0];...
                            cur_obj_loc(1:2)+...
                    cur_obj_loc(3:4)    ];
                    corner_loc_scrn =...
                    [corner_loc(:,1)+fig_pos(1) ...
                            corner_loc(:,2)+fig_pos(2)];
                    scrn_pnt_loc = get(0,'pointerlocation');
                    [dumval,min_ind] = ...
                    min(sum((([corner_loc_scrn-ones(4,1)*...
                        scrn_pnt_loc]).^2)'));
                    if min_ind == 1;
                    FIX_PT = corner_loc(4,:);
                    elseif min_ind ==2;
                    FIX_PT = corner_loc(3,:);
                    elseif min_ind ==3;
                    FIX_PT = corner_loc(2,:);
                    elseif min_ind ==4;
                    FIX_PT = corner_loc(1,:);end
                    set(0,'pointerlocation',...
                    corner_loc_scrn(min_ind,:));

    set(gcf,'windowbuttonupfcn',...
                        'mvrs_obj(''Done'')');
            set(gcf,'windowbuttonmotionfcn',...
                'mvrs_obj(''Resize Object'')');
            set(CUR_OBJ,'selected','on');

            end
        end
    end
elseif strcmp(command_str,'Move Object')
```

```
% CALLBACK FOR WHEN THE POINTER IS MOVED AND THE
% OPERATION GOAL IS TO MOVE AN OBJECT
  if strcmp(CUR_OBJ_TYPE,'text')
        cur_pnt_loc = get(get(CUR_OBJ,'parent'),...
          'currentpoint');
        set(CUR_OBJ,'position',cur_pnt_loc(1,:));
    elseif strcmp(CUR_OBJ_TYPE,'axes')
        cur_obj_loc = get(CUR_OBJ,'position');
        cur_pnt_loc = get(gcf,'currentpoint');
        new_obj_loc = [cur_pnt_loc cur_obj_loc(3:4)];
        set(CUR_OBJ,'position',new_obj_loc);
    end

elseif strcmp(command_str,'Resize Object')
% CALLBACK FOR WHEN THE POINTER IS MOVED AND THE
% OPERATION GOAL IS TO RESIZE AN OBJECT
  if strcmp(CUR_OBJ_TYPE,'axes')
        curr_pnt = get(gcf,'currentpoint');
        relloc = curr_pnt > FIX_PT;
        if all(relloc == [0
0]),set(gcf,'pointer','botl');
        elseif all(relloc == [0
1]),set(gcf,'pointer','topl');
        elseif all(relloc == [1
0]),set(gcf,'pointer','botr');
        elseif all(relloc == [1
1]),set(gcf,'pointer','topr');
        end
        new_pos = [min([curr_pnt ;FIX_PT]),...
              max([abs(curr_pnt-FIX_PT);[1 1]])];
        %keyboard
        set(CUR_OBJ,'position',new_pos);
    end

elseif strcmp(command_str,'Done')
% OPERATION GOAL HAS BEEN COMPLETED SINCE
% USER RELEASED THE MOUSE
  if strcmp(CUR_OBJ_TYPE,'text')
        set(CUR_OBJ,'erasemode','normal');
    elseif strcmp(CUR_OBJ_TYPE,'axes')
        set(CUR_OBJ,'units','normalized');
    end
    set(CUR_OBJ,'selected','off');
    set(gcf,'pointer','arrow');
    set(gcf,'windowbuttonupfcn','');
    set(gcf,'windowbuttonmotionfcn','')
end
```

To test or try out this routine, create a simple plot and a text object like

```
plot(1:10)
text(5,5,'Test String');
```

then activate the moving and resizing feature with

```
mvrs_obj
```

Look at how the resize object feature is incorporated into the program. As an assignment try putting in the functionality that allows you to move uicontrol, line, and patch objects as well. You could also set the *ButtonDownFcn* of the objects (axes and uicontrol objects only) that you want to be able to interactively move to the function **selectmoveresize**. For example, to be able to move all text around, type

```
set(gca,'buttondownfcn','selectmoveresize');
```

Then, after altering the object, set the same property to an empty string so that you don't accidentally move something you don't want to move.

10.8.4.2   Dynamic Boxes Using the RBBOX Function

In some situations, you may want to add a feature that allows the user to click and drag out a dynamic and temporary box for the purpose of selecting objects or identifying regions of a figure. Usually this box is drawn from the location at which the mouse button was first pressed to the current location of the mouse pointer. After the mouse button is released, the box disappears. This is what the code below implements.

M-File

```
function [rect] = dragbox(unitsval)
% DRAGBOX
%
% Usage:
%           [rect] = dragbox(units_string);
% where,
%
% rect:        is the RECT vector over which the
%              drag box is defined ([left bottom width
height]).
% units_string: is a string containing the name
%              of any of the legal units that
%              the figure can have.
%
% Example
%           figure
%           [rect] = dragbox('normalized')
% Wait for mouse button to be pressed.
waitforbuttonpress;

% Determine figure and get its Units.
h_fig = gcf;
original_figunits = get(h_fig,'Units');

% Specify Pixels for units and get location at
% which mouse button is pressed.
set(h_fig,'Units','Pixels');
firstpoint = get(h_fig,'CurrentPoint');

% Create the drag box.
rbbox([firstpoint 0 0],firstpoint);

% Get the location at which button is released.
lastpoint = get(h_fig,'CurrentPoint');
```

```
% Calculate a standard rect vector from two locations.
rect = [min(firstpoint,lastpoint),abs(firstpoint-
lastpoint)];

% Normalize the rect vector to the figure.
figpos = get(h_fig,'Position');
rect = rect./[figpos(3:4) figpos(3:4)];

% Put the rect vector in the specified units.
if nargin == 0
   unitsval = original_figunits;
end

if ~strcmp(lower(unitsval(1)),'n')
  set(h_fig,'Units',unitsval);
  figpos = get(h_fig,'Position');
  rect = rect.*[figpos(3:4) figpos(3:4)];
end

% Put the figure back in the original units.
set(h_fig,'Units',original_figunits);
```

To test out this routine, type

```
figure;
rect = dragbox;
```

and then click and drag in the Figure Window. After you have established the size of the box just dragged out, move it around by typing

```
set(gcf,'buttondownfcn','dragrect(rect)')
```

and once again clicking and dragging in the Figure Window.

10.9   Creating Custom User Interface Components

Power!

With the virtually boundless graphics capabilities of low-level MATLAB programming, if you happen to find that the standard set of MATLAB user interface objects does not suit all of your needs, or if you just want to spice up your GUI and make it more interesting, you can always create your own interface objects. In this section, we will look at creating two custom user interface objects. If you follow along and understand how these operate, you should then be able use your imagination and create just about any type of interface object you want. The key to designing functional interface or display objects is to make use of your knowledge of the available tools (i.e., the properties of the objects and different types of events that can be recognized). There is never one best solution. Therefore, some of the techniques that we have been able to successfully use will be presented. Also bear in mind that these examples are coded for clarity and you might be able to make enhances that use MATLAB more efficiently.

Most of the techniques we will show here rely heavily on the *ButtonDownFcn* of a graphics object. When the user clicks the mouse on the object, that object's *ButtonDownFcn* callback will be evaluated.   Since all graphics objects have this property, you can make an interface with any one you want. For example, you could easily make one out of text object with

```
figure;
axes('xlim',[0 1],'ylim',[0 1],'visible','off');
callback_str = ['set(gco,''position'',[rand(1,2) 0]);'];
text(.5,.5,0,'Click on Me','buttondownfcn',callback_str);
```

and you have a little game that you can play for hours on end. Like trying to swat a fly, every time you click on the text object it will move to some new and random location.

If a 2-dimensional looking flat button suits your needs, create an interface with a patch and text object. However, if that's all you need, you should probably stick with the uicontrol objects since they are easy to program and look a lot better than any patch/text object combination. Also, remember that you can put pictures on your buttons by using the button's *CData* (See Section 10.7.6). Of course, the patch object doesn't need to be a simple square; in fact, you can create any shape you need and turn it into a button. That in itself may already provide the feature you wanted that uicontrols and uimenus could not offer, so that is the first custom component we will demonstrate.

10.9.1 Simulating Buttons with Image Objects

The easiest way to create your own custom 3-dimensional-looking push buttons is to create two images positioned in the same location (i.e., one on top of the other). One represents the appearance of the button in its "off" state and the other represents its appearance in the "on" state (i.e., when the user clicks on the object). Remember that you can achieve similar results by using the *CData* property of a uicontrol object (see Section 10.7.6); however, to create non-standard shaped buttons, you still will need to do something like the method described next.

Usually the hardest or most time-consuming aspect of creating these types of buttons is generating the images. This can be done several ways and you will have to experiment and find the one that works best for you.

1. You can create your own image and color map matrix by typing in the numbers that represent indices to the map of RGB values you have specified. This is probably the most time-consuming and most difficult method because you have to make a mental image of what the numbers represent, but it can be done. Another option is to use an image that is already in RGB format, perhaps created in some image editing software; if you have a color mapped format image you could use something like **makergb** (see section 5.3.3) to create the data for your button.

2. You can piece together axes, patch objects, text, and anything else you want. Size the object the way you want it and then take an image snapshot with F = **getframe**. When used in this manner **getframe** will return a structure in F where F.cdata is an M-by-N-by-3 snapshot. (If your computer does not support truecolor, you will get

data in F.colormap.) Alter the colors to represent the figure in its opposite state and take another snapshot. Combine the two color maps so that both image objects can be shown simultaneously without color distortion (see example below). To use this technique, you must be able to have multiple software packages up on your machine (i.e., MATLAB and your favorite drawing package). Use your favorite software drawing package and create the button icons. Then use X = getframe(reference_fig, capture_rectangle), where reference_fig is the handle to a figure that is purely used as a location reference point and capture_rectangle is a [left bottom width height] vector that defines the region that will be captured with respect to the lower left corner of the reference_fig figure. If you position the drawing package window in a way so that you can see the icons even when you are in MATLAB, you can experimentally determine the correct capture_rectangle vector which captures the portion of the button icons you want. Execute the **getframe** command and then see what you captured by typing

```
image(X);
colormap(X.colormap); %only if X not truecolor (RGB)
```

In the next example, we will use the following code to help generate the two images shown in Figure 10.51. To achieve a 3-dimensional look, we use light and dark shades of a particular color. Decide on a corner from which it should appear that a light source is located with respect to the button. The two sides adjacent to this corner should use light shading for the button's up state image and a dark shading for the button's down state. The opposite two sides can use a thin dark border for the button's up state and a lighter border in the button's down state. Choosing the shaded edge's relative thickness in the two button states is useful, too; your personal preferences and creativity will guide you.

```
h_fig=figure('position',[100 100 50 50],...
'color',[.8 .8 .8],...
'menubar','none');
axes('position',[0 0 1 1],...
    'xlim',[0 1],'ylim',[0 1],...
    'visible','off');
p = patch([.08 .08 1 1],[0 .92 .92 0],...
    [0 0 0 0],[0 0 0 0],'facecolor',[.6 .6 .6],...
    'edgecolor','none') ;
patch(.2*cos(linspace(0,2*pi,4))+.3,...
    .2*sin(linspace(0,2*pi,4))+.3,...
    ones(1,4), ones(1,4),....
    'facecolor',[0 0 0]);% Create triangle
patch(.2*cos(linspace(0,2*pi,6))+.7,...
    .2*sin(linspace(0,2*pi,6))+.4,...
    ones(1,6), ones(1,6),...
    'facecolor',[0 0 0]);  % Create pentagon
patch(.1*cos(linspace(0,2*pi,15))+.5,...
    .1*sin(linspace(0,2*pi,15))+.7,...
    ones(1,15), ones(1,15),...
    'facecolor',[0 0 0]);% Create circle
```

```
l = line([0.08 1 1],[0 0 .92],[2 2 2],'linewidth',2,...
    'color',[.3 .3 .3]);
% This clears the map so that getframe only
% captures what is required.
set(h_fig,'colormap',[]);
Xup=getframe(h_fig);% On State Image
set(gcf,'color',[.2 .2 .2]);
set(p,'facecolor',[.4 .4 .4]);
set(l,'color',[.8 .8 .8]);
Xdw=getframe(h_fig);% Off State Image
```

Combine the two maps so that you will be able to use the two images in the same figure without distorting their colors:

```
mapupdwn = [mapup;mapdw];
```

You will now have the two buttons shown in Figure 10.51.

Figure 10.51   **Up and Down custom buttons using images.**

Now you have developed a pair of images that you can save for future use.

```
save shapeimg Xdw.cdata Xup.cdata mapupdwn
```

To create a button that will toggle between the states represented by the above two images, use something like the following:

```
load shapeimg
figure('position',[100 100 100 100],...
    'menubar','none',...
     'colormap',mapupdwn);
axes('position',[.2 .2 .6 .6],...
    'visible','off','ydir','reverse',...
    'xlim',[0 size(Xup.cdata,2)],'ylim',[1
size(Xup.cdata,1)]);
hold on
image_up = image(Xup.cdata);
image_down = image(Xdw.cdata);
set(image_up,'userdata',image_down,...
    'buttondownfcn',['set(get(gco,''userdata'')'...
        ',''visible'',''on'');' ...
        'set(gco,''visible'',''off'')']);
set(image_down,'userdata',image_up,...
    'visible','off',...
    'buttondownfcn',['set(get(gco,''userdata'')'...
        ',''visible'',''on'');'...
        'set(gco,''visible'',''off'')']);
```

Since the *ButtonDownFcn* is a little difficult to read when it is placed into a string, this code is presented below, as it would exist as actual MATLAB code in a program.

```
set(get(gco,'userdata'),'visible','on');
set(gco,'visible','off');
% Here is where you would tack on any additional
% code that you want executed when this button is
% toggled.
            .
            .
            .
```

   To make the button act more like a push button in the sense that the off state is maintained only while the user holds down the button, all you need to do is add the following line (assuming that the button is in the state in which you want it to normally remain when you set the *WindowButtonUpFcn* property).

```
set(gcf,'windowbuttonupfcn',...
    ['set(gco,''visible'',''on'');' ...
     'set(get(gco,''userdata''),''visible'',''off'')']);
```

This single line will work even if you have several custom buttons in the interface. However, it is usually a good idea to make a button quickly set the *WindowButtonUpFcn* it requires when the button down occurs. Continue with the above example by making the *ButtonDownFcn* set the *WindowButtonUpFcn*. In addition, have the *WindowButtonUpFcn* clear itself after it has been evaluated. This should be done in one of the GUI structures previously discussed; otherwise, you will be endlessly frustrated with errors because a quote or parenthesis is missing.

   When using images to create multiple user interface buttons make sure:

1.   A single color map is applicable for all the images. Create all buttons with a graphics drawing package. Place the images that represent the button's on and off states next to one another and arrange all of the buttons so that a single capture image can be executed. This makes it easy to keep a small-sized color map that works for all the button images. Then break out the individual images by determining which indices of the large captured matrix correspond to the individual button images.

2.   Use one of the GUI programming approaches to make it easy to keep track of all the image object graphics handles and to make the code readable.

3.   When you have many custom buttons, it becomes important that you know what state each object button is in. Consider creating a matrix that has 3-by-M elements for the M custom buttons in your GUI. Each column of the matrix could be dedicated to maintaining information about a particular button. For example, row one of column one could be the handle to the on-state image for a particular button, row 2 could be the handle to the off-state image for that button, and row 3

could be used to indicate the current state of the button (much like the *Value* property of a uicontrol). Keeping track of this kind of information will make it simple to reset, automatically set certain button states, make all your custom buttons mutually exclusive, etc.

4. Make your callbacks as independent as possible so that you don't need to rely on another object's callback, since it may have been changed from what you might expect it to be.

10.9.2 Creating a Dial

Let's step through the process of creating a dial like the one shown in Figure 10.52. We want the user to be able to click on the dial and graphically move the arrow about the arrow's hinge. To move the arrow in a continuous fashion, the user must click down and hold the mouse button while moving about the arrow's hinge and then release when the arrow points to the desired value.

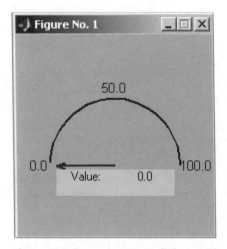

Figure 10.52   **A custom dial control.**

```
function uidial(command_str,Argument1,Argument2)
% UIDIAL
% examples/chap10/uidial.m
% Creates a dial user interface to learn how to
% make a custom GUI object.
% Usage:
%          uidial('initialize',min,max);
%
%  The value of the dial is stored and can be
%  gotten from the current axes UserData property.

if nargin == 0
   command_str = 'initialize';
end
```

```
if ~strcmp(command_str,'initialize')
   handles = get(gcf,'userdata');
   h_arrow = handles(1);
   h_stextval = handles(2);
end

if strcmp(command_str,'initialize')
   % Define default min and max values of dial.
   if nargin == 3
          minval = Argument1;
          maxval = Argument2;
   else
          minval = 0;
          maxval = 100;
   end

   h_fig=figure('Position',[200 200 200 200],...
          'color',[.7 .7 .7],...
          'menubar','none',...
        'resize','off',...
        'Units','normalized');
   h_ax=axes('color',[.7 .7 .7],...
          'xcolor',[.7 .7 .7],...
          'ycolor',[.7 .7 .7],...
          'xtick',[],'ytick',[],...
          'xlim',[-1 1],'ylim',[0 1],...
          'aspect',[NaN 1],...
          'position',[.2 .1 .6 .8]);

   % Draw arrow in its minimum setting.
   arrowx = [0 -1 -.85 NaN -1 -.85];
   arrowy = [0 0 -.05 NaN 0 .05];
   arrowz = [0 0  0  0   0  0];
   % Store a matrix that can be manipulated
   % and used to draw the arrow after a rotation
   % angle has been determined.
   arrowud = [arrowx(:),arrowy(:),...
          arrowz(:),ones(prod(size(arrowx)),1)]';
   h_arrow = line(arrowx,arrowy,...
          'linewidth',2,...
          'clipping','off',...
          'erasemode','background',...
          'userdata',arrowud);

   % Create labels and the radial lines.
   h_stext = uicontrol(h_fig,...
          'style','text',...
          'string','Value:',...
          'position',[.1 .2 .4 .13],...
          'units','norm');
   h_stextval = uicontrol(h_fig,...
          'style','text',...
          'string',sprintf('%2.1f',minval),...
          'position',[.4 .2 .3 .13],...
          'units','norm',...
          'min',minval,'max',maxval);
   h_dialborder = line(1.1*cos(0:.1:pi),...
          1.1*sin(0:.1:pi),...
```

```
            'color',[0 0 0],...
            'clipping','off');
    h_t(1)=text(-1.15,0,sprintf('%2.1f',minval),...
            'horizontalalignment','right');
    h_t(2)=text(1.1,0,sprintf('%2.1f',maxval),...
            'horizontalalignment','left');
    h_t(3)=text(0,1.15,sprintf('%2.1f',...
            (maxval-minval)/2+minval),...
            'horizontalalignment','center',...
            'verticalalignment','bottom');

    % Make sure all the objects that the user might click
on
    % to rotate the arrow with will recognize the initial
    % click.
    set([h_ax;h_t(:);h_dialborder;h_arrow],...
        'buttondownfcn',...
                'uidial(''Set
Calls'');uidial(''Rotate'')');
    set(gcf,'userdata',[h_arrow h_stextval])

elseif strcmp(command_str,'Set Calls')
    % Define when the user clicks on the dial.  Set up
    % the callbacks that should occur when the user moves
or
    % releases the mouse button.
    set(gcf,'windowbuttonupfcn',...
            'set(gcf,''windowbuttonmotion'','''')');
    set(gcf,'windowbuttonmotionfcn','uidial(''Rotate'')');

elseif strcmp(command_str,'Rotate')
    % Define the callback that should occur when the user
    % moves the mouse button.

    % Find out where the mouse pointer is located.
    pt = get(gca,'currentpoint');
    pt = pt(1,1:2);
    % Determine the angle that the pointer is at with
    % respect to the arrow's hinge.
    deg = atan2(pt(2),-pt(1))*180/pi;
    % Make sure the arrow does not swing past limits.
    if deg < 0 & abs(deg) < 90
            deg = 0;
    elseif deg>180 | (deg<0 & abs(deg) > 90)
            deg = 180;
    end

    % Scale angle linearly between dial's minimum
    % and maximum values.
    minval = get(h_stextval,'min');
    maxval = get(h_stextval,'max');
    val = (deg/(180-0)*((maxval-minval)))+minval;

    % Store the value in the current axes UserData
    % where it can be retrieved by an application.
    set(gca,'userdata',val);

    % Create transformed coordinate points for the
```

```
      % arrow.
      arrowud = get(h_arrow,'userdata');
      A = viewmtx(deg,90);
      newarrow = A*arrowud;
      set(h_arrow,'xdata',newarrow(1,:),...
          'ydata',newarrow(2,:));

      % Update the value indicator.
      set(h_stextval,'string',sprintf('%2.1f',val));

end
```

This program may be altered so that the dial could be used as a means of displaying values from an application, rather than just as an application input device. In addition, it would also be nice to be able to specify the position that the dial should occupy within any given figure so that multiple dials could be created as part of a GUI. Adding these features would be a good exercise.

APPENDIX : QUICK REFERENCES

The purpose of this appendix is to give you a convenient set of quick references. Included here is a short bibliography of texts that we feel will give you a solid background in the graphical representation of information, GUI development, and MATLAB. Also, here are the graphics commands and graphics objects properties in MATLAB.

Bibliographic References

1. Tufte, E. R., *The Visual Display of Quantitative Information*, Graphics Press, Cheshire, CT, 1990.

2. Tufte, E. R., *Envisioning Information*, Graphics Press, Cheshire, CT, 1990.

3. Thalmann, D., *Scientific Visualization and Graphics Simulation*, John Wiley and Sons, Inc., Chichester West Sussex, England, 1990.

4. Weinshenk, S., Jamar, P., Yeo, S. C., *GUI Design Essentials*, John Wiley and Sons, Inc., New York, NY, 1997.

Graphics Commands and Object Properties

MATLAB Data Formats – Section 3.2.1

Data Formats	Command	Returns
MAT - MATLAB workspace	LOAD	Variables in file
CSV - Comma separated numbers	CSVREAD	Double array
TXT – Formatted data in a text file	TEXTREAD	Double array
DAT - Formatted text	IMPORTDATA	Double array
DLM - Delimited text	DLMREAD	Double array
TAB - Tab separated text	DLMREAD	Double array

Spreadsheet Formats	Command	Returns
XLS - Excel worksheet	XLSREAD	Double array and cell array
WK1 - Lotus 123 worksheet	WK1READ	Double array and cell array

Scientific Data Formats	Command	Returns
CDF - Common Data Format	CDFREAD	Cell array of CDF records
FITS - Flexible Image Transport System	FITSREAD	Primary or extension table data
HDF - Hierarchical Data Format	HDFREAD	HDF or HDF-EOS data set

Image Formats	Command	Returns
TIFF - TIFF image	IMREAD	Truecolor, grayscale or indexed image(s).
PNG - PNG image	IMREAD	Truecolor, grayscale or indexed image
HDF - HDF image	IMREAD	Truecolor or indexed image(s)
BMP - BMP image	IMREAD	Truecolor or indexed image

continued on next page

Audio Formats	Command	Returns
AU – Next/Sun Sound	AUREAD	Sound data and sample rate
SND – Next/Sun Sound	AUREAD	Sound data and sample rate
WAV – Microsoft Wave Sound	WAVREAD	Sound data and sample rate

Movie Formats	Command	Returns
AVI - Movie	AVIREAD	MATLAB movie

Line Color, Marker Style, and Line Style Strings – Section 3.3.1

Line Color		Marker Style	
character	*creates*	*character*	*creates*
b or blue	blue line	.	point
g or green	green line	o	circle
r or red	red line	x	x-mark
c or cyan	cyan line	+	plus
m or magenta	magenta line	*	star
y or yellow	yellow line	s	square
k or black	black line	d	diamond
		v	triangle down
Line Style		^	triangle up
character	*creates*	<	triangle left
-	solid	>	triangle right
:	dotted	p	pentagram
-.	dashdot	h	hexagram
--	dashed		

TeX Characters Available in MATLAB – Section 3.4.5

TeX Characters	Result	TeX Characters	Result	TeX Characters	Result
\alpha	α	\upsilon	υ	\sim	~
\beta	β	\phi	φ	\leq	≤
\gamma	γ	\chi	χ	\infty	∞
\delta	δ	\psi	ψ	\clubsuit	♣
\epsilon	ε	\omega	ω	\diamondsuit	♦
\zeta	ζ	\Gamma	Γ	\heartsuit	♥
\eta	η	\Delta	Δ	\spadesuit	♠
\theta	θ	\Theta	Θ	\leftrightarrow	↔
\vartheta	ϑ	\Lambda	Λ	\leftarrow	←
\iota	ι	\Xi	Ξ	\uparrow	↑
\kappa	κ	\Pi	Π	\rightarrow	→
\lambda	λ	\Sigma	Σ	\downarrow	↓
\mu	μ	\Upsilon	Υ	\circ	∘
\nu	ν	\Phi	Φ	\pm	±
\xi	ξ	\Psi	Ψ	\geq	≥
\pi	π	\Omega	Ω	\propto	∝
\rho	ρ	\forall	∀	\partial	∂
\sigma	σ	\exists	∃	\bullet	•
\varsigma	ς	\ni	∋	\div	÷
\tau	τ	\cong	≅	\neq	≠
\equiv	≡	\approx	≈	\aleph	ℵ
\Im	ℑ	\Re	ℜ	\wp	℘
\otimes	⊗	\oplus	⊕	\oslash	∅
\cap	∩	\cup	∪	\supseteq	⊇
\supset	⊃	\subseteq	⊆	\subset	⊂
\int	∫	\in	∈	\o	
\rfloor	⌋	\lceil	⌈	\nabla	∇
\lfloor	⌊	\cdot	·	\dots	...
\perp	⊥	\neg	¬	\prime	′
\wedge	∧	\times	×	\0	∅
\rceil	⌉	\surd	√	\mid	|
\vee	∨	\varpi	ϖ	\copyright	©
\langle	⟨	\rangle	⟩		

TeX Stream Modifiers – Section 3.4.5

TeX Stream Modifier	*Description*
\bf	Bold font.
\it	Italics font.
\sl	Oblique font (rarely used).
\rm	Normal font.
^	Make part of string superscript.
_	Make part of string subscript.
\fontname{*fontname*}	Specify the font family to use.
\fontsize{*fontsize*}	Specify the font size in FontUnits.

Projection Types – Section 4.2.1

Projection Type	How to Interpret	How to Use
Orthographic Projection	Think of the "viewing volume" as a box whose opposite sides are parallel, so the distance from the camera does not affect the size of surfaces in the plot.	Used to maintain the actual size of objects and the angle between objects. This works well for data plots. Real-world objects look unnatural.
Perspective Projection	The "viewing volume" is the projection of a pyramid where the apex has been cut off parallel to the base. Objects farther from the camera appear smaller.	Used to create more "realistic" views of objects. This works best for real-world objects. Data plots may look distorted.

Summary of the Axis Function – Section 4.2.2

Syntax	Affect
`axis([xmin xmax ymin ymax])`	Sets the x- and y-axis limits.
`axis([xmin xmax ymin ymax zmin zmax cmin cmax])`	Sets the x-, y-, and z-axis limits and the color scaling limits.
`v = axis`	Returns a row vector containing the x-, y-, and z-axis limits, i.e., scaling factors for the x-, y-, and z-axis.
`axis auto`	Computes the current axes' limits automatically, based on the minimum and maximum values of x, y, and z data.
`axis 'auto x'` `"    "'auto y'` `"    "'auto x'` `"    "'auto xz'` `"    "'auto yz'` `"    "'auto xy'`	Computes the indicated axis limit automatically.
`axis manual`	Freezes scaling of the current limits. Used with **hold** forces subsequent plots to use the same limits.
`axis tight` or `axis fill`	Sets the axis limits to the range of the data.
`axis ij`	Sets the origin of the coordinate system to the upper left corner. The i-axis is vertical, increasing from top to bottom. The j-axis is horizontal, increasing from left to right.
`axis xy`	This is the default coordinate system with the origin at the lower left corner. The x-axis is horizontal increasing from left to right, and the y-axis is vertical increasing from bottom to top.

Scalar Volume Computation Functions – Section 4.3.1

Function	Action
FVC = isocaps(X,Y,Z,V,ISOVALUE)	Computes an isosurface end cap geometry for data V at isosurface value ISOVALUE and returns a structure containing the faces, vertices, and colors of the end cap which can be passed directly to the **patch** function.
NC = isocolors(X,Y,Z,C,VERTICES)	Computes the colors of isosurface vertices VERTICES using color values C and returning them in the array NC.
N = isonormals(X,Y,Z,V,VERTICES)	Computes the normals (N) of isosurface vertices VERTICES by using the gradient of the data in V.
FV = isosurface(X,Y,Z,V,ISOVALUE)	Extracts an isosurface at ISOVALUE in the volume V, returning the structure FV containing the faces and vertices of the isosurface, suitable for use with the **patch** function.
NFV = reducepatch(P,R)	Reduces the number of faces in a patch P by a fraction R of the original faces. It returns the structure NFV containing the new faces and vertices.
[NX, NY, NZ, NV] = reducevolume(X,Y,Z,V,[Rx Ry Rz])	Reduces the number of elements in a volume by only keeping every Rx, Ry, Rz element in the corresponding x, y, or z direction.
NFV = shrinkfaces(P,SF)	Reduces the size of patch P by shrink factor SF, returning a structure NFV containing the new faces and vertices.
W = smooth3(V,'gaussian', SIZE) W = smooth3(V,'box', SIZE)	Smooths the data in V according to the convolution kernel of size SIZE specified by the given string.
FVC = surf2patch(S)	Converts a surface object S into a patch object. FVC is a structure containing the faces, vertices, and colors of the new patch.
[NX, NY, NZ, NV] = subvolume(X,Y,Z,V,LIMITS)	Extracts a subset of volume data from V using limits LIMITS = [xmin xmax ymin ymax zmin zmax].
contourslice(X,Y,Z,V,Sx,Sy,Sz)	Draws contours in a volume slice plane at the points in the vectors Sx, Sy, and Sz.
patch(x,y,z,C)	Creates a patch in the 3-D space of color defined by C.
slice(X,Y,Z,V,Sx,Sy,Sz)	Draws a slice plane described by the vectors Sx, Sy, Sv, through the volume V.

Graphics Objects Creation Functions – Section 7.1

Graphics Object	Low-Level Creation Function	Description
Figure	figure or figure(H)	A window to show other graphics objects.
Axes	axes, axes(H), or axes('position',RECT)	The axes for showing graphs in a figure.
UIcontrol	Uicontrol	The user interface control is used to execute a function in response to the user.
UImenu	Uimenu	User defined menus in the figure.
UIcontextmenu	uicontextmenu('PropertyName1',value1,...)	A pop-up menu that appears when a user right-clicks on a graphics object.
Image	image(C) or image(x,y,C)	A 2-D bitmap.
Light	light('PropertyName','PropertyValue',...)	Light sources that affect the coloring of patch and surface objects.
Line	line(x,y) or line(x,y,z)	A line in 2-D or 3-D plots.
Patch	patch(x,y,c) or patch(x,y,z,c)	A polygon that is filled with some color or texture and has *edges*.
Rectangle	rectangle, rectangle('Position',[x,y,w,h]), or rectangle('Curvature',[x,y],...)	A 2-D shape; can be rectangle or oval created within an axes object.
Surface	surface(X,Y,Z,C), surface(X,Y,Z), surface(Z,C), surface(Z)	3-D representation of data plotted as heights above the x-y plane.
Text	text(x,y,text_string) or text(x,y,z,text_string)	Character strings used in a figure.

Universal Object Properties – Section 7.4.3

Property	Read Only	ValueType/Options	Format
BusyAction	No	[{queue} | cancel]	row
ButtonDownFcn	No	string	row
Children	No*	handle(s)	column
Clipping	No	[{on} | off]	row
CreateFcn	No	string	row
DeleteFcn	No	string	row
HandleVisibility	No	[{on} | callback | yes]	row
HitTest	No	[{on} | off]	row
Interruptible	No	[no | {yes} | off | {on}]	row
Parent	No	handle	one element
Selected	No	[{off} | on]	row
SelectionHighlight	No	[{no} | yes | {off} | on]	row
Tag	No	string	row
Type	Yes	string	row
UserData	No	number(s) or string	matrix
Visible	No	[{on} | off]	row

* Although you cannot create new handles in the *Children* property, you can change the order of the handles and so change the stacking order of the objects.

Root Properties – Section 7.5.1

Property	Read-Only	ValueType/Options	Format
Display Related			
FixedWidthFontName	No	string	row
ScreenDepth	Yes	integer	1 element
ScreenSize	Yes	[left bottom width height]	4-element row
Related to the State of MATLAB			
CallbackObject	Yes	handle	1 element
CurrentFigure	No	handle	1 element
ErrorMessage	No	string	row
PointerLocation	No	[x-coordinate,y-coordinate]	2-element row
PointerWindow	Yes	handle	1 element
ShowHiddenHandles	Yes	[on | {off}]	row
Behavior Related			
Diary	No	[on | {off}]	row
DiaryFile	No	string	row
Echo	No	[on | {off}]	1 element
Format	No	[short | long | {shortE} | longE | hex | bank | + | rat]	row
FormatSpacing	No	[{loose} | compact]	row
Language	No	string	row
RecursionLimit	No	integer	1 element
Units	No	[inches | centimeters | normalized | points | {pixels}]	row

Figure Properties – Section 7.5.2

Property	Read Only	ValueType/Options	Format
		Positioning the Figure	
Position	No	[left bottom height width]	4-element row
Units	No	[inches | centimeters | normalized | points | {pixels}]	row
		Style & Appearance	
Color	No	[Red Green Blue] or color string	RGB vector
MenuBar	No	[{figure} | none]	1 element
Name	No	string	row
NumberTitle	No	[{on} | off]	row
Resize	No	[{on} | off]	row
WindowStyle	No	[{normal} | modal]	row
		Colormap Controls	
Colormap	No	M RGB number triplets	M-by-3 matrix
Dithermap	No	N RGB number triplets	M-by-3 matrix
Dithermapmode	No	[auto | {manual}]	row
FixedColors	No	N RGB number triplets	N-by-3 matrix
MinColormap	No	number	1 element
ShareColors	No	[no | {yes}]	row
		Transparency	
Alphamap	No	default is 64 values progression from 0 to 1	M-by-1 vector
		Renderer	
BackingStore	No	[{on} | off]	row
DoubleBuffer	No	[on | {off}]	row
Renderer	No	[{patinters} | zbuffer | OpenGL]	row
RendererMode	No	[{auto} | manual]	row
		Current State	
CurrentAxes	No	handle	1 element
CurrentCharacter	No	character	1 element
CurrentObject	No	handle	1 element
CurrentPoint	No	[x-coordinate, y-coordinate]	2-element row
SelectionType	Yes	[normal | extended | alt | open]	row

continued on next page

Property	Read Only	ValueType/Options	Format
Properties that Affect the Pointer			
Pointer	No	[crosshair | fullcrosshair | {arrow} | ibeam | watch | topl | topr | botl | botr | left | top | right | bottom | circle | cross | fleur | custom]	string
PointerShapeHotSpot	No	(row index, column index)	2-element row
PointerShapeCData	No	1s where black, 2s where white, NaNs where transparent	16-by-16
Callback Execution			
CloseRequestFcn	No	string, function handle, or cell-array {'closereq'}	string, 1-element, cell-array
KeyPressFcn	No	string	string
ResizeFcn	No	string	string
UIContextMenu	No	Number	1 element
WindowButtonDownFcn	No	string	string
WindowButtonMotionFcn	No	string	string
WindowButtonUpFcn	No	string	string
Controlling Access to Objects			
IntegerHandle	No	[{on} | off]	string
NextPlot	No	[{add} | replace | replacechildren]	string
Properties that Affect Printing			
InvertHardcopy	No	[{on} | off]	string
PaperOrientation	No	[{portrait} | landscape]	string
PaperPosition	No	[left bottom width height]	4-element row
PaperPositionMode	No	[{auto} | manual]	string
PaperSize	No	[width height]	2-element row
PaperType	No	[{usletter} | uslegal | A0 | A1 | A2 | A3 | A4 | A5 | B0 | B1 | B2 | B3 | B4 | B5 | arch-A | arch-B | arch-C | arch-D | arch-E | A | B | C | D | E | tabloid | <custom>]	string
PaperUnits	No	[{inches}|centimeters| normalized | points]	string
General			
FileName	No	A name of a FIG-File to be used with GUIDE; see Chapter 10.	string

Axis Properties – Section 7.5.3

Property	Read Only	ValueType/Options	Format
Properties Affecting Transparency and Lighting			
ALim	No		
ALimMode	No	[{auto} | manual]	
AmbientLightColor	No		
Properties Controlling Boxes and Tick Marks			
Box	No	[on | {off}]	row
TickLength	No	[2-Dticklength 3-Dticklength]	2-element row
TickDir	No	[{in} | out]	
TickDirMode	No	[{auto} | manual]	
XMinorTick	No	[on | {off}]	row
XTick	No	numbers	
XTickLabel	No	string	matrix
XTickLabelMode	No	[{auto} | manual]	row
XTickMode	No	[{auto} | manual]	row
Properties Affecting Character Formats			
FontAngle	No	[{normal} | italic | oblique]	
FontName	No	name of desired font	string
FontSize	No	number	1 element
FontUnits	No	[inches | centimeters | normalized | {points} | pixels]	string
FontWeight	No	[light | {normal} | demi | bold]	string

continued on next page

Property	Read Only	ValueType/Options	Format
Properties Determining Axis Location and Position			
Position	No	[left bottom width height]	4-element row
Units	No	[inches | centimeters | {normalized} | points | pixels | characters]	
XAxisLocation	No	[top | {bottom}]	string
YaxisLocation	No	[{left} | right]	row
CurrentPoint	No	mouse click near and far x, y, z axis locations	2-by-3 matrix
Title	No	handle of text object	1 element
Properties Affecting Grids, Lines, and Color			
Color	No	[Red Green Blue] or color string	
ColorOrder	No	M RGB number triplets	M-by-3 matrix
CLim	No	[cmin cmax]	2-element row
CLimMode	No	[{auto} | manual]	string
DrawMode	No	[{normal} | fast]	
XGrid	No	[on | {off}]	
GridLineStyle	No	[- | – | {:} | -. | none]	string
Layer	No	[top | {bottom}]	string
LineStyleOrder	No	string array of linestyle symbol(s)	matrix
LineWidth	No	number	1 element
MinorGridLineStyle	No	[- | – | {:} | -. | none]	
XColor	No	[Red Green Blue] or color string	row
Xform	No	4 x 4 Perspective Transformation	4 x 4 matrix
XLabel	No	Handle of text object	1 element
XMinorGrid	No	[on | {off}]	row
NextPlot	No	[add | {replace} | replacechildren]	string
Properties Affecting Axis Limits			
DataAspectRatio	No	[x y z] relative ratio of axis lengths	2-element row
DataAspectRatioMode	No	[{auto} | manual]	string
PlotBoxAspectRatio	No	[x y z] relative ratios of box lengths	3-element row
PlotBoxAspectRatioMode	No	[{auto} | manual]	
XDir	No	[{normal} | reverse]	row
XLim	No	[xmin xmax]	2-element row
XLimMode	No	[{auto} | manual]	row
XScale	No	[{linear} | log]	row

continued on next page

Property	Read Only	ValueType/Options	Format
Axes Properties Related to Viewing Perspective			
CameraPosition	No	[x y z] numbers	3-element row
CameraPositionMode	No	[{auto} | manual]	string
CameraTarget	No	[x y z] numbers	3-element row
CameraTargetMode	No	[{auto} | manual]	string
CameraUpVector	No	[x y z] numbers	3-element row
CameraUpVectorMode	No	[{auto} | manual]	string
CameraViewAngle	No	number	1 element
CameraViewAngleMode	No	[{auto} | manual]	string
Layer	No	[top | {bottom}]	string
Projection	No	[{orthographic} | perspective]	
View	No	[DegreesAzimuth DegreesElevation]	2-element row

Line Properties – Section 7.5.4

Property	Read Only	ValueType/Options	Format
Color	No	[Red Green Blue] or color string	RGB row
EraseMode	No	[{normal} | background | xor | none]	row
LineStyle	No	[{-} | -- | : | -. | none]	row
LineWidth	No	number	1 element
Marker	No	[+ | o | * | . | x | square | diamond | v | ^ | > | < | pentagram | hexagram | {none}]	row
MarkerSize	No	number	1 element
MarkerEdgeColor	No	[none | {auto}] -or- a ColorSpec	row
MarkerFaceColor	No	[{none} | auto] -or- a ColorSpec	row
XData	No	numbers	vector
YData	No	numbers	vector
ZData	No	numbers	vector

Rectangle Properties – Section 7.5.5

Property	Read Only	ValueType/Options	Format
Curvature	No	[x, y]	1 or 2 element
EraseMode	No	[{normal} | background | xor | none]	row
FaceColor	No	ColorSpec | {none}	row
EdgeColor	No	{ColorSpec} | none	row
LineStyle	No	[{-} | -- | : | -. | none]	row
LineWidth	No	number	1 element
Position	No	[x,y,width,height]	vector

Patch Properties – Section 7.5.6

Property	Read Only	ValueType/Options	Format
Properties Defining Patch Objects			
Faces	No	permutation of 1:M	N-by-V matrix
Vertices	No	numbers x-, y-, z-coordinates	M-by-3 matrix
XData	No	coordinates of the points at the vertices	vector or matrix
YData	No	coordinates of the points at the vertices	vector or matrix
ZData	No	coordinates of the points at the vertices	vector or matrix
Properties Specifying Lines, Color, and Markers			
CData	No	numbers	vector
CDataMapping	No	[direct | {scaled}]	row
EdgeColor	No	[none | {flat} | interp] or [Red Green Blue] or color string	row
FaceColor	No	[none | {flat} | interp] or [Red Green Blue] or color string	row
FaceVertexCData	No	RGB per patch, face, or vertex	matrix
LineStyle	No	[{'-'} | '—' | '-.' | ':' | 'none']	row
LineWidth	No	number	1 element
Marker	No	['square' | 'diamond' | 'v' | '^' | '>' | '<' | '.' | 'pentagram' | 'hexagram' | 'o' | 'x' | '+' | '*' | {none}]	row
MarkerEdgeColor	No	[none | {auto} | [R G B] | color_string]	row
MarkerFaceColor	No	[{none} | auto | [R G B] | color_string]	row
MarkerSize	No	number	1 element
Properties Affecting Lighting and Transparency			
AmbientStrength	No	numbers	vector
BackFaceLighting	No	[unlit | lit | {reverselit}]	row
FaceLighting	No	[none | {flat} | gouraud | phong]	row
DiffuseStrength	No	number	1 element
EdgeLighting	No	[{none} | {flat} | gouraud | phong]	row
SpecularColorReflectance	No	number ranging from 0 to 1	1 element
SpecularExponent	No	number > or = to 1	1 element
SpecularStrength	No	number ranging from 0 to 1	1 element
VertexNormals	No	numbers	M-by-3 matrix
NormalMode	No	[{auto} | manual]	row
EraseMode	No	[{normal} | none | xor | background]	row
AlphaDataMapping	No	[none |direct | {scaled}]	row
EdgeAlpha	No	[{scalar = 1} | flat | interp]	1 element or string
FaceAlpha	No	[{scalar = 1} | flat | interp]	1 element or string
FaceVertexAlphaData	No	transparency data	1 element or M-by-1 matrix

Surface Properties – Section 7.5.7

Property	Read Only	ValueType/Options	Format
Properties that Define a Surface			
XData	No	coordinates of the points at the vertices	vector or matrix
YData	No	coordinates of the points at the vertices	vector or matrix
ZData	No	coordinates of the points at the vertices	vector or matrix
Properties that Specify Lines, Colors, and Markers			
CData	No	numbers	vector
CDataMapping	No	[direct | {scaled}]	row
LineStyle	No	[{'-'} | '--' | '-.' | ':' | 'none']	row
LineWidth	No	number	1 element
EdgeColor	No	[none | {flat} | interp] or [Red Green Blue] or color string	row
FaceColor	No	[none | {flat} | interp | texturemap] or [Red Green Blue] or color string	row
Marker	No	['square' | 'diamond' | 'v' | '^' | '>' | '<' | '.' | 'pentagram' | 'hexagram' | 'o' | 'x' | '+' | '*' | {none}]	row
MarkerEdgeColor	No	[none | {auto} | [R G B] | color_string]	row
MarkerFaceColor	No	[{none} | auto | [R G B] | color_string]	row
MarkerSize	No	number	1 element
Properties Affecting Lighting and Transparency			
AmbientStrength	No	numbers	vector
BackFaceLighting	No	[unlit | lit | {reverselit}]	row
DiffuseStrength	No	number	1 element
EdgeLighting	No	[{none} | {flat} | gouraud | phong]	row
FaceLighting	No	[none | {flat} | gouraud | phong]	row
NormalMode	No	[{auto} | manual]	row
SpecularColorReflectance	No	number ranging from 0 to 1	1 element
SpecularExponent	No	number > or = to 1	1 element
SpecularStrength	No	number ranging from 0 to 1	1 element
VertexNormals	No	numbers	M-by-3 matrix
AlphaData	No	default = 1 (opaque)	M-by-N matrix of double or uint8
AlphaDataMapping	No	[none |direct | {scaled}]	row
EdgeAlpha	No	[{scalar = 1} | flat | interp]	1 element or string
FaceAlpha	No	[{scalar = 1} | flat | interp]	1 element or string

Image Properties – Section 7.5.8

Property	Read Only	ValueType/Options	Format
General Properties of the Image Object			
CData	No	numbers	matrix or M-by-N-by-3 array
CDataMapping	No	[{direct}	scaled]	row
XData	No	[min, max] default = [1, size(CData,2)]	2-element vector
YData	No	[min max] default = [1, size(CData, 1)]	2-element vector
Properties Affecting Transparency			
AlphaData	No	default = 1 (opaque)	M-by-N matrix of double or uint8
AlphaDataMapping	No	[{none} |direct	scaled]	row

Text Properties – Section 7.5.9

Property	Read Only	ValueType/Options	Format
Color	No	[Red Green Blue] or color string	RGB row
Editing	No	[{off}	on]	row
EraseMode	No	[{normal}	none	xor	background]	row
Extent	Yes	[left bottom width height]	4-element row
FontAngle	No	[{normal}	italic	oblique]	1 element
FontName	No	string	row
FontSize	No	numbers	1 element
FontUnits	No	[inches	centimeters	normalized	points	pixels	{data}]	row
FontWeight	No	[light	{normal}	demi	bold]	row
HorizontalAlignment	No	[{left}	center	right]	row
Interpreter	No	[{tex}	none]	row
Position	No	[x y z] coordinates	row
Rotation	No	[AngleInDegrees]	1 element
String	No	string	row
Units	No	[inches	centimeters	normalized	points	pixels	{data}]	row
VerticalAlignment	No	[top	cap	{middle}	baseline	bottom]	row

Alpha Properties – Section 8.4.1

Property	Read Only	ValueType/Options	Format
AlphaData	No	m-by-n matrix of transparency data for image and surface objects	matrix
AlphaDataMapping	No	none | direct | scaled none = default for images scaled = default for patches	row
FaceAlpha	No	Transparency for faces	row or scalar
EdgeAlpha	No	Transparency for edges	row or scalar
FaceVertexAlphaData	No	Alpha data property for patches	row or scalar
ALim	No	Alpha axis limits	vector
ALimMode	No	Alpha axis limits mode	row
Alphamap	No	Figure Alphamap	matrix

·Uicontrol Properties – Section 10.3.2

Property	Read Only	ValueType/Options	Format
BackgroundColor	No	[Red Green Blue] or color string	RGB row
ButtonDownFcn	No	string	row
CData	No		matrix
CallBack	No	string	row
Enable	No	[on | {off} | inactive]	row
Extent	Yes	[0,0,width,height]	row
FontAngle	No	[{normal} | italic | oblique]	row
FontName	No	string	row
FontSize	No	number	1 element
FontUnits	No	{points} | normalized | inches | centimeters | pixels	row
FontWeight	No	[light | {normal} | demi | bold]	row
ForegroundColor	No	[Red Green Blue] or color string	RGB row
HorizontalAlignment	No	[left | {center} | right]	row
Interruptible	No	{on} | off	row
ListBoxTop	No	number	1 element
Max	No	number	1 element
Min	No	number	1 element
Position	No	[left bottom width height]	4-element row
String	No	string	string matrix
Style	No	[{pushbutton} | radiobutton | togglebutton | checkbox | edit | text | slider | frame | popupmenu | list box]	row
SliderStep	No	number	1 element
TooltipString	No	string	row
Units	No	[inches | centimeters | normalized | points | {pixels}]	row
UIContextMenu		handle	1 element
Value	No	number	1 element
Tag	No	string	row
UserData	No	string(s) or number(s)	matrix
Visible	No	[{on} | off]	row

Uimenu Properties – Section 10.4.1

Property	Read Only	ValueType/Options	Format
Accelerator	No	string	row
CallBack	No	string	row
Checked	No	[on | {off}]	row
Children	Yes	object_handles	column
Enable	No	[on | {off} | inactive]	row
ForegroundColor	No	[Red Green Blue] or color string	RGB row
Label	No	string	row
Position	No	[left bottom width height]	4-element row
Separator	No	[on {off}]	row
Interruptible	No	[{on} | off]	row
Tag	No	string	row
UserData	No	string(s) or number(s)	matrix
Visible	No	[{on} | off]	row

INDEX

2

2-D plot(s)
  elementary, 25

A

area plots, 83
aspect ratio, 46
  demo, 261
axes
  children, 196
  creating multiple, 63
  CurrentAxes property, 227
  object, 63, 194, 208
  Position property, 291
  properties, 236–66
axis
  auto, 45, 139, 498
  command, 41
  equal, 140, 173
  normal, 42, 140
  square, 42, 43, 140
  tight, 42, 139, 498
  xy, 42, 139, 498

B

bar graph(s), 67
  3-D, 71
  bar3h, 71
  clustered, 70
  histogram, 72
  stacked, 70
boxes. *See also* properties,axis,box
  Box property, 238
  dynamic, 483
built-in surfaces, 132
  cylinder, 134
  ellipsoid, 133
  peaks, 132
  sphere, 132

C

callback
  and the event queue, 474
  defined, 217
  figure properites affecting execution, 230
  function prototypes, 456
  in GUIDE, 452
  interrupting, 478
  object property, 217
  uicontrol, 400

uimenu, 424
check boxes, 391
clabel, 116, 197
Clipping property, 208
color
  data values, 279
  edge, 277
  face, 196, 277, 286, 310
  invisible, 323
color map, 94, 106
combination plot(s), 122–27
  contour and quiver, 123
  surface and contour, 125
combination plots, 122
command
  axis, 41, 89, 261
  bar, 67
  bar3h, 71
  clear, 155
    accidental in a GUI, 441
    with CallBack, 439
    with UserData, 408
  clf, 37
  colorbar, 328
  compass, 87
  drawnow, 374
  figure, 199, 374
  fopen, 24
  fread, 24
  fscanf, 24
  get
    with Clim, 306
  hold, 37, 63
    with contourslice, 150
  input, 475, 479
  legend, 56
  line, 253, 374
    color, 495
  loglog, 26
  movie, 366
  pause, 215, 374, 479
  refresh, 226, 319, 475
  rotate, 273
  rotate3d, 142
  semilogx, 26
  semilogy, 26
  shading
    faceted, 112
    flat, 112
    interp, 112
  sphere, 132
  surf, 132, 312
  uicontrol, 409

Command History, 4, 5, 8
contour
    plot
        3-D, 120
        color, 116
        filled, 119
        using clabel with, 116
    plots, 115
        specifying, 115

D

darkening color maps, 319
data
    function, 19
    importing, 21
    importing high-level data, 22
    importing low-level-data, 24
    low-level I/O commands, 25
    measured, 20
    non-uniform, 3-D plotting of, 110
    quantity and dimension, 14
    sources of, 19
    standard formats, 21
    using fread, 24
    using fscanf, 24
    visualizing, 12
default
    data file extension, 23
Desktop, 5
dialog boxes
    errordlg, 414
    helpdlg, 414
    inputdlg, 414
    msgbox, 415
    questdlg, 231, 414
    warndlg, 414

E

editable text, 392, 403, 406, 452, 465
    with pop-up menus, 466
    with sliders, 464
ellipsoid
    equation for, 133
    function, 133
erase mode, 375
error bar, 77

F

figure
    properties affecting style and appearance, 223
    properties affecting the colormap, 224
    properties that affect position, 222
Figure Window, 5
function

area, 83
axis('ij'), 259
brighten, 319
clabel, 116
closereq, 230
contourf, 119
copyobj, 210
cylinder, 134
delete, 202
drawnow, 374
ellipsoid, 133
errordlg. *See* dialog boxes
feather, 86
fill, 93
findobj, 297, 408
fplot, 91
gco, 228
get, 148, 202, 204, 287
getframe, 215, 361, 383, 473
gradient, 106
helpdlg, 414. *See* dialog boxes
hidden, 114
hist, 72
hold, 122
    with axis, 139
legend, 57
loglog, 35
mesh, 106, 108, 129, 134
meshc, 124
meshgrid, 104, 158
movie, 362, 367
movie2avi, 371
moviein, 361, 368
msgbox. *See* dialog boxes
orient, 235
patch, 132, 146, 313
pcolor, 329, 330
plot
    general, 26–35
plot3, 101–4
polardb, 89
print, 186
questdlg. *See* dialog boxes
quiver, 121
rose, 90
semilogx, 35
semilogy, 35
set, 149, 202, 204
    multiple windows, 218
sidetext, 51
specular, 347
sphere, 132
stairs, 74
str2mat, 241
subplot, 63
trimesh, 129

trisurf, 129
view, 4
functions
  plotting, 91
    ezplot, 92
    fplot, 91

H

handle graphics, 40, 199
  graphics object handles, 198–202
help
  on-line, 7
  pull-down, 6
hidden lines
  removing, 113
hierarchy
  graphics objects, 193
  tree, 195
histogram
  angle with rose, 90
  in image processing, 72
  of normal random data, 200

I

icons, on buttons, 486
image object, 197, 208, 304, 306
  with color map, 315
Image object, 287

L

labels
  using cell arrays, 79
labels, on scatter plots, 55
light, 301
light objects, 193, 196, 286
  creation, 197
line
  color, 33
  creating color varying, 350
  width, controlling, 269
linestyle
  surface property, 284
list box, 395
load command, 21 - 23, 155, 160, 163, 174–77,
  182, 317, 335–37, 487, 494
  importing high level data, 22

M

mapping
  texture, 334
  with imagesc, 181
  with mesh, 106
memory, preallocating for movies, 367
mesh

plot, 105, 108
surfaces, creating 3-D, 104
modal
  dialog box, 231
  WindowStyle, 224
movie
  allocating memory, 361
  converting to avi, 371
  format, 22

N

NaN. *See* Not-a-Number
Not-a-Number, 163, 230, 276, 323

O

object(s)
  axes, 63
  hierarchy, 98
    tree, 195
objects
  in the Property Editor, 97
OpenGL, 189, 226
output, 220
  formatted, 220

P

patch, 273
plot(s)
  area, 83
  command, 26
  shaded surface, 112
  waterfall, 109
pointer
  custom, 230
  location, 218
  types, 229
polar plot, 88
print
  preview, 183
  pull-down, 183
  setup window, 183
properties
  manipulation, 204
properties, axis, 236–66
  adding plots while keeping current axis limits,
    259
  Box, 238, 251
  CameraPosition, 265
  CameraTarget, 266
  CameraUpVector, 266
  CLim, 255
  Color, 248
  ColorOrder, 252, 253, 255
  DataAspectRatio, 261

DrawMode, 251
FontAngle, 245
FontName, 245
FontSize, 245
FontUnits, 245
FontWeight, 245
GridLineStyle, 248
Layer, 251
limits
   XLim, 256
   XLimMode, 256
   YLim, 256
   YLimMode, 256
   ZLim, 256
   ZLimMode, 256
LineStyleOrder, 253
LineWidth, 248
NextPlot, 253
PlotBoxAspectRatio, 261
Position, 245, 261
TickDir, 239
TickDirMode, 239
TickLength, 239
Units, 245
View, 265
XaxisLocation, 246
XColor, 249
Xform, 265
XGrid, 248
Xtick, 239
XTickLabel, 239
XTickLabelMode, 243
XTickMode, 243
YaxisLocation, 246
YColor, 249
YGrid, 248
ZColor, 249
ZGrid, 248
properties, default
  overriding, 239
  setting, 295
properties, figure, 220–36
  Alphamap, 225
  BackingStore, 225
  CloseRequestFcn, 230
  Color, 223
  CurrentAxes, 227
  CurrentCharacter, 227, 232
  CurrentObject, 227
  CurrentPoint, 228
  Dithermap, 225
  DithermapMode, 225
  DoubleBuffer, 226
  FixedColors, 224
  IntegerHandle, 234
  InvertHardCopy, 235

MenuBar, 224
Name, 224
NumberTitle, 224
PaperOrientation, 235
PaperPosition, 235
PaperSize, 235
PaperType, 235
PaperUnits, 235
Pointer, 229
PointerShapeCData, 230
PointerShapeHotSpot, 230
Position, 222
Renderer, 226
RenderMode, 226
ResizeFcn, 224, 233
SelectionType, 228
UIContextMenu, 233
Units, 223
WindowStyle, 224
properties, image
  AlphaData, 319, 353
  AlphaDataMapping, 319
  Alphamap, 353
  CData, 287
  XData, 288
  YData, 288
properties, line, 266–71
  Color, 268
  EraseMode, 271
  LineStyle, 268
  LineWidth, 268
  Marker, 268
  MarkerEdgeColor, 270
  MarkerFaceColor, 270
properties, patch, 273
  AlphaDataMapping, 281
  AmbientStrength, 280
  BackFaceLighting, 281
  CData, 278
  CDataMapping, 279
  DiffuseStrength, 280
  EdgeAlpha, 281
  EdgeColor, 277, 278
  EdgeLighting, 280
  FaceAlpha, 281
  FaceColor, 278
  FaceLighting, 280
  Faces, 275
  FaceVertexAlphaData, 281
  FaceVertexCData, 278
  LineStyle, 277
  LineWidth, 277
  Marker, 277
  NormalMode, 280
  SpecularColorReflectance, 280
  SpecularExponent, 280

SpecularStrength, 280
VertexNormals, 280, 281
Vertices, 275
XData, 275
YData, 275
ZData, 275
properties, rectangle, 272
  Curvature, 272
  EdgeColor, 272
  FaceColor, 272
  LineStyle, 272
  LineWidth, 272
  Position, 272
properties, root, 215
  Callback, 217
  Diary, 219
  DiaryFile, 219
  Echo, 219
  ErrorMessage, 218
  FixedWidthFontName, 216
  Format, 219
  FormatSpacing, 219
  Language, 220
  PointerWindow, 219
  RecursionLimit, 220
  ScreenDepth, 217
  ScreenSize, 217
  ShowHiddenHandles, 219
  Units, 220
properties, surface, 281–87
  AmbientStrength, 286
  BackFaceLighting, 286
  CData, 283
  CDataMapping, 286
  DiffuseStrength, 286
  EdgeColor, 284
  EdgeLighting, 286
  FaceColor, 283, 286
  FaceLighting, 286
  LineStyle, 284
  LineWidth, 284
  MarkerEdgeColor, 284
  MarkerSize, 284
  MeshStyle, 284
  NormalMode, 286
  SpecularColorReflectance, 286
  SpecularExponent, 286
  SpecularStrength, 286
  VertexNormals, 286
  XData, 282
  YData, 282
  ZData, 283
properties, text
  Editing, 294
  EraseMode, 294
  Extent, 294

FontName, 292, 294
FontSize, 294
FontWeight, 294
HorizontalAlignment, 293
Interpreter, 295
Position, 291, 294
Rotation, 291
String, 294
Units, 291
VerticalAlignment, 293
properties, uicontrol
  BackgroundColor, 400
  ButtonDownFcn, 400
  CallBack, 400, 409
  CData, 400
  Children, 409
  Enable, 401
  Extent, 402
  FontAngle, 403
  FontName, 403
  FontSize, 403
  FontUnits, 403
  FontWeight, 403
  ForegroundColor, 402
  HorizontalAlignment, 403
  Interruptible, 408
  ListBoxTop, 407
  Max, 404
  Min, 404
  Parent, 409
  Position, 405
  SliderStep, 404
  String, 406
  Style, 406
  Tag, 408
  TooltipString, 405
  Type, 409
  Units, 408
  UserData, 408
  Value, 404
  Visible, 409
properties, universal
  BusyAction, 208
  ButtonDownFcn, 207
  Children, 208
  Clipping, 208
  CreateFcn, 210
  DeleteFcn, 211
  HandleVisibility, 211
  HitTest, 212
  Interruptible, 208
  Parent, 208
  Selected, 212
  SelectionHighlight, 213
  Tag, 213
  Type, 213

UserData, 214
Visible, 214
property editor, 203
pseudocolor
caxis, 306
pcolor, 329
with lines, 350

Q

quiver plots, 121

R

radial units, 89
red-green-blue, 94
refresh, 376

S

sidetext, 51
slider(s), 385, 397, 404
Specular Lighting Model, 347
stacking order, 207, 227, 251
stairstep graphs, 74
stem plots, 75–76
3-D, 127
stretchable GUI, 412
subplot
making odd axes, 66
symbols
Greek alphabet, 242
plot, 31, 34
pointer, 229

T

text files
loading, 22
saving, 22
texture mapping, 334

V

viewpoint, 265
visualization
and motion, 17
characteristics of good, 12
interacting with, 17
reasons to visualize, 11
with color, light, and shading, 14

W

warning dialog, 414
Waterfall Plots. *See* plot(s) waterfall

X

xdata
image property, 288
in animation, 376
line property, 267

Z

Z-buffering, 226